Frontiers of Supercomputing II

Published Titles in the Los Alamos Series
in Basic and Applied Sciences
Edited by David H. Sharp and L. M. Simmons, Jr.

1. Wildon Fickett and William C. Davis, *Detonation*
2. Charles L. Mader, *Numerical Modeling of Detonation*
3. Robert D. Cowan, *The Theory of Atomic Structure and Spectra*
4. Ben R. Finney and Eric M. Jones, eds., *Interstellar Migration and the Human Experience*
5. Wildon Fickett, *Introduction to Detonation Theory*
6. Grant Heiken and Kenneth Wohletz, *Volcanic Ash*
7. N. Metropolis, D. H. Sharp, W. J. Worlton, and K. R. Ames, eds., *Frontiers of Supercomputing*
8. Charles L. Mader, *Numerical Modeling of Water Waves*
9. S. Kass, J. Patera, R. Moody, and R. Slansky, *Affine Lie Algebras, Weight Multiplicities, and Branching Rules*
10. S. M. Ulam, *Analogies between Analogies: The Mathematical Reports of S. M. Ulam and His Los Alamos Collaborators*
11. Torlief E. O. Ericson, Vernon W. Hughes, Darragh E. Nagel, and John C. Allred, *The Meson Factories*
12. Karyn R. Ames and Alan Brenner, eds., *Frontiers of Supercomputing II: A National Reassessment*

Frontiers of Supercomputing II

A National Reassessment

Edited by
Karyn R. Ames
and
Alan Brenner

UNIVERSITY OF CALIFORNIA PRESS
Berkeley   Los Angeles   London

University of California Press
Berkeley and Los Angeles, California

University of California Press
London, England

Copyright © 1994 by The Regents of the University of California

Library of Congress Cataloging-in-Publication Data

Frontiers of supercomputing II: a national reassessment / edited by
Karyn R. Ames and Alan Brenner.
    p. cm.—(Los Alamos series in basic and applied sciences; 12)
    Papers from the 2nd Frontiers of Supercomputing Conference held at Los Alamos National Laboratory, 8/20–24/90.
    Includes bibliographical references.
    ISBN 0–520–08401–2 (acid-free paper)
    1. Supercomputers—Congresses. I. Ames, Karyn R. II. Brenner, Alan. III. Frontiers of Supercomputing Conference (2nd: 1990: Los Alamos National Laboratory). IV. Title: Frontiers of supercomputing two. V. Title: Frontiers of supercomputing 2. VI. Series.
QA76.88.F76   1994
338.4'700411'0973—dc20   93-29197
                                                                                   CIP

Printed in the United States of America

1 2 3 4 5 6 7 8 9

The paper used in this publication meets the minimum requirements of American National Standard for Information Sciences—Permanence of Paper for Printed Library Materials, ANSI Z39.48–1984 ∞

Contents

Preface ix

Acknowledgments xi

1 Opening, Background, and Questions Posed for This Conference

Welcome 3
  Sig Hecker
Supercomputing as a National Critical Technologies Effort 7
  Senator Jeff Bingaman
Goals for Frontiers of Supercomputing II and Review of
Events since 1983 15
  Kermith Speierman
Current Status of Supercomputing in the United States 21
  Erich Bloch

2 Technology Perspective

Overview 35
  Robert Cooper
Supercomputing Tools and Technology 37
  Tony Vacca
High-Performance Optical Memory Technology at MCC 41
  John Pinkston
Digital Superconductive Electronics 47
  Fernand Bedard
Enabling Technology: Photonics 55
  Alan Huang

3 Vector Pipeline Architecture

Vector Architecture in the 1990s 65
  Les Davis
In Defense of the Vector Computer 69
  Harvey Cragon
Market Trends in Supercomputing 73
  Neil Davenport
Massively Parallel SIMD Computing on Vector Machines
Using PASSWORK 77
  Ken Iobst
Vectors Are Different 85
  Steven J. Wallach

4 Scalable Parallel Systems

Symbolic Supercomputing 95
  Alvin Despain
Parallel Processing: Moving into the Mainstream 101
  H. T. Kung
It's Time to Face Facts 115
  Joe Brandenburg
Large-Scale Systems and Their Limitations 123
  Dick Clayton
A Scalable, Shared-Memory, Parallel Computer 129
  Burton Smith
Looking at All of the Options 135
  Jerry Brost

5 Systems Software

Parallel Software 141
  Fran Allen
Supercomputer Systems-Software Challenges 147
  David L. Black
Future Supercomputing Elements 155
  Bob Ewald
Compiler Issues for TFLOPS Computing 163
  Ken Kennedy
Performance Studies and Problem-Solving Environments 169
  David Kuck
Systems and Software 177
  George Spix

6 User-Interface Software

Parallel Architecture and the User Interface 183
    Gary Montry
Object-Oriented Programming, Visualization, and
User-Interface Issues 191
    David Forslund
Software Issues at the User Interface 199
    Oliver McBryan
What Can We Learn from Our Experience with Parallel
Computation up to Now? 221
    Jack T. Schwartz

7 Algorithms for High-Performance Computing

Parallel Algorithms and Implementation Strategies on
Massively Parallel Supercomputers 227
    R. E. Benner
The Interplay between Algorithms and Architectures: Two
Examples 239
    Duncan Buell
Linear Algebra Library for High-Performance Computers 243
    Jack Dongarra
Design of Algorithms 259
    C. L. Liu
Computing for Correctness 267
    Peter Weinberger

8 The Future Computing Environment

Interactive Steering of Supercomputer Calculations 275
    Henry Fuchs
A Vision of the Future at Sun Microsystems 287
    Bill Joy
On the Future of the Centralized Computing Environment 293
    Karl-Heinz A. Winkler
Molecular Nanotechnology 299
    Ralph Merkle
Supercomputing Alternatives 311
    Gordon Bell

9 Industrial Supercomputing

Overview of Industrial Supercomputing 333
  Kenneth W. Neves
Shell Oil Supercomputing 347
  Patric Savage
Government's High Performance Computing Initiative
Interface with Industry 353
  Howard E. Simmons
An Overview of Supercomputing at General Motors
Corporation 359
  Myron Ginsberg
Barriers to Use of Supercomputers in the Industrial
Environment 373
  Robert Hermann

10 Government Supercomputing

Planning for a Supercomputing Future 379
  Norm Morse
High-Performance Computing at the National
Security Agency 387
  George Cotter
The High Performance Computing Initiative: A Way to Meet
NASA's Supercomputing Requirements for Aerospace 393
  Vic Peterson
The Role of Computing in National Defense Technology 399
  Bob Selden
NSF Supercomputing Program 403
  Larry Smarr

11 International Activity

A Look at Worldwide High-Performance Computing and Its
Economic Implications for the U.S. 413
  Robert Borchers
  Seymour Goodman
  Michael Harrison
  Alan McAdams
  Emilio Millán
  Peter Wolcott
Economics, Revelation, Reality, and Computers 435
  Herbert E. Striner

12 Experience and Lessons Learned

Supercomputing since 1983 449
  Lincoln Faurer
Lessons Learned 453
  Ben Barker
The John von Neumann Computer Center: An Analysis 469
  Al Brenner
Project THOTH: An NSA Adventure in Supercomputing,
1984–88 481
  Larry Tarbell
The Demise of ETA Systems 489
  Lloyd Thorndyke
FPS Computing: A History of Firsts 497
  Howard Thrailkill

13 Industry Perspective: Policy and Economics for High-Performance Computing

Why Supercomputing Matters: An Analysis of the Economic
Impact of the Proposed Federal High Performance
Computing Initiative 505
  George Lindamood
Government as Buyer and Leader 515
  Neil Davenport
Concerns about Policies and Economics for High-Performance
Computing 517
  Steven J. Wallach
High-Performance Computing in the 1990s 521
  Sheryl L. Handler
A High-Performance Computing Association to Help the
Expanding Supercomputing Industry 525
  Richard Bassin
The New Supercomputer Industry 529
  Justin Rattner
The View from DEC 535
  Sam Fuller
Industry Perspective: Remarks on Policy and Economics for
High-Performance Computing 541
  David Wehrly

14 What Now?

Conference Summary 549
  David B. Nelson
The High Performance Computing Initiative 557
  Eugene Wong
Government Bodies as Investors 563
  Barry Boehm
Realizing the Goals of the HPCC Initiative: Changes Needed 569
  Charles Brownstein
The Importance of the Federal Government's Role in
High-Performance Computing 575
  Sig Hecker
Legislative and Congressional Actions on High-Performance
Computing and Communications 579
  Paul G. Huray
The Federal Role as Early Customer 589
  David B. Nelson
A View from the Quarter-Deck at the National Security
Agency 593
  Admiral William Studeman
Supercomputers and Three-Year-Olds 599
  Al Trivelpiece
NASA's Use of High-Performance Computers: Past, Present,
and Future 603
  Vice Admiral Richard H. Truly
A Leadership Role for the Department of Commerce 607
  Robert White
Farewell 611
  Senator Pete Domenici

Contributors 615

Preface

In 1983, Los Alamos National Laboratory cosponsored the first Frontiers of Supercomputing conference and, in August 1990, cosponsored Frontiers of Supercomputing II: A National Reassessment, along with the National Security Agency, the Defense Advanced Research Projects Agency, the Department of Energy, the National Aeronautics and Space Administration, the National Science Foundation, and the Supercomputing Research Center.

Continued leadership in supercomputing is vital to U.S. technological progress, to domestic economic growth, to international industrial competitiveness, and to a strong defense posture. In the seven years that passed since the first conference, the U.S. was able to maintain this lead, although that lead has significantly eroded in several key areas. To help maintain and extend a leadership position, the 1990 conference aimed to facilitate a national reassessment of U.S. supercomputing and of the economic, technical, educational, and governmental barriers to continued progress. The conference addressed events and progress since 1983, problems in the U.S. supercomputing industry today, R&D priorities for high-performance computing in the U.S., and policy at the national level.

The challenges in 1983 were to develop computer hardware and software based on parallel processing, to build a massively parallel computer, and to write new schemes and algorithms for such machines. In the 1990s, the dream of computers with parallel processors is being realized. Some computers, such as Thinking Machines Corporation's Connection Machine, have more than 65,000 parallel processors and thus are massively parallel.

Participants and speakers at the 1990 conference included senior managers and policy makers, chief executive officers and presidents of companies, computer vendors, industrial users, U.S. senators, high-level federal officials, national laboratory directors, and renowned academicians.

The discussions published here incorporate much of the widely ranging, often spontaneous, and invariably lively exchanges that took place among this diverse group of conferees.

Specifically, *Frontiers of Supercomputing II* features presentations on the prospects for and limits of hardware technology, systems architecture, and software; new mathematical models and algorithms for parallel processing; the structure of the U.S. supercomputing industry for competition in today's international industrial climate; the status of U.S. supercomputer use; and highlights from the international scene. The proceedings conclude with a session focused on government initiatives necessary to preserve and extend the U.S. lead in high-performance computing.

Conferees faced a new challenge—a dichotomy in the computing world. The supercomputers of today are huge, centrally located, expensive mainframes that "crunch numbers." These computers are very good at solving intensive calculations, such as those associated with nuclear weapons design, global climate, and materials science. Some computer scientists consider these mainframes to be dinosaurs, and they look to the powerful new microcomputers, scientific workstations, and minicomputers as the "supercomputers" of the future. Today's desktop computers can be as powerful as early versions of the Cray supercomputers and are much cheaper than mainframes.

Conference participants expressed their views that the mainframes and the powerful new microcomputers have complementary roles. The challenge is to develop an environment in which the ease and usefulness of desktop computers are tied to the enormous capacity and performance of mainframes. Developments must include new user interfaces, high-speed networking, graphics, and visualization. Future users may sit at their desktop computers and, without knowing it, have their work parceled out to mainframes, or they may access databases around the world.

Los Alamos National Laboratory and the National Security Agency wish to thank all of the conference cosponsors and participants. The 1990 conference was a tremendous success. When the next Frontiers of Supercomputing conference convenes, the vision of a seamless, comprehensive computing environment may then be a reality. The challenge now is to focus the energies of government, industry, national laboratories, and universities to accomplish this task.

Acknowledgments

The second Frontiers of Supercomputing conference held at Los Alamos National Laboratory, Los Alamos, New Mexico, August 20–24, 1990, was a tremendous success, thanks to the participants. As colleagues in high-performance computing, the conference participants avidly interacted with each other, formed collaborations and partnerships, and channeled their talents into areas that complemented each other's activities. It was a dynamic and fruitful conference, and the conference organizers extend special thanks to all of the participants.

Lawrence C. Tarbell, Jr., of the National Security Agency (NSA) was one of the conference organizers. The other conference organizer was William L. "Buck" Thompson, Special Assistant to the Director of Los Alamos National Laboratory. Members of the organizing committee from Los Alamos were Andy White and Gary Doolen. The organizing committee members from the NSA were Norman Glick and Byron Keadle; from the Supercomputing Research Center, Harlow Freitag; from the National Science Foundation, Tom Weber; from the Department of Energy, Norm Kreisman; from the Defense Advanced Research Projects Agency, Stephen Squires; and from the National Aeronautics and Space Administration, Paul Smith.

The success of this conference was in no small measure due to Donila Martinez of Los Alamos National Laboratory. She became the nerve center of northern New Mexico in finding places for conference participants to stay and in taking care of myriad conference preparation details.

Thanks also go to Kermith Speierman from NSA. He was the inspiration for the first Frontiers of Supercomputing conference in 1983 and was to a great extent the inspiration for this second conference, as well.

Nick Metropolis can clearly be called one of the true fathers of computing. He was in Los Alamos in the very early days, during the Manhattan Project, and he became the person in charge of building the MANIAC computer. He can tell you about the dawn of parallel processing.

You might think we are just entering that era. It actually began in Los Alamos about 50 years ago, when teams of people were operating mechanical calculators in parallel.

All recording and transcription of the conference was done by Steven T. Brenner, a registered professional reporter. Kyle T. Wheeler of the Computing and Communications Division at Los Alamos National Laboratory provided guidance on computing terminology.

Lisa Rothrock, an editor with B. I. Literary Services, in Los Alamos, New Mexico, gave much-needed editorial assistance for the consistency, clarity, and accuracy of these proceedings. Page composition and layout were done by Wendy Burditt, Chuck Calef, and Kathy Valdez, compositors at the Los Alamos National Laboratory Information Services Division. Illustrations were prepared for electronic placement by Linda Gonzales and Jamie Griffin, also of the Los Alamos National Laboratory Information Services Division.

1

Opening, Background, and Questions Posed for This Conference

Sig Hecker, Director of Los Alamos National Laboratory, welcomed attendees to the conference and introduced Senator Bingaman for a welcome speech. Kermith Speierman of the National Security Agency reviewed events since the last Frontiers of Supercomputing conference (1983), set the goals of the current conference, and charged the participants to meet those goals. The keynote address was given by Erich Bloch, who presented his perspective on the current status of supercomputing in the United States.

Session Chair

Larry Tarbell, National Security Agency

Buck Thompson, Los Alamos National Laboratory

Welcome

Sig Hecker

Siegfried S. Hecker is the Director of Los Alamos National Laboratory, in Los Alamos, New Mexico, a post he has held since January 1986. Dr. Hecker joined the Laboratory as a Technical Staff Member in the Physical Metallurgy Group in 1973 and subsequently served as Chairman of the Center for Materials Science and Division Leader of Materials Science and Technology. He began his professional career at Los Alamos in 1968 as a Postdoctoral Appointee. From 1970 to 1973, he worked as a Senior Research Metallurgist at General Motors Research Laboratories. He earned his Ph.D. in metallurgy from Case Western Reserve University in 1968.

Dr. Hecker received the Department of Energy's E. O. Lawrence Award for Materials Science in 1984. In 1985, he was cited by Science Digest as one of the year's top 100 innovators in science. In October of 1989, he delivered the Distinguished Lecture in Materials and Society for the American Society for Metals. The American Institute of Mining, Metallurgical, and Petroleum Engineers awarded him the James O. Douglas Gold Medal in 1990.

Among the scientific organizations in which Dr. Hecker serves is the Leadership/Applications to Practice Committee of the Metallurgical Society, the Board of Directors of the Council on Superconductivity for American Competitiveness, and the Board of Advisors of the Santa Fe Institute. Public-service agencies in which he is active include the

University of New Mexico Board of Regents, the Board of Directors of Carrie Tingley Hospital in Albuquerque, the Los Alamos Area United Way Campaign, and the Los Alamos Ski Club, of which he is President.

Welcome to Los Alamos and to New Mexico. I think most of you know that it was in 1983—in fact, seven years ago this week—that we held the first Frontiers of Supercomputing conference here at Los Alamos under the sponsorship of Los Alamos National Laboratory and the National Security Agency (NSA) to assess the critical issues that face supercomputing. Today we are here to make a national reassessment of supercomputing. The expanded number of sponsors alone, I think, reflects the increased use of supercomputing in the country. The sponsors of this conference are NSA, Los Alamos National Laboratory, the Defense Advanced Research Projects Agency, the Department of Energy, the National Science Foundation, and the Supercomputing Research Center.

I want to make a few brief remarks, both about the conference, as well as computing at the Laboratory. I found it very interesting to go back and look through the first *Frontiers of Supercomputing* book. Several things haven't changed at all since the last conference. K. Speierman, in his conference summary, pointed out very nicely that increased computational power will allow us to make significant advances in science, particularly in nonlinear phenomena. Supercomputing, we pointed out at the first conference, also will improve our technology and allow us to build things more efficiently. That certainly remains ever so true today. Indeed, leadership in high-performance computing is obviously vital to U.S. military and economic competitiveness.

In the preface to *Frontiers of Supercomputing* (Metropolis et al. 1986), the conference participants indicated that it will take radical changes in computer architecture, from single to massively parallel processors, to keep up with the demand for increased computational power. It was also fascinating that the authors at that time warned that the importance of measures to more effectively use available hardware cannot be overemphasized, namely measures such as improved numerical algorithms and improved software. Once again, these comments remain ever so true today.

However, there are a number of things that have changed since 1983. I think we have seen a substantial increase in parallel processing. At the Laboratory today, the CRAY Y-MPs are the workhorses for our computations. We have also made great progress in using the massively parallel

Connection Machines, from Thinking Machines Corporation, to solve demanding applications problems.

I think all the way around, in the country and in the world, we have seen a revolution in the computing environment, namely, that the personal computer has come into its own—to the tune of about 50 million units in the decade of the 1980s. That number includes one user, my eight-year-old daughter, who now has computational power at her fingertips that scientists wish they would have had a decade or two ago. Also, the trend toward high-power scientific workstations, networking, and ultra-high-speed graphics will forever change the way we do computing.

Another thing that hasn't changed, however, is the insatiable appetite of scientists who want more and more computing power. Seven years ago we had a few CRAY-1s at Los Alamos, and, just to remind you, that was only seven years after Seymour Cray brought serial number 1 to Los Alamos back in 1976. Today we have about 65 CRAY-1 equivalents, plus a pair of Connection Machine 2s. Nevertheless, I constantly hear the cry for more computational horsepower. At Los Alamos, that need is not only for the defense work we do but also for many other problems, such as combustion modeling or enhanced oil recovery or global climate change or how to design materials from basic principles.

However, a fundamental change has occurred. I think today, to remain at the forefront of computing, we can't simply go out and buy the latest model of supercomputer. We clearly will have to work smarter, which means that we'll have to work much more in conjunction with people at universities and with the computer and computational equipment manufacturers.

Therefore, I look forward to this reassessment in Frontiers of Supercomputing II, and I think it will be an interesting week. Typically, it's the people who make a conference. And as I look out at the audience, I feel no doubt that this will be a successful conference.

It is my pleasure this morning to introduce the person who will officially kick off the conference. We are very fortunate to have Senator Jeff Bingaman of New Mexico here. Senator Bingaman also played a similar role at the conference in 1983, shortly after he was elected to the United States Senate.

Senator Bingaman grew up in Silver City, a little town in the southern part of the state. He did his undergraduate work at Harvard and received a law degree from Stanford University. He was Attorney General for the State of New Mexico before being elected to the United States Senate.

I have had the good fortune of getting to know Senator Bingaman quite well in the past five years. He certainly is one of the greatest advocates for science and technology in the United States Congress. He serves on the Senate Armed Services Committee and also on the Senate Energy and Natural Resources Committee. On the Armed Services Committee, he heads the Subcommittee on Defense Industry and Technology. In both of those committees, he has been a strong advocate for science and technology in the nation, and particularly in Department of Defense and Department of Energy programs. In the Armed Services subcommittee, he spearheaded an effort to focus on our critical technologies and competitiveness, both from a military, as well as an economic, standpoint. And of course, there is no question that supercomputing is one of those critical technologies.

Thus, it is most appropriate to have Senator Bingaman here today to address this conference, and it's my honor and pleasure to welcome him to Los Alamos.

Reference

Frontiers of Supercomputing, N. Metropolis, D. H. Sharp, W. J. Worlton, and K. R. Ames, Eds., University of California Press, Berkeley, California (1986).

… *Opening, Background, and Questions Posed for This Conference* … 7

Supercomputing as a National Critical Technologies Effort

Senator Jeff Bingaman

Senator Jeff Bingaman (D-NM) began his law career as Assistant New Mexico Attorney General in 1969. In 1978 he was elected Attorney General of New Mexico. Jeff was first elected to the United States Senate in 1982 and reelected in 1988. In his two terms, Jeff has focused on restoring America's economic strength, preparing America's youth for the 21st century, and protecting our land, air, and water for future generations.

Jeff was raised in Silver City, New Mexico, and attended Harvard University, graduating in 1965 with a bachelor's degree in government. He then entered Harvard University Law School, graduating in 1968. Jeff served in the Army Reserves from 1968 to 1974.

It is a pleasure to be here and to welcome everyone to Los Alamos and to New Mexico.

I was very fortunate to be here seven years ago, when I helped to open the first Frontiers of Supercomputing conference on a Monday morning in August, right here in this room. I did look back at the remarks I made then, and I'd like to cite some of the progress that has been made since then and also indicate some of the areas where I think we perhaps are still in the same ruts we were in before. Then I'll try to put it all in a little broader context of how we go about defining a rational technology policy for the entire nation in this post-Cold War environment.

Back in 1983, I notice that my comments then drew particular attention to the fact that Congress was largely apathetic and inattentive to the challenge that we faced in next-generation computing. The particular fact or occurrence that prompted that observation in 1983 was that the Defense Advanced Research Projects Agency's (DARPA's) Strategic Computing Initiative, which was then in its first year, had been regarded by some in Congress as a "bill payer"—as one of those programs that you can cut to pay for supposedly higher-priority strategic weapons programs. We had a fight that year while I worked with some people in the House to try to maintain the $50 million request that the Administration had made for funding the Strategic Computing Program for DARPA.

Today, I do think that complacency is behind us. Over the past seven years, those of you involved in supercomputing/high-performance supercomputing have persuasively made the case both with the Executive Branch and with the Congress that next-generation computers are critical to the nation's security and to our economic competitiveness. More importantly, you have pragmatically defined appropriate roles for government, industry, and academia to play in fostering development of the key technologies needed for the future and—under the leadership of the White House Science Office, more particularly, of the Federal Coordinating Committee on Science, Engineering, and Technology (FCCSET)—development of an implementation plan for the High Performance Computing Initiative.

That initiative has been warmly received in Congress. Despite the fact that we have cuts in the defense budget this year and will probably have cuts in the next several years, both the Senate Armed Services Committee and the House Armed Services Committee have authorized substantial increases in DARPA's Strategic Computing Program. In the subcommittee that I chair, we increased funding $30 million above the Administration's request, for a total of $138 million this next year. According to some press reports I've seen, the House is expected to do even better.

Similarly, both the Senate Commerce Committee and the Senate Energy Committee have reported legislation that provides substantial five-year authorizations for NSF at $650 million, for NASA at $338 million, and for the Department of Energy (DOE) at $675 million, all in support of a national high-performance computing program. Of course, the National Security Agency and other federal agencies are also expected to make major contributions in the years ahead.

Senator Al Gore deserves the credit for spearheading this effort, and much of what each of the three committees that I've mentioned have done follows the basic blueprint laid down in S. B. 1067, which was a bill introduced this last year that I cosponsored and strongly supported. Mike Nelson, of Senator Gore's Commerce Committee staff, will be spending the week with you and can give you better information than I can on the prospects in the appropriations process for these various authorizations.

One of the things that has struck me about the progress in the last seven years is that you have made the existing institutional framework actually function. When I spoke in 1983, I cited Stanford University Professor Edward Feigenbaum's concern (expressed in his book *The Fifth Generation*) that the existing U.S. institutions might not be up to the challenge from Japan and his recommendation that we needed a broader or bolder institutional fix to end the "disarrayed and diffuse indecision" he saw in this country and the government. I think that through extraordinary effort, this community, that is, those of you involved in high-performance supercomputing, have demonstrated that existing institutions can adapt and function. You managed to make FCCSET work at a time when it was otherwise moribund. You've been blessed with strong leadership in some key agencies. I'd like to pay particular tribute to Craig Fields at DARPA and Erich Bloch at NSF. Erich is in his last month of a six-year term as the head of NSF, and I believe he has done an extraordinary job in building bridges between the academic world, industry, and international laboratories. His efforts to establish academic supercomputer centers and to build up a worldwide high-data-rate communications network are critical elements in the progress that has been made over the last seven years. Of course, those efforts were not made and those successes were not accomplished without a lot of controversy and complaints from those who felt their own fiefdoms were challenged.

On the industrial side, the computer industry has been extraordinarily innovative in establishing cooperative institutions. In 1983, both the Semiconductor Research Cooperative (SRC) and Microelectronics and Computer Technology Corporation (MCC) were young and yet unproved. Today SRC and MCC have solid track records of achievement, and MCC has had the good sense to attract Dr. Fields to Austin after his dismissal as head of DARPA, apparently for not pursuing the appropriate ideological line.

More recently, industry has put together a Computer Systems Policy Project, which involves the CEOs of our leading computer firms, to think through the key generic issues that face the industry. Last month, the R&D directors of that group published a critical technologies report outlining the key success factors that they saw to be determinative of U.S. competitiveness in the 16 critical technologies for that industry.

As I see it, all of these efforts have been very constructive and instructive for the rest of us and show us what needs to be done on a broader basis in other key technologies.

The final area of progress I will cite is the area I am least able to judge, namely, the technology itself. My sense is that we have by and large held our own as a nation vis-à-vis the rest of the world in competition over the past seven years. I base this judgment on the Critical Technology Plan—which was developed by the Department of Defense (DoD), in consultation with DOE—and the Department of Commerce's Emerging Technologies Report, both of which were submitted to Congress this spring. According to DoD, we are ahead of both Japan and Europe in parallel computer architectures and software producibility. According to the Department of Commerce report, we are ahead of both Japan and Europe in high-performance computing and artificial intelligence. In terms of trends, the Department of Commerce report indicates that our lead in these areas is accelerating relative to Europe but that we're losing our lead in high-performance computing over Japan and barely holding our lead in artificial intelligence relative to Japan.

Back in 1983, I doubt that many who were present would have said that we'd be as well off as we apparently are in 1990. There was a great sense of pessimism about the trends, particularly relative to Japan. The Japanese Ministry of International Trade and Industry (MITI) had launched its Fifth Generation Computer Project by building on their earlier national Superspeed Computer Project, which had successfully brought Fujitsu and Nippon Electric Corporation to the point where they were challenging Cray Research, Inc., in conventional supercomputer hardware. Ed Feigenbaum's book and many other commentaries at the time raised the specter that this technology was soon to follow consumer electronics and semiconductors as an area of Japanese dominance.

In the intervening years, those of you here and those involved in this effort have done much to meet that challenge. I'm sure all of us realize that the challenge continues, and the effort to meet it must continue. While MITI's Fifth Generation Project has not achieved its lofty goals, it has helped to build an infrastructure second only to our own in this critical field. Japanese industry will continue to challenge the U.S. for first place. Each time I've visited Japan in the last couple of years, I've made it a point to go to IBM Japan to be briefed on the progress of Japanese industry, and they have consistently reported solid progress being made there, both in hardware and software.

I do think we have more of a sense of realism today than we had seven years ago. Although there is no room for complacency in our nation about the efforts that are made in this field, I think we need to put aside the notion that the Japanese are 10 feet tall when it comes to developing technology. Competition in this field has helped both our countries. In multiprocessor supercomputers and artificial intelligence, we've spawned a host of new companies over the past seven years in this country. Computers capable to 10^{12} floating-point operations per second are now on the horizon. New products have been developed in the areas of machine vision, automatic natural-language understanding, speech recognition, and expert systems. Indeed, expert systems are now widely used in the commercial sector, and numerous new applications have been developed for supercomputers.

Although we are not going to be on top in all respects of supercomputing, I hope we can make a commitment to remain first overall and to not cede the game in any particular sector, even those where we may fall behind.

I have spent the time so far indicating progress that has been made since the first conference. Let me turn now to just a few of the problems I cited in 1983 and indicate some of those that still need to be dealt with.

The most fundamental problem is that you in the supercomputing field are largely an exception to our technology policy-making nationwide. You have managed through extraordinary effort to avoid the shoals of endless ideological industrial-policy debate in Washington. Unfortunately, many other technologies have not managed to avoid those shoals.

Let me say up front that I personally don't have a lot of patience for these debates. It seems to me our government is inextricably linked with industry through a variety of policy mechanisms—not only our R&D policy but also our tax policy, trade policy, anti-trust policy, regulatory policy, environmental policy, energy policy, and many more. The sum total of these policies defines government's relationship with each industry, and the total does add up to an industrial policy. This is not a policy for picking winners and losers among particular firms, although obviously we have gone to that extent in some specific cases, like the bailouts of Lockheed and Chrysler and perhaps in the current debacle in the savings and loan industry.

In the case of R&D policy, it is clearly the job of research managers in government and industry to pick winning technologies to invest in. Every governor in the nation, of both political parties, is trying to foster winning technologies in his or her state. Every other industrialized nation is doing the same. I don't think anybody gets paid or promoted for picking losing technologies.

Frankly, the technologies really do appear to pick themselves. Everyone's lists of critical technologies worldwide overlap to a tremendous degree. The question for government policy is how to insure that some U.S. firms are among the world's winners in the races to develop supercomputers, advanced materials, and biotechnology applications—to cite just three examples that show up on everybody's list.

In my view, the appropriate role for government in its technology policy is to provide a basic infrastructure in which innovation can take place and to foster basic and applied research in critical areas that involve academia, federal laboratories, and industry so that risks are reduced to a point where individual private-sector firms will assume the remaining risk and bring products to market. Credit is due to Allan D. Bromley, Assistant to the President for Science and Technology, for having managed to get the ideologues in the Bush Administration to accept a government role in critical, generic, and enabling technologies at a precompetitive stage in their development. He has managed to get the High Performance Computing Initiative, the Semiconductor Manufacturing Technology Consortium, and many other worthwhile technology projects covered by this definition.

Frankly, I have adopted Dr. Bromley's vocabulary—"critical, generic, enabling technologies at a precompetitive stage"—in the hope of putting this ideological debate behind us. In Washington we work studiously to avoid the use of the term "industrial policy," which I notice we used very freely in 1983. My hope is that if we pragmatically go about our business, we can get a broad-based consensus on the appropriate roles for government, industry, and academia in each of the technologies critical to our nation's future. You have, as a community, done that for high-performance supercomputing, and your choices have apparently passed the various litmus tests of a vast majority of members of both parties, although there are some in the Heritage Foundation and other institutions who still raise objections.

Now we need to broaden this effort. We need to define pragmatically a coherent, overall technology policy and tailor strategies for each critical technology. We need to pursue this goal with pragmatism and flexibility, and I believe we can make great headway in the next few years in doing so.

Over the past several years, I have been attempting to foster this larger, coherent national technology policy in several ways. Initially, we placed emphasis on raising the visibility of technology issues within both the Executive Branch and the Congress. The Defense Critical Technology Plan and the Emerging Technologies Report have been essential parts of raising the visibility of technological issues. Within industry I have tried to encourage efforts to come up with road maps for critical technologies, such as those of the Aerospace Industries Association, John Young's Council on Competitiveness, and the Computer Systems Policy Project. It is essential that discussion among government, industry, and academia be fostered and that the planning processes be interconnected at all levels, not just at the top.

At the top of the national critical technologies planning effort, I see the White House Science Office. Last year's Defense Authorization Bill established a National Critical Technologies Panel under Dr. Bromley, with representation from industry, the private sector, and government. They recently held their first meeting, and late this year they will produce the first of six biennial reports scheduled to be released between now and the year 2000. In this year's defense bill, we are proposing to establish a small, federally funded R&D center under the Office of Science and Technology Policy, which would be called the Critical Technologies Institute. The institute will help Dr. Bromley oversee the development of interagency implementation plans under FCCSET for each of the critical technologies identified in the national critical technologies reports (much like the plan on high-performance computing issued last year). Dr. Ed David, when he was White House Science Advisor under President Nixon, suggested to me that the approach adopted by the Federally Funded Research and Development Centers was the only way to insure stability and continuity in White House oversight of technology policy. After looking at various alternatives, I came to agree with him.

Of course, no structure is a substitute for leadership. I believe that the policy-making and reporting structure that we've put in place will make the job of government and industry leaders easier. It will ensure greater visibility for the issues, greater accountability in establishing and pursuing technology policies, greater opportunity to connect technology policy with the other government policies that affect the success or failure of U.S. industry, and greater coherence among research efforts in government, industry, and academia. That is the goal that we are pursuing.

I think we will find as we follow this path that no single strategy will be appropriate to each technology or to each industry. What worked for high-performance supercomputing will not transfer readily to advanced materials or to biotechnology. We will need to define appropriate roles in each instance in light of the existing government and industry structure in that technology. In each instance, flexibility and pragmatism will need to be the watchwords for our efforts.

My hope is that if another conference like this occurs seven years from now, we will be able to report that there is a coherent technology policy in place and that you in this room are no longer unique as having a White House-blessed implementation plan.

You may not feel you are in such a privileged position at this moment compared to other technologies, and you know better than I the problems that lie ahead in ensuring continued American leadership in strategic computing. I hope this conference will identify the barriers that remain in the way of progress in this field. I fully recognize that many of those barriers lie outside the area of technology policy. A coherent technology strategy on high-performance computing is necessary but clearly not sufficient for us to remain competitive in this area.

I conclude by saying I believe that you, and all others involved in high-performance supercomputing, have come a great distance in the last seven years and have much to be proud of. I hope that as a result of this conference you will set a sound course for the next seven years.

Thank you for the opportunity to meet with you, and I wish you a very productive week.

Opening, Background, and Questions Posed for This Conference   15

Goals for Frontiers of Supercomputing II and Review of Events Since 1983

Kermith Speierman

> *At the time of the first Frontiers of Supercomputing conference in 1983, Kermith H. "K." Speierman was the chief scientist at the National Security Agency (NSA), a position he held until 1990. He has been a champion of computing at all levels, especially of supercomputing and parallel processing. He played a major role in the last conference. It was largely through his efforts that NSA developed its parallel processing capabilities and established the Supercomputing Research Center.*

I would like to review with you the summary of the last Frontiers of Supercomputing conference in 1983. Then I would like to present a few representative significant achievements in high-performance computing over this past seven years. I have talked with some of you about these achievements and I appreciate your help. Last, I'd like to talk about the goals of this conference and share with you some questions that I think are useful for us to consider during our discussions.

1983 Conference Summary

In August of 1983, at the previous conference, we recognized that *there is a compelling need for more and faster supercomputers. The Japanese*, in fact, have shown that they *have a national goal in supercomputation and can achieve effective cooperation between government, industry, and academia* in their country. I think the Japanese shocked us a little in 1983, and we were a bit complacent then. However, I believe we are now guided more by our needs, our capabilities, and the idea of having a consistent, balanced program with other sciences and industry. So I think we've reached a level of maturity that is considerably greater than we had in 1983. I think U.S. vendors are now beginning, as a result of events that have gone on during this period, to be very serious about massively parallel systems, or what we now tend to call scalable parallel systems.

The only evident approach to achieve large increases over current supercomputer speeds is through massively parallel systems. However, there are some interesting ideas in other areas like optics that are exciting. But I think for this next decade we do have to look very hard at the scalable parallel systems.

We don't know how to use parallel architectures very well. The step from a few processors to large numbers is a difficult problem. It is still a challenge, but we now know a great deal more about using parallel processors on real problems. It is still very true that *much work is required on algorithms, languages, and software to facilitate the effective use of parallel architectures.*

It is also still true that *the vendors need a larger market for supercomputers to sustain an accelerated development program.* I think that may be a more difficult problem now than it was in 1983 because the cost of developing supercomputers has grown considerably. However, the world market is really not that big—it is approximately a $1 billion-per-year market. In short, the revenue base is still small.

Potential supercomputer applications may be far greater than current usage indicates. In fact, I think that the number of potential applications is enormous and continues to grow.

U.S. computer companies have a serious problem buying fast, bipolar memory chips in the U.S. We have to go out of the country for a lot of that technology. I think our companies have tried to develop U.S. sources more recently, and there has been some success in that. Right now, there is considerable interest in fast bipolar SRAMs. It will be interesting to see if we can meet that need in the U.S.

Packaging is a major part of the design effort. As speed increases, you all know, packaging gets to be a much tougher problem in almost a nonlinear way. That is still a very difficult problem.

Supercomputers are systems consisting of algorithms, languages, software, architecture, peripherals, and devices. They should be developed as systems that recognize the critical interaction of all the parts. You have to deal with a whole system if you're going to build something that's usable.

Collaboration among government, industry, and academia on supercomputer matters is essential to meet U.S. needs. The type of collaboration that we have is important. We need to find collaboration that is right for the U.S. and takes advantage of the institutions and the work patterns that we are most comfortable with. As suggested by Senator Jeff Bingaman in his presentation during this session, the *U.S. needs national supercomputer goals and a strategic plan to reach those goals.*

Events in Supercomputing since 1983

Now I'd like to talk about representative events that I believe have become significant in supercomputing since 1983. After the 1983 conference, the National Security Agency (NSA) went to the Institute for Defense Analyses (IDA) and said that they would like to establish a division of IDA to do research in parallel processing for NSA. We established the Supercomputing Research Center (SRC), and I think this was an important step.

Meanwhile, NSF established supercomputing centers, which provided increased supercomputer access to researchers across the country. There were other centers established in a number of places. For instance, we have a Parallel Processing Science and Technology Center that was set up by NSF at Rice University with Caltech and Argonne National Laboratory. NSF now has computational science and engineering programs that are extremely important in computational math, engineering, biology, and chemistry, and they really do apply this new paradigm in which we use computational science in a very fundamental way on basic problems in those areas.

Another event since 1983, scientific visualization, has become a really important element in supercomputing.

The start up of Engineering Technology Associates Systems (ETA) was announced at the 1983 banquet speech by Bill Norris. Unfortunately, ETA disbanded as an organization in 1989.

In 1983, Denelcor was a young organization that was pursuing an interesting parallel processing structure. Denelcor went out of business, but their ideas live on at Tera Computer Company, with Burton Smith behind them.

Cray Research, Inc., has trifurcated into three companies since 1983. One of those, Supercomputing Systems, Inc., is receiving significant technological and financial support from IBM, which is a very positive direction.

At this time, the R&D costs for a new supercomputer chasing very fast clock times are $200 or $300 million. I'm told that's about 10 times as much as it was 10 years ago.

Japan is certainly a major producer of supercomputers now, but they haven't run away with the market. We have a federal High Performance Computing Initiative that was published by the Office of Science and Technology Policy in 1989, and it is a result of the excellent interagency cooperation that we have. It is a good plan and has goals that I hope will serve us well.

The Defense Advanced Research Projects Agency's Strategic Computing Program began in 1983. It has continued on and made significant contributions to high-performance computing.

We now have the commercial availability of massively parallel machines. I hope that commercial availability of these machines will soon be a financial success.

I believe the U.S. does have a clear lead in parallel processing, and it's our job to take advantage of that and capitalize on it. There are a significant number of applications that have been parallelized, and as that set of applications grows, we can be very encouraged.

We now have compilers that produce parallel code for a number of different machines and from a number of different languages. The researchers tell me that we have a lot more to do, but there is good progress here. In the research community there are some new, exciting ideas in parallel processing and computational models that should be very important to us.

We do have a much better understanding now of interconnection nets and scaling. If you remember back seven years, the problem of interconnecting all these processors was of great concern to all of us.

There has been a dramatic improvement in microprocessor performance, I think primarily because of RISC architectures and microelectronics for very-large-scale integration. We have high-performance workstations now that are as powerful as CRAY-1s. We have special accelerator boards that perform in these workstations for special functions at very high rates. We have minisupercomputers that are both vector and scalable parallel machines. And UNIX is certainly becoming a standard for high-performance computing.

We are still "living on silicon." As a result, the supercomputers that we are going to see next are going to be very hot. Some of them may be requiring a megawatt of electrical input, which will be a problem.

I think there is a little flickering interest again in superconducting electronics, which provides a promise of much smaller delay-power products, which in turn would help a lot with the heat problem and give us faster switching speeds.

Conference Goals

Underlying our planning for this conference were two primary themes or goals. One was the national reassessment of high-performance computing—that is, how much progress have we made in seven years? The other was to have a better understanding of the limits of high-performance computing. I'd like to preface this portion of the discussion by saying that not all limits are bad. Some limits save our lives. But it is very important to understand limits. By limits, I mean speed of light, switching energy, and so on.

The reassessment process is one, I think, of basically looking at progress and understanding why we had problems, why we did well in some areas, and why we seemed to have more difficulties in others. Systems limits are questions of architectural structures and software. Applications limits are a question of how computer architectures and the organization of the system affect the kinds of algorithms and problems that you can put on those systems. Also, there are financial and business limits, as well as policy limits, that we need to understand.

Questions

Finally, I would like to pose a few questions for us to ponder during this conference. I think we have to address in an analytical way our ability to remain superior in supercomputing. Has our progress been satisfactory? Are we meeting the high-performance computing needs of science, industry, and government? What should be the government's role in high-performance computing?

Do we have a balanced program? Is it consistent? Are there some show-stoppers in it? Is it balanced with other scientific programs that the U.S. has to deal with? Is the program aggressive enough? What benefits will result from this investment in our country?

The Gartner report addresses this last question. What will the benefits be if we implement the federal High Performance Computing Initiative?

Finally, 1 want to thank all of you for coming to this conference. I know many of you, and 1 know that you represent the leadership in this business. I hope that we will have a very successful week.

Current Status of Supercomputing in the United States

Erich Bloch

Erich Bloch serves as a Distinguished Fellow at the Council on Competitiveness. Previously, he was the Director of the National Science Foundation. Early in his career, in the 1960s, Erich worked with the National Security Agency as the Program Manager of the IBM Stretch project, helping to build the fastest machine that could be built at that time for national security applications. At IBM, Erich was a strong leader in high-performance computing and was one of the key people who started the Semiconductor Research Cooperative.

Eric is chairman of the new Physical Sciences, Math, and Engineering Committee (an organ of the Federal Coordinating Committee on Science, Engineering, and Technology), which has responsibility for high-performance computing. He is also a member of the National Advisory Committee on Semiconductors and has received the National Medal of Technology from the President.

I appreciate this opportunity to talk about supercomputing and computers and technology. This is a topic of special interest to you, the National Science Foundation, and the nation.

But it is also a topic of personal interest to me. In fact, the Los Alamos Synchrotron Laboratory has special meaning for me. It was my second home during the late fifties and early sixties, when I was manager of IBM's Stretch Design and Engineering group.

How the world has changed! We had two-megabit—not megabyte—core memories, two circuit/plug-in units with a cycle time of 200 nanoseconds. Also, in pipelining, we had the first "interrupt mechanisms" and "look-ahead mechanisms."

But some things have stayed the same: cost overruns, not meeting specs, disappointing performance, missed schedules! It seems that these are universal rules of supercomputing.

But enough of this. What I want to do is talk about the new global environment, changes brought about by big computers and computer science, institutional competition, federal science and technology, and policy issues.

The Global Imperative

Never before have scientific knowledge and technology been so clearly coupled with economic prosperity and an improved standard of living. Where access to natural resources was once a major source of economic success, today access to technology—which means access to knowledge—is probably more important. Industries based primarily on knowledge and fast-moving technologies—such as semiconductors, biotechnology, and information technologies—are becoming the new basic industries fueling economic growth.

Advances in information technologies and computers have revolutionized the transfer of information, rendering once impervious national borders open to critical new knowledge. As the pace of new discoveries and new knowledge picks up, the speed at which knowledge can be accessed becomes a decisive factor in the commercial success of technologies.

Increasing global economic integration has become an undeniable fact. Even large nations must now look outward and deal with a world economy. Modern corporations operate internationally to an extent that was undreamed of 40 years ago. That's because it would have been impossible to operate the multinational corporations of today without modern information, communications, and transportation technologies.

Moreover, many countries that were not previously serious players in the world economy are now competitors. Global economic integration has been accompanied by a rapid diffusion of technological capability in the form of technically educated people. The United States, in a dominant position in nearly all technologies at the end of World War II, is now only one producer among many. High-quality products now come from countries that a decade or two ago traded mainly in agricultural products or raw materials.

Our technical and scientific strength will be challenged much more directly than in the past. Our institutions must learn to function in this environment. This will not be easy.

Importance of Computers—The Knowledge Economy

Amid all this change, computing has become a symbol for our creativity and productivity and a barometer in the effort to maintain our competitive position in the world arena. The development of the computer, and its spread through industry, government, and education, has brought forth the emergence of knowledge as the critical new commodity in today's global economy. In fact, computers and computer science have become the principal enabling technology of the knowledge economy.

Supercomputers, in particular, are increasingly important to design and manufacturing processes in diverse industries: oil exploration, aeronautics and aerospace, pharmaceuticals, energy, transportation, automobiles, and electronics, just to name the most obvious examples. They have become an essential instrument in the performance of research, a new tool to be used alongside modeling, experimentation, and theory, that pushes the frontiers of knowledge, generates new ideas, and creates new fields. They are also making it possible to take up old problems—like complex-systems theory, approaches to nonlinear systems, genome mapping, and three-dimensional modeling of full aircraft configurations—that were impractical to pursue in the past.

We are only in the beginning of a general exploitation of supercomputers that will profoundly affect academia, industry, and the service sector. During the first 30 years of their existence, computers fostered computer science and engineering and computer architecture. More recently, we have seen the development of computational science and engineering as a means of performing sophisticated research and design tasks. Supercomputer technology and network and graphics technology, coupled with mathematical methods for algorithms, are the basis for this development.

Also, we have used the von Neumann architecture for a long time. Only recently is a new approach in massive parallelism developing. The practical importance of supercomputers will continue to increase as their technological capabilities advance, their user access improves, and their use becomes more simple.

Computers—A Historic Perspective

Let's follow the development of computing for a moment. The computer industry is an American success story—the product of our ingenuity and of a period of unquestioned market and technological leadership in the first three and a half decades after World War II.

What did we do right?

First, we had help from historical events. World War II generated research needs and a cooperative relationship among government, academia, and the fledgling computer industry. Government support of computer research was driven by the Korean War and the Cold War. Federal funding was plentiful, and it went to commercially oriented firms capable of exploiting the technology for broader markets.

But we had other things going for us as well. There were important parallel developments and cross-feeding between electronics, materials, and electromechanics. There was a human talent base developed during the war. There was job mobility, as people moved from government labs to industry and universities, taking knowledge of the new technologies with them.

There was also a supportive business climate. U.S. companies that entered the field—IBM, Sperry Corporation, National Cash Register, Burroughs—were able to make large capital investments. And there was an entrepreneurial infrastructure eager to exploit new ideas.

Manufacturing and early automation attempts had a revolutionary impact on the progress of computer development. It's not fully appreciated that the mass production of 650s, 1401s, and later, 7090s and 360s set the cost/performance curve of computers on its precipitous decline and assured technology preeminence.

Industry leaders were willing to take risks and play a hunch. Marketing forecasts did not justify automation; IBM proceeded on faith and demonstrated that the forecasts were consistently on the low side. A typical assessment of the time was that "14 supercomputers can satisfy the world demand."

We had another thing going for us—our university research enterprise. Coupling research and education in the universities encouraged human talent at the forefront of the computer field and created computer departments at the cutting edge of design and construction: Illinois, MIT, IAS, and the University of Pennsylvania.

Clearly, it was the right mix of elements. But there was nothing inevitable about our successful domination of the field for the last 30 years. That was partly attributable to the failures of our competitors.

England provides a good case study of what can go wrong. It had the same basic elements we had:
- the right people (Turing, Kilburn);
- good universities (Manchester, Cambridge, Edinburgh); and
- some good companies (Ferranti, Lyons).

So why did it not compete with us in this vital industry? One reason, again, is history. World War II had a much more destructive effect on Britain than on us. But there were more profound reasons. The British government was not aggressive in supporting this new development. As Kenneth Flam points out, the British defense establishment was less willing than its American counterpart to support speculative and risky high-tech ventures.

The British government did not assume a central role in supporting university research. British industry was also more conservative and the business climate less favorable. The home market was too small; industry was unable to produce and market a rapidly changing technology, and it did not recognize the need to focus on manufacturability. Finally, there was less mobility of talented people between government, industry, and universities. In fact, there was more of a barrier to educating enough people in a new technological world than in the U.S.

Why bring up this old history? Because international competition in computing is greater, and the stakes higher, than ever before. And it is not clear that we are prepared to meet this competition or that our unique advantages of the 1950s exist today:
- Government policy toward high-risk, high-technology industries is less clear than in the 1950s. The old rationale for close cooperation—national defense—is no longer as compelling. Neither is defense the same leading user of high technology it once was.
- The advantage of our large domestic market is now rivaled by the European Economic Community (EEC) and the Pacific Rim countries.
- Both Japan and the EEC are mounting major programs to enhance their technology base, while our technology base is shrinking.
- Japan, as a matter of national policy, is enhancing cooperation between industry and universities—not always their own universities but sometimes ours.
- Industry is less able and willing to take the risk that IBM and Sperry did in the 1950s. The trend today is toward manipulating the financial structure for short-term profits.
- Finally, although the stakes and possible gains are tremendous, the costs of developing new generations of technology have risen beyond the ability of all but the largest and strongest companies, and sometimes of entire industries, to handle.

Corrective Action

What should we do so that we do not repeat the error of Great Britain in the 1950s? Both the changing global environment and increasing foreign competition should focus our attention on four actions to ensure that our economic performance can meet the competition.

First, we must make people—including well-educated scientists and engineers and a technically literate work force and populous—the focus of national policy. Nothing is more important than developing and using our human resources effectively.

Second, we must invest adequately in research and development.

Third, we must learn to cooperate in developing precompetitive technology in cases where costs may be prohibitive or skills lacking for individual companies or a even an industry.

Fourth, we must have access to new knowledge, both at home and abroad.

Let me discuss each of these four points.

Human Resources

People are the crucial resource. People generate the knowledge that allows us to create new technologies. We need more scientists and engineers, but we are not producing them.

In the last decade, employment of scientists and engineers grew three times as fast as total employment and twice as fast as total professional employment. Most of this growth was in the service sector, in which employment of scientists and engineers rose 5.7 per cent per year for the last decade. But even in the manufacturing sector, where there was no growth at all in total employment, science and engineering employment rose four per cent per year, attesting to the increasing technical complexity of manufacturing.

So there is no doubt about the demand for scientists and engineers. But there is real doubt that the supply will keep up. The student population is shrinking, so we must attract a larger proportion of students into science and engineering fields just to maintain the current number of graduates.

Unfortunately, the trend is the other way. Freshman interest in engineering and computer sciences decreased during the 1980s, but it increased for business, humanities, and the social sciences. Baccalaureates in mathematics and computer science peaked in 1986 and have since declined over 17 per cent. Among the physical and biological sciences, interest has grown only marginally.

In addition, minorities and women are increasingly important to our future work force. So we must make sure these groups participate to their fullest in science and engineering. But today only 14 per cent of female students, compared to 25 per cent of male students, are interested in the natural sciences and engineering in high school. By the time these students receive their bachelor's degrees, the number of women in these fields is less than half that of men. Only a tiny fraction of women go on to obtain Ph.Ds.

The problem is even worse among Blacks, Native Americans, and Hispanics at every level—and these groups are a growing part of our population. Look around the room and you can see what I mean.

To deal with our human-resources problem, NSF has made human resources a priority, with special emphasis on programs to attract more women and minorities. At the precollege level, our budget has doubled since 1984, with many programs to improve math and science teachers and teaching. At the undergraduate level, NSF is developing new curricula in engineering, mathematics, biology, chemistry, physics, computer sciences, and foreign languages. And we are expanding our Research for Undergraduates Program.

My question to you is, how good are our education courses in computer science and engineering? How relevant are they to the requirements of future employers? Do they reflect the needs of other disciplines for new computational approaches?

R&D Investment

In the U.S., academic research is the source of most of the new ideas that drive innovation. Entire industries, including semiconductors, biotechnology, computers, and many materials areas, are based on research begun in universities.

The principal supporter of academic research is the federal government. Over the last 20 years, however, we have allowed academic research to languish. As a per cent of gross national product, federal support for academic research declined sharply from 1968 to 1974 and has not yet recovered to the 1968 level. Furthermore, most of the recent growth has occurred in the life sciences. Federal investment in the physical sciences and engineering, the fields that are most critical for competitive technologies, has stagnated. As a partial solution to this problem, NSF and the Administration have pressed for a doubling of the NSF budget by 1993. This would make a substantial difference and is essential to our technological and economic competitiveness.

We must also consider the balance between civilian and defense R&D. Today, in contrast to the past, the commercial sector is the precursor of leading-edge technologies, whereas defense research has become less critical to spawning commercial technology.

But this shift is not reflected in federal funding priorities. During the 1980s, the U.S. government sharply increased its investment in defense R&D as part of the arms buildup. Ten years ago, the federal R&D investment was evenly distributed between the defense and civilian sectors. Today the defense sector absorbs about 60 per cent. In 1987 it was as high as 67 or 68 per cent.

In addition to the federal R&D picture, we must consider the R&D investments made by industry, which has the prime responsibility for technology commercialization. Industry cannot succeed without strong R&D investments, and recently industry's investment in R&D has declined in real terms. It's a moot point whether the reason was the leveraged buyout and merger binge or shortsighted management action or something else. The important thing is to recognize the problem and begin to turn it around.

Industry must take advantage of university research, which in the U.S. is the wellspring of new concepts and ideas. NSF's science and technology centers, engineering research centers, and supercomputer centers are designed with this in mind, namely, multidisciplinary, relevant research with participation by the nonacademic sector.

But on a broader scale, the High Performance Computing Initiative developed under the direction of the Office of Science and Technology Policy requires not only the participation of all concerned agencies and industry but everybody's participation, especially that of the individuals and organizations here today.

Technology Strategy

Since World War II the federal government has accepted its role as basic research supporter. But it cannot be concerned with basic research, only. The shift to a world economy and the development of technology has meant that in many areas the scale of technology development has grown to the point where, at least in some cases, industry can no longer support it alone.

The United States, however, has been ambivalent about the government role in furthering the generic technology base, except in areas such as defense, in which government is the main customer. In contrast, our foreign competitors often have the advantage of government support, which reduces the risk and assures a long-term financial commitment.

Nobody questions the government's role of ensuring that economic conditions are suitable for commercializing technologies. Fiscal and monetary policies, trade policies, R&D tax and antitrust laws, and interest rates are all tools through which the government creates the financial and regulatory environment within which industry can compete. But this is not enough. In addition, government and industry, together, must cooperate in the proper development of generic precompetitive technology in areas where it is clear that individual companies or private consortia are not able to do the job.

In many areas, the boundary lines between basic research and technology are blurring, if not overlapping completely. In these areas, generic technologies at their formative stages are the base for entire industries and industrial sectors. But the gestation period is long; it requires the interplay with basic science in a back-and-forth fashion. Developing generic technologies is expensive and risky, and the knowledge diffuses quickly to competitors.

If, at one time, the development of generic technology was a matter for the private sector, why does it now need the support of government?

First, it is not the case that the public sector was not involved in the past. For nearly 40 years, generic technology was developed by the U.S. in the context of military and space programs supported by the Department of Defense and the National Aeronautics and Space Administration. But recent developments have undermined this strategy for supporting generic technology:

- As I already said, the strategic technologies of the future will be developed increasingly in civilian contexts rather than in military or space programs. This is the reverse of the situation that existed in the sixties and seventies.
- American industry is facing competitors that are supported by their governments in establishing public/private partnerships for the development of generic technologies, both in the Pacific Rim and in the EEC.
- What's more, the cost of developing new technologies is rising. In many key industries, U.S. companies are losing their market share to foreign competitors—not only abroad but at home, as well. They are constrained in their ability to invest in new, risky technology efforts. They need additional resources.

But let's be clear . . .

The "technology strategy" that I'm talking about is not an "industrial policy." Cooperation between government and industry does not mean a centrally controlled, government-coordinated plan for industrial development. It is absolutely fundamental that the basic choices concerning which products to develop and when must remain with private industry, backed by private money and the discipline of the market. But we can have this and also have the government assume a role that no longer can be satisfied by the private sector.

Cooperation is also needed between industry and universities in order to get new knowledge moving smoothly from the laboratory to the market. Before World War II, universities looked to industry for research support. During and after the war, however, it became easier for universities to get what they needed from the government, and the tradition slowly grew that industry and universities should stay at arm's length. But this was acceptable only when government was willing to carry the whole load, and that is no longer true. Today, neither side can afford to remain detached.

Better relations between industry and universities yield benefits to both sectors. Universities get needed financial support and a better vantage point for understanding industry's needs. Industry gets access to the best new ideas and the brightest people and a steady supply of the well-trained scientists and engineers it needs.

Cooperation also means private firms must learn to work together. In the U.S., at least in this century, antitrust laws have forced companies to consider their competitors as adversaries. This worked well to ensure competition in the domestic market, but it works less well today, when the real competition is not domestic, but foreign. Our laws and public attitudes must adjust to this new reality. We must understand both that cooperation at the precompetitive level is not a barrier to fierce competition in the marketplace and that domestic cooperation may be the prerequisite for international competitive success.

The evolution of the Semiconductor Manufacturing Technology Consortium is a good example of how government support and cooperation with industry leads to productive outcomes.

International Cooperation

Paradoxically, we must also strengthen international cooperation in research even as we learn to compete more aggressively. There is no confining knowledge within national or political boundaries, and no nation can afford to rely on its own resources for generating new knowledge. Free access to new knowledge in other countries is necessary to remain competitive, but it depends on cooperative relationships.

In addition, the cost and complexity of modern research has escalated to the point where no nation can do it all—especially in "big science" areas and in fields like AIDS, global warming, earthquake prediction, and nuclear waste management. In these and other fields, sharing of people and facilities should be the automatic approach of research administrators.

Summary

My focus has been on the new global environment; the changes brought about by computers and computer science; international competition, its promise and its danger; and the role of government. But more important is a sustained commitment to cooperation and to a technical work force—these are the major determinants of success in developing a vibrant economy.

In the postwar years, we built up our basic science and engineering research structure and achieved a commanding lead in basic research and most strategic technologies. But now the focus must shift to holding on to what we accomplished and to building a new national technology structure that will allow us to achieve and maintain a commanding lead in the technologies that determine economic success in the world marketplace.

During World War II, the freedom of the world was at stake. During the Cold War, our free society was at stake. Today it is our standard of living and our leadership of the world as an economic power that are at stake.

Let me leave you with one thought: computers have become a symbol of our age. They are also a symbol and a barometer of the country's creativity and productivity in the effort to maintain our competitive position in the world arena. As other countries succeed in this area or overtake us, computers can become a symbol of our vulnerability.

2

Technology Perspective

This session focused on technology for supercomputing—its current state, projections, limitations, and foreign dependencies. The viability of the U.S. semiconductor industry as a source of parts was considered. The possible roles of gallium arsenide, silicon, superconductive, and electro-optical technologies in supercomputers were discussed. Packaging, cooling, computer-aided design, and circuit simulation were also discussed.

Session Chair

Robert Cooper,
Atlantic Aerospace Electronics Corporation

Overview

Robert Cooper

Robert Cooper is currently the President, CEO, and Chairman of the Board of Atlantic Aerospace Electronics Corporation. Previously, he served simultaneously as Assistant Secretary of Defense for Research and Technology and Director of the Defense Advanced Research Projects Agency (DARPA). Under his directorship, DARPA moved into areas such as programs in advanced aeronautical systems, gallium arsenide microelectronic circuits, new-generation computing technology, and artificial intelligence concepts. Bob has also been the Director of the NASA Goddard Space Flight Center and the Assistant Director of Defense Research at MIT's Lincoln Laboratory. Bob holds a doctorate from MIT in electrical engineering and mathematics.

When I was at Goddard, we started the first massively parallel processor that was built, and it subsequently functioned at Goddard for many, many years. Interestingly enough, as I walked into this room to be on this panel, one of the folks who was on that program sat down next to me and said that he remembered those days fondly.

I'm really quite impressed by this group, and I subscribe to the comment that I heard out in the hallway just before the first session. One person was talking to another and said that he had never seen such a high concentration of computing genius in one place since 1954 at the Courant Institute, when John von Neumann dined alone. Be that as it may, I am nevertheless confident that if anything can be made to happen in the high-end computer industry in this country, this group can play a key role in making it happen.

That comment also goes for the panel today, which is going to attack the problems of technology and perspectives for the future. We actually are starting this conference from a technical perspective by looking at the future—considering the prospects for computation—rather than looking toward the past, as we did in the first session.

Before we get started with our first speaker, I'd like to say a couple of words about what I see happening to the technology of high-end computing in the U.S. and in the world. Basically, the enabling technologies for high-end computing are the devices themselves. The physical constraints are the things that you will hear a lot about in this session: the logic devices; the memory devices; the architectural concepts, to a certain extent, which are determined by how you can fit these things together; and the interconnect technologies.

The main issue with technology developments in this area in this country is that we are somehow unable to take advantage of all of these things at the scale required to put large-scale systems together, and that is one of the reasons why we started the Strategic Computing Initiative back in 1983 at the Defense Advanced Research Projects Agency (DARPA), and that is why I think we are all hanging so much hope on the High Performance Computing Initiative that has come out of the study activity at DARPA and at the Office of Science and Technology Policy since about 1989.

I think it is the technology transition problem that we have to face. There is a role for government and a role for industry in the transition. I have been associated with some companies recently who have tried to take technology that they developed or that was somewhat common in the industry and make products out of it. I think that before we finish this particular session, we should talk about the issue of technology transition.

Supercomputing Tools and Technology

Tony Vacca

> *Tony Vacca is the Vice President of Technology at Cray Research, Inc., and has responsibility for product and technology development beyond Cray's C90 vector processor. Tony has had over 20 years' experience with circuit design, packaging, and storage. He began his career working at Raytheon Company as a design engineer, thereafter joining Control Data Corporation. From 1981 to 1989, he was the leader of the technology group at Engineering Technology Associates Systems. Tony has a bachelor of science degree in electrical engineering from the Michigan Technological Institute and has done graduate work at Northeastern and Stanford Universities.*

The supercomputer technologies, or more generally, high-performance computer technologies, cover a broad spectrum of requirements that have to be looked at simultaneously at any given time to meet the goals, which are usually schedule-driven.

From a semiconductor perspective, the technologies fall into four classes: silicon, gallium arsenide, the superconductor, and the optical. In parallel, we have to look simultaneously at such things as computer-aided design tools, under which is a category of elements that get increasingly important as microminiaturization and scaling of integration rise.

Also, we have to look at the packaging issues, and there are a lot of computer-aided design tools that are helping us in that area. As was discussed earlier, the issue of thermal management at all levels is very crucial, but the need for performance still dominates; we have to keep that in perspective.

Silicon is a very resilient technology, and every time it gets challenged it appears to respond. There are a lot of challenges to silicon, but I don't see many candidates in the near future that are more promising in the area of storage, especially dynamic storage, and possibly in some forms of logic.

Gallium arsenide has struggled over the last 10 years and is finally coming out as a "real" technology. Gallium arsenide has sent some false messages in some forms because some of the technology has focused not on performance but on power consumption. When it focuses on both, it will be much more effective for us. Usually when we are applying these technologies, we have to focus on the power and the speed simultaneously, especially because we are putting more processors on the floor.

The optical technology, from our viewpoint, has been used a lot in the communications between various mediums. When people talk about multigigahertz operations, I have some difficulty because I'm fighting to get 500-megahertz, single-bit optics in production from U.S. manufacturers. When people talk about the ability of 20-, 50-,100-, and 500-gigabit-per-second channels, I believe that is possible in some form, but I don't know how producible the concept is.

Cryogenic technology was fairly successful several years ago. Cryogenic technology is a superconductive Josephson junction technology that also needs significant help to bring it to viable production; to apply it, one needs a catalyst that the other two technologies are beginning to have.

Interestingly, there may be some people that believe that if you escape an architecture that is massively parallel, you can escape advancing technology. I think switching does not change the focus at all to a need for high-performance technology because the massively parallel entry points are the points of entrance and not the points of completion. If we lose focus across a 10-year span, we will have fooled ourselves into believing that we have defocused semiconductor technology, logic, packaging, or interconnect technology. I think we will have to have high-performance technology to stay in the leadership position that we are in.

In the interests of our position of leadership, I have been keeping what I call a U.S.-Japan technology "score card" (Figure 1). Design tools, packaging, integrated-circuit chips, and semiconductor capital equipment are the primary technology categories. The boxes under those categories in Figure 1 indicate where I think we are, relative to Japan, in

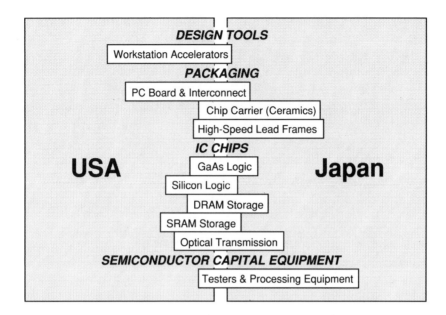

Figure 1. USA-Japan technology "score card."

these particular technological areas. These are key areas that I think we sometimes avoid developing. However, we must concentrate on these areas and areas of technology like these because they are the basis for developing the technologies we can build products from.

We cannot select technologies "à la carte" and discount other technologies. Technologies must be selected and balanced against one another. In the past, we didn't focus so much on supercomputer technologies because supercomputers a few years ago were alive and well, and a few companies were carrying the ball. A few years ago we didn't focus on semiconductor technologies because they were alive and well, and we were doing such a good job.

Now there is the capital issue that I think is very significant. If you consider the extent to which major semiconductor suppliers in the U.S. today depend on foreign capital equipment for getting their jobs done, then you appreciate that we are facing a very crucial issue.

High-Performance Optical Memory Technology at MCC

John Pinkston

John Pinkston was the Vice President and Director of the Exploratory Initiatives Program at the Microelectronics and Computer Technology Corporation. Currently, he is a Research and Development Fellow at the National Security Agency. Dr. Pinkston received a bachelor's degree in electrical engineering from Princeton University and a Ph.D., also in electrical engineering, from MIT.

During this session we are going to hear about high-speed devices for logic, memory, and packaging, which are necessary and critical to build any high-performance supercomputing system. I would like to talk about a high-performance bulk-storage technology that we have been working on at the Microelectronics and Computer Technology Corporation (MCC), which, if it is successful, could impact very significantly the performance of supercomputer systems that deal with data-intensive projects.

Specifically, I am talking about volume holographic storage. Picture in your mind a disk that has a latency time of 10 microseconds and an I/O transfer of about a gigabit per second. That is the kind of technology that would impact environments where "solid-state disks" are used today.

Basically what we are working on is optical technology—storing a hologram for an array of bits in a photosensitive crystal in the form of a two-dimensional page. The motivation for MCC's involvement with

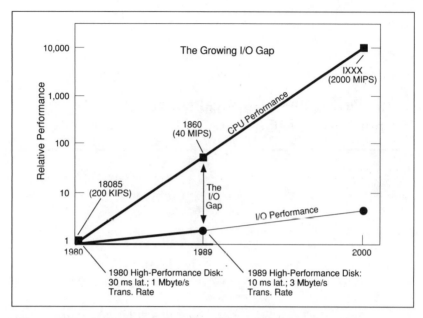

Figure 1. Why MCC got involved with holographic storage.

holographic storage is the widening gap between CPU and I/O performance (Figure 1). In areas where the problem is data intensive and data limited, the I/O performance is the limiting factor of the performance of the overall system.

The concept is shown in Figure 2. Data are brought in and stored in a spatial light modulator, which is essentially a square array of spots that are transparent or opaque. The modulator is illuminated by a light from a laser and is then imaged onto a photosensitive crystal, with about a one-millimeter-square area. The pattern is interfered with by a reference beam from the same laser brought in at an angle, which creates an interference pattern or a hologram in this area of the crystal material.

The crystal is a photorefractive material that stores an image. The interference pattern, which has areas of high intensity and low intensity, creates a local change in the index of refraction where the light intensity is high (Figure 3). Essentially, electrons get excited into mobile states and settle back down where the light intensity is not so great. If the light is taken away, the electrons freeze in some trapping states, and you are left with essentially a charge grating written in the material that persists and contains the hologram of the image.

Technology Perspective 43

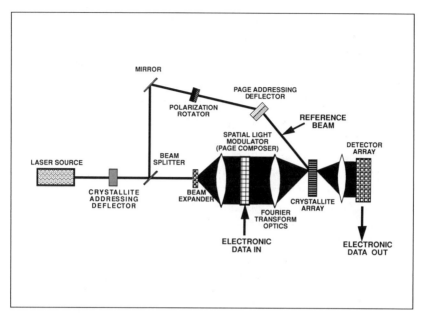

Figure 2. Optical configuration of the hologram concept.

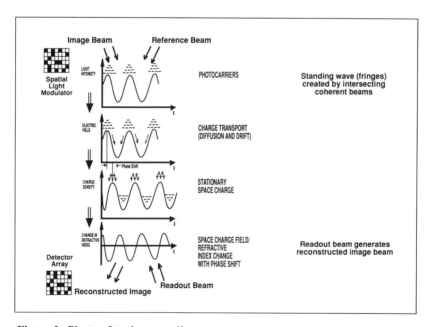

Figure 3. Photorefractive recording.

To read data, the crystallite array is illuminated with the reference beam, which scatters off the diffraction grating, reconstructs the object beam, and is imaged onto a detector array such as a charge-coupled device (CCD). One can store many pages into the same volume of material, as with holograms, by varying the angle of the incoming beam and therefore varying the spacing of the grating. You can think of it as spatial frequency division multiplexing of signals superimposed in the same module.

We have seen storage in the range of 30 to 50 pages in a region that we call a stack, and you can have multiple stacks in nonoverlapping volumes of the crystal.

Readout can occur in the 10-microsecond time frame. Writing takes a little longer—in the 100-microsecond time frame.

This technology offers potential storage density in the gigabyte range. This density is not really going to compete with very large archival disks but is very competitive with high-performance disks today. This idea has been around for a while, for probably 20 years, but recent developments have made it more attractive than before.

The problems in the past had been that, first, the material was very difficult to work with. We use a strontium barium niobate. One can also use a bismuth silicon oxide. These are both very hard materials to obtain in sufficient purity and quality.

Second, there was a problem that no one had been able to overcome. Both reads and writes illuminate the crystal and cause some fading of the holograms in that stack.

Basically, in our lab we have developed a way of making the crystal by stacking up a bunch of fibers, which can be grown much more easily than a large bulk crystal, thereby getting around the material-availability problem. Further, we've produced a nondestructive readout technique. Figure 4 lists the innovations MCC has patented in the fields of crystallite-array and nondestructive-readout technology.

The technology is now quite promising. Performance projections are targeted in the $1 to $20 per megabyte range, with multiple hundreds of megabits per second I/O rates and read latency in the microsecond time frame (Figure 5). Capacity is somewhere in the gigabyte range.

We feel we have to be able to beat what semiconductors can do at the system level by probably about five to 10 times in terms of cost per bit in the year in which our product becomes available.

We have built up a small functioning unit that has potential. It is about a foot on a side and has a laser, two acousto-optic deflectors, several mirrors and lenses, a spatial light modulator, a storage crystal, and a CCD

Technology Perspective 45

ARRAY OF CRYSTALLITES (FIBERS)

- Scalable storage capacity (larger array)
- Larger capacity (more pages per stack)
- Low crosstalk
- Lends itself to low-cost production

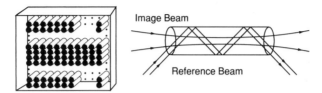

NONDESTRUCTIVE READOUT TECHNIQUE

- Allows prolonged readout in photorefractive material
- Billions of reads without signal-to-noise degradation
- May result in archival storage

Figure 4. MCC-patented innovations in two key Holostore technologies.

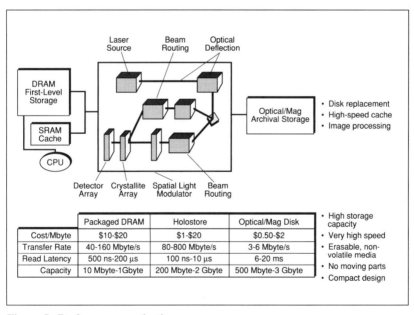

	Packaged DRAM	Holostore	Optical/Mag Disk
Cost/Mbyte	$10-$20	$1-$20	$0.50-$2
Transfer Rate	40-160 Mbyte/s	80-800 Mbyte/s	3-6 Mbyte/s
Read Latency	500 ns-200 µs	100 ns-10 µs	6-20 ms
Capacity	10 Mbyte-1Gbyte	200 Mbyte-2 Gbyte	500 Mbyte-3 Gbyte

- High storage capacity
- Very high speed
- Erasable, non-volatile media
- No moving parts
- Compact design

Figure 5. Performance projections.

- Project jointly funded by a number of MCC shareholders and associates
- Current project aimed at prototype demonstration by end of 1992

	Prototype Targets	Future Targets
Page Size	64 Kbit	1 Mbit
Pages per Stack	30	100
Stacks per Array	900	90,000
Storage Array		
Size	10 cm	?
Capacity	200 Mbyte	> 10 Gbyte
Media	Fixed array of crystallites	Removable module
Avg. Page Read Time	10 microseconds	100 nanoseconds
Avg. Page Write Time	100 microseconds	10 microseconds
Transfer Rate	80 Mbyte/second	> 10 Gbyte/second

Figure 6. Bobcat II project.

detector array. The unit is hooked up to a PC and is operating. We plan to develop a prototype (Bobcat II) that we hope to have available by the end of 1992. The capabilities of Bobcat II are outlined in Figure 6.

Applications for the unit will include the following:
- disk drive replacement,
- high-speed cache memory,
- high-speed storage with direct optical interfaces to fiber-optic communications networks,
- high-speed image-acquisition processing,
- survivable mass storage for demanding environments, and
- optical computing.

Digital Superconductive Electronics

Fernand Bedard

Fernand D. Bedard graduated magna cum laude from Fordham University with a B.S. degree in physics and mathematics and received his Ph.D. in physics from Johns Hopkins University, where he held an NSF fellowship. He subsequently taught physics at the University of Cincinnati and, since coming to the Washington, DC, area, has taught at American University and the University of Maryland. He has authored or coauthored 25 publications in areas of microwave spectroscopy, optical pumping, superconductivity, and semiconductors.

He is currently a Fellow at the National Security Agency (NSA) Research and Engineering organization and is a Special Assistant to the Chief of Research at that organization. Immediately before taking up these posts, he served as Program Manager for Technology Base Research and Development, which provides for anticipating NSA's future mission requirements. Prior assignments included the directorship of both the NSA Office of Research Physics Division and of the Office of Research, itself.

One of the major ingredients in the continual improvement of high-performance computers has been the increase in clock rate of the machines (Figure 1). The upper symbols show the clock interval of representative computers as they have evolved. Below each of these points is shown the gate delay of the logic devices, 10 to 20 times smaller

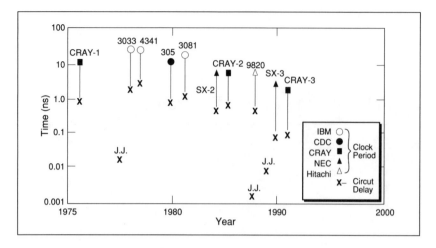

Figure 1. Trends in logic speed.

than the clock interval, to allow multiple logic levels, package delay, and skew to be accommodated. At the lower left is a data point, circa 1978, of an exploratory superconductive device whose fully loaded gate delay was roughly 50 picoseconds at that time; obviously today there is no computer that reflects that device's performance. A major effort to bring that about in the U.S. was terminated several years later.

At just about that time, the Ministry of International Trade and Industry (MITI) in Japan established a Superspeed Project—of which superconductive devices were an element—whose goal was to demonstrate an advanced computer that used nonsilicon technology. Out of this work came some very impressive results from the major participants, Hitachi, Fujitsu, Nippon Electric Corporation (NEC), and MITI's Electro Technical Laboratory. Fujitsu's work is particularly noteworthy. They demonstrated astounding chip-level performance by first building the functionality of an AMD 2901 on a chip that operated with the characteristics shown in Table 1. They proudly pointed out that the *chip* power dissipation, five milliwatts, was equal to the power of a *single gate* of the fastest semiconductor competitor. The 2.5-micrometer feature size was reduced to 1.5 micrometers to demonstrate gate performance (Figure 2)—near one-picosecond delay. Using 1.5-micrometer lithography, they then shrank the microprocessor onto approximately one-third of the five-millimeter-square chip, added more memory, and repeated the demonstration, achieving the results shown in Table 2. Notice that the gate complexity is roughly 3000 gates with a six-milliwatt power consumption—about two microwatts per gate. The next chip demonstration

Table 1. Performance of 4-Bit Microprocessor

Device	Si[a]	GaAs[b]	Josephson
Maximum Clock (MHz)	30	72	770
Power (W)	1.4	2.2	0.005

[a] AMD, 1985 data book
[b] Vitesse, 1987 GaAs IC Symposium

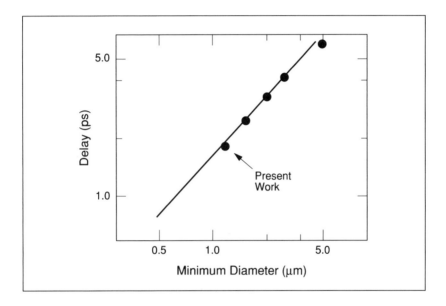

Figure 2. Gate delay versus junction diameter (Fujitsu, August 1988).

Table 2. Performance of Subnanosecond 4-Bit Josephson Processor

Instruction ROM Access Time	100 ps
Bit-Slice Microprocessor Clock Frequency	1.1 GHz
Multiplier-Accumulator Multiplication Time	200 ps
Power Dissipation	6.1 mW (1.9 µW/gate)
Number of Gates	3,056
Number of Junctions	24,000

was of a digital signal-processor chip, again using 1.5-micrometer feature size. This time the gate count was 6300, the clock rate was one gigahertz, and once more the power consumption was low—12 milliwatts, or again about two microwatts per gate (Tables 3 and 4).

If you look at the usual delay-versus-power plot (Figure 3) to size up the comparisons, you find that silicon, as represented by NEC's SX-3, and gallium arsenide, as planned by the CRAY-3, are in the 70–80-picosecond unloaded-gate-delay regime and 250-picosecond loaded-gate-delay regime. The gate power consumption is in the milliwatts-per-gate domain, whereas the Fujitsu demonstrations are in the microwatts-per-gate domain for power while providing sub-10-picosecond loaded gate delay.

Table 3. Fujitsu's Specifications for Digital Signal Processor Chip

Gate Count	6,300
Josephson Junction Count	23,000
Minimum Junction Size	1.5 μm
ROM Instruction	64w × 24b
Coefficient	16w × 8b
Data RAM	16w × 8b × 2
Multiplier	8b × 8b
ALU	13b, 16 functions
Chip Size	5.0 × 5.0 mm
Power	12 mW

Table 4. Fujitsu Circuit Performance

Instruction ROM Access Time	200 ps
Data RAM Access Time	130 ps
Multiplication	240 ps
Adding in ALU	410 ps
Internal Machine Clock Cycle	1 GHz
Second-Order IIR Filter	7 ns

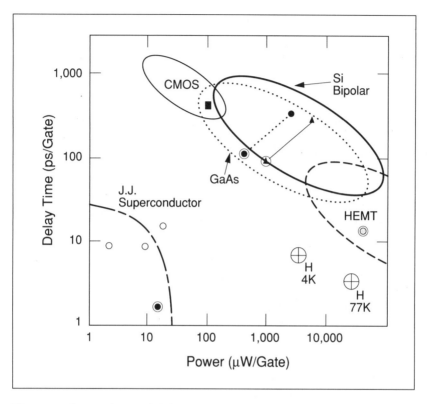

Figure 3. Comparisons of delay versus power for digital superconducting, semiconducting gate, and circuit.

What about factoring in the refrigerator power, approximately 500 to 1000 times the required computer logic power consumption? First, even accounting for that produces a faster chip performance-per-watt total and, more importantly, puts the power consumption where it is easily dealt with—at the unregulated power line, not at the tightly packed (for high-speed clocks) logic engine. Furthermore, the cooling and power supply requirements of conventional technology are rarely referred to and factored in at the system level.

There is an effort under way presently to demonstrate a switching network, a crossbar, using superconductive devices in such a way as to exploit their high speed and very low power, along with the advantage of zero-resistance transmission lines. The prototype, a 128-×-128 crossbar (Figure 4), is designed to switch two gigabits per second of data per serial channel, with room temperature inputs and outputs. The power dissipation at 4K should be 20–40 milliwatts and, even with a refrigerator

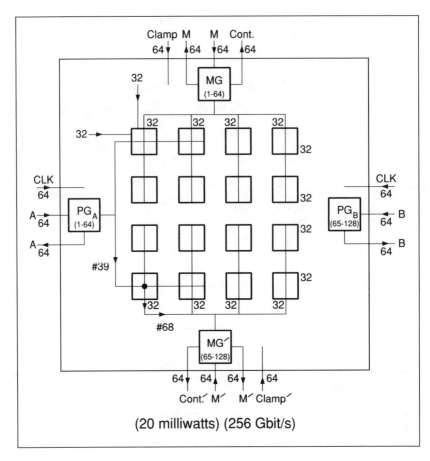

Figure 4. Pathway through a 128-×-128 crossbar (32-×-32 chip).

"penalty," would be a small fraction of the room-temperature electronics it services and would be much lower than any semiconductor competitor of lesser performance. The round trip "request-acknowledge" time should be approximately 10 nanoseconds, including address and arbitration time (Figure 5). If successful, the architecture, which depends intrinsically upon the devices, should allow the building of a 1024-×-1024 crossbar (Figure 6) with substantially the same access times as the 128-×-128 crossbar. The system's speed limitation is determined by the speed of light and the inability of semiconductors to keep up.

Technology Perspective 53

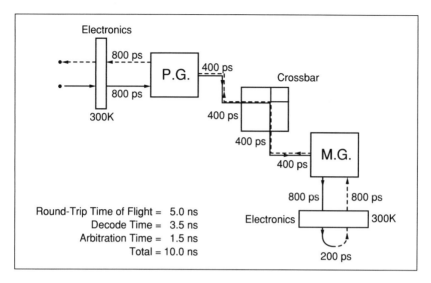

Figure 5. Timing through crossbar, from processor to memory and back.

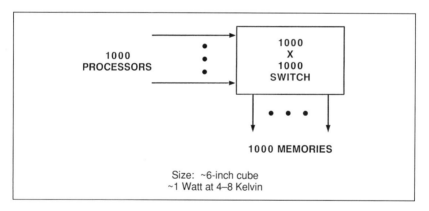

Figure 6. A 1024-×-1024 crossbar switch.

Enabling Technology: Photonics

Alan Huang

Alan Huang is head of the Digital Optics Research Department in the Communications Systems Research Laboratory at AT&T Bell Laboratories. He has been interested in optical computing for almost 20 years. Dr. Huang is also known for his contributions to very-large-scale integration and broadband communications networks. He received his Ph.D. in electrical engineering from Stanford University, Palo Alto, California. He has published over 50 papers and is the holder of 20 patents.

Introduction

Computers, as we know today, will be just be one component of an *intellectual power grid* in which computation and storage will become commodities traded over optical fiber "power lines." Success will hinge on the successful integration of computers, communications, and their associated technologies—electronics and photonics at both a macro and micro level.

At the micro level, the parallelism of optics is the most important factor. Architecturally, this connectivity can be used to transparently extend the name space and simplify the coordination of thousands of microprocessors into a unified micro-distributed computer. The goal is a thousand interconnections, each at one gigabit per second.

At the macro level, the bandwidth of optics is the most important parameter. Architecturally, this connectivity can be used to transparently extend the name space and simplify the coordination of thousands of computers into a unified macro-distributed computer. Our goal is one connection at a terabit per second.

A Thousand Interconnections, Each at One Gigabit per Second

One of the main reasons for trying to use optics is its connectivity. It is relatively easy for a lens to convey a 100-by-100 array of channels, each with the bandwidth of an optical fiber. This is shown in Figure 1. One thousand twenty-four optical connections can be implemented in the same space it takes to make one electronic connection.

One of the fundamental technologies that makes all of these optical interconnects possible is molecular beam epitaxy (MBE). This technology gives us the ability to grow crystals atom by atom with the precision of plus or minus one atomic layer over a two-inch wafer. See Figure 2. What good is this? By varying the thickness and elemental composition,

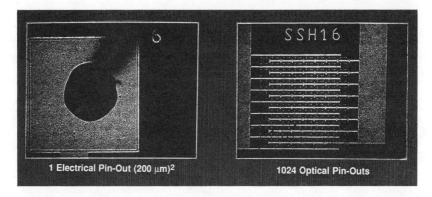

Figure 1. One thousand twenty-four optical connections contained within the same area as one electronic connection.

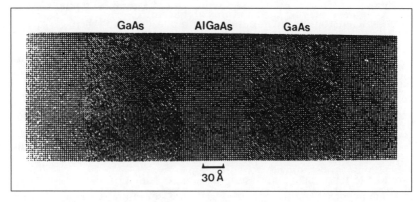

Figure 2. Plus or minus one atomic layer precision of molecular beam epitaxy.

we can grow optical components such as mirrors. If we change the recipe, we can grow quantum wells, which give the material unusual optical properties. We can also grow *p-n* junctions to make electronics. This process of MBE gives us a way of integrating optics, materials, and electronics at an atomic level, which blurs the traditional distinction between electronics and optics.

One of the devices developed on the basis of this technology is the SEED device (Prise et al. 1991), a light-controlled mirror that we can toggle between 10 and 60 per cent reflectivity. These devices function as flip-flop with optical inputs and outputs. We have fabricated arrays of up to 32K devices and have run some of these devices at one gigahertz.

A second device based on MBE is the microlaser (Jewell et al. 1991). MBE was used to grow a mirror, a quantum well, and then a second mirror. We can then fabricate millions of lasers by etching the wafer. This is shown in Figure 3. Our yield is over 95 per cent, and the raw cost is approximately $0.0001 per laser. The yields and cost of this process will dramatically affect the availability of lasers. This technology is useful in terms of the connectivity of optics because it demonstrates that thousands of lasers can be fabricated in a very small area.

A second reason for using optics is the bandwidth. An optical channel has over one terahertz of bandwidth. A thousand channels, each at one gigabit per second, can also be accomplished by using wavelength division multiplexing techniques to break this bandwidth into thousands of individual channels. The microlasers shown in Figure 3 can also be used in this manner. These wafers can be grown on a slight slant. This technique would make each of the microlasers function at a slightly different wavelength.

One of the problems with trying to achieve a thousand interconnects, each at one gigabit per second, is the optical packaging. In electronics the circuit boards, sockets, etc., are quite standardized. Optical setups have usually been one of a kind and quite large, with many micrometer adjustments. We have directed a large part of our effort at miniaturizing and simplifying this packaging. Our first system took three optical benches, each 4 by 12 feet, to interconnect three optical logic gates. The next year, we were able to reduce this to a 1- by 1-foot module that interconnected 32 gates. A year later, we interconnected four of these 1- by 1-foot modules to build a simple optical pipelined processor (Prise et al. 1991). See Figure 4. Six months later, another group hooked three 8-by-8 arrays of optical logic gates together with a 2- by 3-foot setup. A year later, they interconnected six arrays, each 32 by 32, with a 1- by 1-foot system. We have since managed to reduce most of the optics in our

Figure 3. An array of surface emitting microlasers.

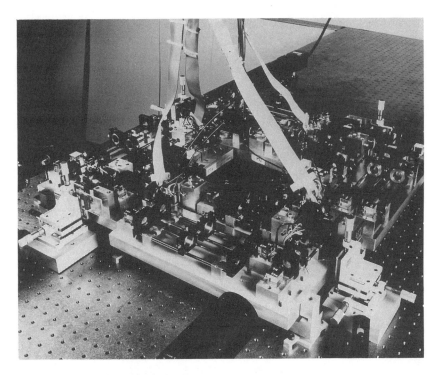

Figure 4. A simple optical pipelined processor.

original 1- by 1-foot module to a module 1 by 2 by 3 inches in size (Figure 5).

We are now trying to reduce most of the optics in our original 1- by 1-foot module so that it fits onto the surface of a quarter. This technology, three-dimensional planar optics (Streibl et al. 1989), basically replaces the lenses with holograms of the lenses and fabricates these holograms with photolithography (Figure 6). We have demonstrated complex optical systems with lenses capable of conveying a 32 by 32 array of spots with greater than a 90 per cent diffraction efficiency.

One Connection at One Terabit per Second

Another reason for trying to use optics is for its speed. Optical nonlinearities have been measured down to the femtosecond (10^{-15} s), whereas electronics, because of the mobility of electrons in a semiconductor, has a built-in limit at around 10 picoseconds (10^{-12} s). The large bandwidths also allow us to go ultrafast. It frees us from the inductive and capacitive limitations of electronics. We have recently demonstrated an all-optical fiber logic AND gate, a NOT gate, a XOR gate, a 1.6-terahertz

Figure 5. A miniaturized version of one of the modules of the simple optical processor.

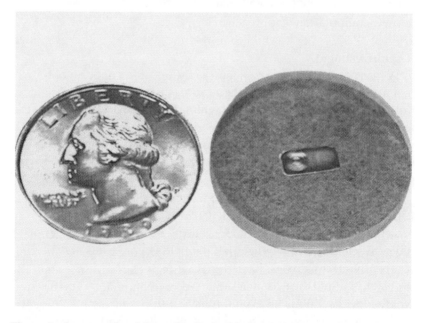

Figure 6. An example of three-dimensional planar optics.

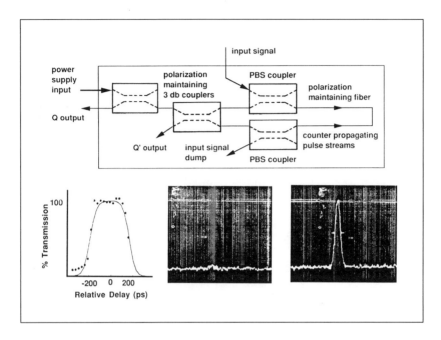

Figure 7. An ultrafast all-optical fiber logic gate.

optical clock, a 2.5-gigabits-per-second multiplexer (Whitaker et al. 1991), and a 254-bit optical dynamic RAM (Figure 7). This is the beginning of a new digital technology that has the potential of working in the terahertz regime.

References

J. L. Jewell, J. P. Harbison, A. Scherer, Y. H. Lee, and L. T. Florez, "Vertical-Cavity Surface-Emitting Lasers: Design, Growth, Fabrication, Characterization," *IEEE Quantum Electronics* 27, 1332–1346 (1991).

M. E. Prise, N. C. Craft, M. M. Downs, R. E. LaMarche, L. A. D'Asaro, L. M. Chirovsky, and M. J. Murdocca, "Optical Digital Processor Using Arrays of Symmetric Self-Electro-optic Effect Devices," *Applied Optics* 30, 2287–2296 (1991).

N. Streibl, K. Brenner, A. Huang, J. Jahns, J. Jewell, A. W. Lohmann, D. A. B. Miller, M. Murdocca, M. E. Prise, and T. Sizer, "Digital Optics," *Proceedings of the IEEE* 77 (12), 1954–1969 (1989).

N. A. Whitaker, Jr., H. Avramopoulos, P. M. W. French, M. C. Gabriel, R. E. LaMarche, D. J. DiGiovanni, and H. M. Presby, "All-Optical Arbitrary Demultiplexing at 2.5 Gbits/s with Tolerance to Timing Jitter," *Optics Letters* **16**, 1838–1840 (1991).

3

Vector Pipeline Architecture

This session focused on the promise and limitations of architectures featuring a moderate number of tightly coupled, powerful vector processors—the limitations, dependencies, sustained-performance potential, processor performance, interconnection topologies, and applications domains for such architectures. How fast do things have to be to eliminate vectors and use only scalar processors?

Session Chair

Les Davis, Cray Research, Inc.

Vector Architecture in the 1990s

Les Davis

> *Les Davis has been with Cray Research, Inc., since its founding in 1972. Initially, he was the chief engineer for the CRAY-1 project. Today, Mr. Davis is the Executive Vice President and a member of the Technology Council, which formulates the company's strategic direction. Before joining Cray Research, Mr. Davis was Director of Electrical Engineering and General Manager of the Chippewa Laboratory for Control Data Corporation.*

As the title of this session suggests, we will be interested in pursuing the discussion on vector processing. It is interesting to note, when you go to a meeting like this, how many people—the same people—have been in this business for 20 years or longer. I don't know if that's a good thing or a bad thing, but at least it attests to your persistence in sticking with one kind of work despite the passage of time and several companies. It is also interesting to note, now, how openly we discuss high-performance computing. I can remember in the early days when my kids would ask me what I did, I'd kind of mumble something about scientific computing, hoping they wouldn't ask what that was. Yet, it's surprising today that we talk about it quite openly.

Nevertheless, it is the politicians that I have trouble communicating with about high-performance computing. I am hoping that after a series of meetings like this, we will be able to convince the politicians of the importance of high-performance computing.

I believe that the vector architecture in the 1980s played almost the same role that the first transistorized computers played in the 1960s. Vector architecture really did offer the researchers an opportunity to do things that in the late 1960s and late 1970s we were unable to achieve with the machines that were available at that time.

I think the vector machines were characterized by several positive things. One was the introduction of very large memories, high-bandwidth memories to support those large memories, and very efficient vectorizing compilers. As a result of those combinations, we saw several orders of magnitude improvement in performance over what previous architectures offered.

On the negative side, scalar processing did not move along quite as rapidly because it was restrained by slow clock rates. If you looked at the performance improvements, you only saw a factor of perhaps 2.5 from 1975 through 1990. On the other side of the coin, if you looked at the ability to incorporate not only vectorization but also large or reasonably large numbers of vector processors tightly coupled to the memory, you saw, in many cases, several orders of magnitude improvement in performance.

I also think the multiprocessor vector machines were another significant step that we took in the 1980s, and now we are able to couple up to 16 processors in a very tight fashion. Interprocessor communication and memory communication actually allow us to make very efficient use of those machines.

The other important thing is that we have allowed our compiler developers to move along and take advantage of these machines. I think a lot of that work will take and be transportable when we look at some of the newer architectures that we are examining in the research and development areas.

I think the importance of the U.S. retaining its leadership in the supercomputer industry has been stated many times. For us to retain that leadership in the high-performance computing area, we must be able to maintain our lead in the manufacturing, as well as in the design of the systems. That is something that was touched on in the pevious session, but I think it has much more importance than a lot of people attach to it.

We need now to be able to compete in the world markets because in many cases, that is one of the few ways in which we not only can get research and development dollars but also can perfect our manufacturing capabilities. If we are not able to do that, I don't think we're going to be able to capitalize on some of the new technologies and new developments that are taking place today.

I think the vector architectures are going to be the backbone of our high-performance computing initiative throughout the 1990s. This is not to say that there will not be newer software and hardware architectures that will be coming along. However, if we are not able to take and maintain the leadership with our current types of architectures, I know very well of a group of people that are located overseas that would just love to be able to do that.

My commitment here is to make sure that we not only are looking ahead and trying to make sure that we move very aggressively with new architectures, both in hardware and software, but also that we are not giving up and losing sight of the fact that we have quite a commitment to a large number of people today that have invested in these vector-type architectures.

In Defense of the Vector Computer

Harvey Cragon

Harvey G. Cragon has held the Ernest Cockrell, Jr., Centennial Chair in Engineering at the University of Texas, Austin, since 1984. Previously he was employed at Texas Instruments for 25 years, where he designed and constructed the first integrated-circuit computer, the first transistor-transistor logic computer, and a number of other computers and microprocessors. His current interests center upon computer performance and architecture design. He is a fellow of the Institute of Electrical and Electronics Engineers (IEEE) and a member of the IEEE Computer Society, the National Academy of Engineering, and the Association for Computing Machinery (ACM). Professor Cragon received the IEEE Emanuel R. Piore Award in 1984 and the ACM-IEEE Eckert-Mauchly Award in 1986. He is also a trustee of The Computer Museum in Boston.

As several have said this morning, parallel computers, as an architectural concept, were talked about and the research was done on them before the advent of vector machines. The vector machine is sort of the "new kid on the block," not the other way around.

Today I am going to defend the vector computer. I think that there are some reasons why it is the workhorse of the industry and why it has been successful and will continue to be successful.

The first reason is that in the mid-1960s, about 1966, it suddenly dawned on me, as it had on others, that the Fortran DO loop was a direct invocation of a vector instruction. Everything would not be vectorizable, but just picking out the Fortran DO loops made it possible to compile programs for the vector machine. That is, I think, still an overwhelming advantage that the vector machines have—that the arrayed constructs of languages such as Ada are vectorizable.

The second reason is that there is a natural marriage between pipelining and the vector instruction. Long vectors equal long pipelines, short clock periods, and high performance. Those items merge together very well.

Now, back to the programming point of view and how vectors more or less got started. Erich Bloch, in Session 1, was talking about the Stretch computer. I remember reading in a collection of papers on Stretch that there was a flow chart of a vector subroutine. I looked at that and realized that's what ought to be in the hardware. Therefore, we were taking known programming constructs and mapping them into hardware.

Today it strikes me that we are trying to work the parallel computer problem the other way. We are trying to find the programming constructs that will work on the hardware. We come at it from the programming point of view.

I believe that vector pipeline machines give a proper combination of the space-time parallelism that arises in many problems. The mapping of a problem to perform pipeline and vector instructions is more efficient and productive than mapping the same type of problem to a fixed array—to an array that has fixed dimensionality.

We worked at Illinois on the ILLIAC-IV. It was a traumatic decision to abandon that idea because a semiconductor company would love to have something replicated in large numbers. However, we did not know how to program ILLIAC-IV, but we did know how to program the vector machine.

Looking to the future, I think that vector architecture technology is fairly mature, and there are not a whole lot of improvements to make. We are going to be dependent in large measure on the advances in circuit technology that we will see over the next decade. A factor of 10 is probably still in the works for silicon.

Will the socioeconomic problems of gallium arsenide and Josephson junctions overcome the technical problems? Certainly, as Tony Vacca (Session 2) said, we need high-performance technology just as much as the parallel computer architects need it.

I have made a survey recently of papers in the International Solid State Circuits Conference, and it would appear that over the last 10 years, clock rates in silicon have improved about 25 per cent per year. This would translate into a 10^9-type clock rate in another eight or 10 years. At those clock rates, the power problems would become quite severe. If we try to put one of these things on a chip, we have got real power problems that have got to be solved.

I also perceive another problem facing us—that we have not paid as much attention to scalar processing as we should. Given that either pipeline vector machines stay dominant or that the multiprocessors become dominant, we still have to have higher-performance scalar machines to support them. I think that we need research in scalar machines probably as much as, if not more than, we need research in vector machines or parallel machines.

I tend to believe that the RISC theology is going the wrong way and that what we really need to do is raise rather than lower the level of abstraction so that we can get the proper computational rates out of scalar processing that we really need to support vector or parallel machines.

In conclusion, there is a saying from Texas: if it's not broke, don't fix it. There is also another saying: you dance with the one that "brung" you. Well, the one that "brung" us was the vector machine, so let's keep dancing.

Market Trends in Supercomputing

Neil Davenport

Neil Davenport is the former President and CEO of Cray Computer Corporation. Before the spinoff of the company from Cray Research, Inc. (CRI), in November 1989, Neil served from 1981 to 1988 as the Cray Research Ltd. (UK) Managing Director for Sales, Support, and Service for Northern Europe, the Middle East, India, and Australia; from 1988 to November 1989, he was Vice President of Colorado Operations, with responsibility for the manufacture of the CRAY-3. Before joining CRI, he worked 11 years for ICL in England, the last three managing the Education and Research Region, which had marketing responsibility for the Distributed Array Processor program.

Since 1976 and the introduction of the CRAY-1, which for the purpose of this paper is regarded as the start of the supercomputer era, the market for large-scale scientific computers has been dominated by machines of one architectural type. Today, despite the introduction of a number of new architectures and despite the improvement in performance of machines at all levels in the marketplace, most large-scale scientific processing is carried out on vector pipeline computers with from one to eight processors and a common memory. The dominance of this architecture is equally strong when measured by the number of machines installed or by the amount of money spent on purchase and maintenance.

As with every other level of the computer market, the supply of software follows the dominant hardware. Accordingly, the library of application software for vector pipeline machines has grown significantly. The investment by users of the machines and by third-party software houses in this architecture is considerable.

The development of vector pipeline hardware since 1976 has been significant, with the prospect of machines with 100 times the performance of the CRAY-1 being delivered in the next year or two. The improvement in performance of single processors has not been sufficient to sustain this growth. Multiple processors have become the norm for the highest-performance offerings from most vendors over the past few years. The market leader, Cray Research, Inc., introduced its first multi-processor system in 1982.

Software development for single processors, whether part of a larger system or not, has been impressive. The proportion of Fortran code that is vectorized automatically by compilers has increased continuously since 1976. Several vendors offer good vectorization capabilities in Fortran and C. For the scientist, vectorization has become transparent. Good code runs very well on vector pipeline machines. The return for vectorization remains high for little or no effort on the part of the programmer. This improvement has taken the industry 15 years to accomplish.

Software for multiprocessing a single task has proved to be much more difficult to write. Preprocessors to compilers to find and highlight opportunities for parallel processing in codes are available, along with some more refined structures for the same function. As yet, the level of multitasking single programs over multiple processors remains low. There are exceptional classes of problems that lend themselves to multitasking, such as weather models. Codes for these problems have been restructured to take advantage of multiple processors, with excellent results. Overall, however, the progress in automatic parallelization and new parallel-application programs has been disappointing but not surprising. The potential benefits of parallel processing and massively parallel systems have been apparent for some time. Before 1980, a number of applications that are well suited to the massively parallel architecture were running successfully on the ICL Distributed Array Processor. These included estuary modeling, pattern recognition, and image processing. Other applications that did not map directly onto the machine architecture did not fare so well, including oil reservoir engineering, despite considerable effort.

The recent improvements in performance and the associated lowering in price of microprocessors has greatly increased the already high level of attraction to massively parallel systems. A number of vendors have introduced machines to the market, with some success. The hardware issues seem to be manageable, with the possible exception of common memory. The issues for system and application software are still formidable. The level of potential reward and the increase in the numbers of players will accelerate progress, but how quickly? New languages and new algorithms do not come easily, nor are they easily accepted.

In the meantime, vector pipeline machines are being enhanced. Faster scalar processing with cycle times down to one nanosecond are not far away. Faster, larger common memories with higher bandwidth are being added. The number of processors will continue to increase as slowly as the market can absorb them. With most of the market momentum—also called user friendliness, or more accurately, user familiarity—still being behind such machines, it would seem likely that the tide will be slow to turn.

In summary, it would appear that the increasing investment in massive parallelism will yield returns in some circumstances that could be spectacular; but progress will be slow in the general case. Intermediate advances in parallel processing will benefit machines of 16 and 64 processors, as well as those with thousands. If these assumptions are correct, then the market share position in 1995 by type of machine will be similar to that of today.

Massively Parallel SIMD Computing on Vector Machines Using PASSWORK

Ken Iobst

Kenneth Iobst received a B.S. degree in electrical engineering from Drexel University, Philadelphia, in 1971 and M.S. and Ph.D. degrees in electrical engineering/computer science from the University of Maryland in 1974 and 1981, respectively. Between 1967 and 1985, he worked as an aerospace technologist at the NASA Langley Research Center and the NASA Goddard Space Flight Center and was actively involved in the Massively Parallel Processor Project. In 1986 he joined the newly formed Supercomputing Research Center, where he is currently employed as a research staff member in the algorithms group. His current research interests include massively parallel SIMD computation, SIMD computing on vector machines, and massively parallel SIMD architecture.

When I first came to the Supercomputing Research Center (SRC) in 1986, we did not yet have a SIMD research machine—i.e., a machine with single-instruction-stream, multiple-data-streams capability. We did, however, have a CRAY-2. Since I wanted to continue my SIMD research started at NASA, I proceeded to develop a simulator of my favorite SIMD machine, the Goodyear MPP (Massively Parallel Processor), on the CRAY-2.

This SIMD simulator, called PASSWORK (PArallel SIMD Simulation WORKbench), now runs on seven different machines and represents a truly machine-independent SIMD parallel programming environment. Initially developed in C, PASSWORK is now callable from both C and

Fortran. It has been used at SRC to develop bit-serial parallel algorithms, solve old problems in new ways, and generally achieve the kind of performance one expects on "embarrassingly" parallel problems.

As a result of this experience, I discovered something about the equivalence between a vector machine and a real SIMD machine that I would now like to share with you. In general, the following remarks apply to both the Goodyear MPP and Thinking Machines Corporation's CM-2.

There are two basic views of a vector machine like the CRAY-2. In the traditional vector/scalar view, the CRAY-2 has four processors, each with 16K words of local memory, 256 megawords of globally shared memory, and a vector processing speed of four words per 4.1 nanoseconds. From a massively parallel point of view, the CRAY-2 has a variable number of bit-serial processors (4K per vector register) and a corresponding amount of local memory per processor equal to 2^{34} processor bits.

Given an understanding of SIMD computing, one can see how the broadcast of a single instruction to multiple processors on a SIMD machine is analogous to the pipelined issue of vector instructions on a vector machine. There is a natural sort of equivalence here between these two seemingly different machine architectures.

As can be seen in Figure 1, there are two basic computing domains—a vector/scalar domain and a bit-serial domain. In the vector/scalar domain, we do things conventionally. In the bit-serial domain, we are more able to trade space for time and to solve the massively parallel parts of problems more efficiently. This higher performance results from operating on small fields or kernels with linear/logarithmic bit-serial computational complexity. In this bit-serial domain, we are operating on fully packed words, where the bits of a word are associated with single-bit processors, not with a physical numeric representation.

If you take a single problem and break it up into a conventional and a bit-serial part, you may find that a performance synergy exists. This is true whenever the whole problem can be solved in less time across two domains instead of one. This capability may depend heavily, however, on an efficient mechanism to translate between the computing domains. This is where the concept of corner-turning becomes very important.

The concept of corner-turning allows one to view a computer word of information sometimes as containing spatial information (one bit per processor) and at other times as containing numeric information, as is depicted in Figure 2. Corner-turning is the key to high-performance SIMD computing on vector machines and is best implemented in hardware with a separate vector functional unit in each CPU. With this support,

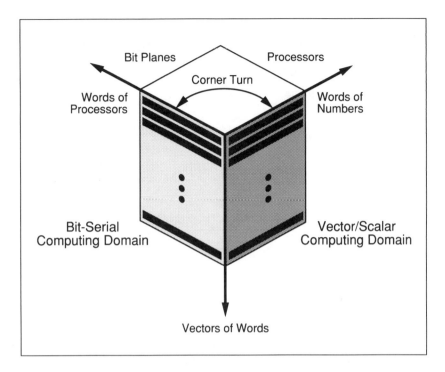

Figure 1. Vector corner-turning operation.

vector machines would be used much more extensively for SIMD computing than they are today.

To give you an idea of how things might be done on such a machine, let's look at the general routing problem on a SIMD machine. Suppose we have a single bit of information in each of 4K processors and wish to arbitrarily route this information to some other processor. To perform this operation on a real SIMD machine requires some sort of sophisticated routing network to handle the simultaneous transmissions of data, given collisions, hot spots, etc. Typically, the latencies associated with parallel routing of multiple messages are considerably longer than in cases where a single processor is communicating with one other processor.

On a vector machine, this routing is pipelined and may suffer from bank conflicts but in general involves very little increased latency for multiple transmissions. To perform this kind of routing on a vector machine, we simply corner-turn the single bit of information across 4K processors into a 4K vector, permute the words of this vector with hardware scatter/gather, and then corner-turn the permuted bits back into the original processors.

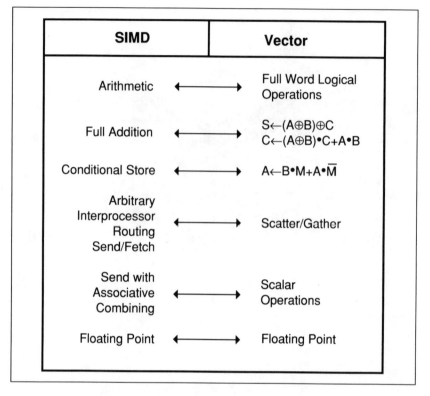

Figure 2. SIMD/vector equivalences through corner turning.

Using this mechanism for interprocessor SIMD communication on a vector machine depends heavily on fast corner-turning hardware but in general is an order of magnitude faster than the corresponding operation on a real SIMD machine. For some problems, this routing time dominates, and it becomes very important to make corner-turning as fast as possible to minimize this "scalar part" of this parallel SIMD problem. This situation is analogous to minimizing the scalar part of a problem according to Amdahl's Law.

Figure 2 shows some other equivalences between SIMD computing and vector/scalar computing. Some of these vector/scalar operations do not require corner turning but suffer from a different kind of overhead—the large number of logical operations required to perform basic bit-serial arithmetic. For example, bit-serial full addition requires five logical operations to perform the same computation that a real SIMD machine performs in a single tick. Fortunately, a vector machine can sometimes hide this latency with multiple logical functional units. Conditional store, which is frequently used on a SIMD machine to enable or disable computation across a subset of processors, also suffers from this same overhead.

There are some other "SIMD operations," however, that are actually performed more effectively on a vector/scalar machine than on a real SIMD machine. This seems like a contradiction, but the reference to "SIMD operation" here is used in the generic sense, not the physical sense. Operations in this class include single-bit tallies across the processors and the global "or" of all processors that is frequently used to control SIMD instruction issue.

Single-bit tallies across the processors are done much more efficiently on a vector machine using vector popcount hardware than on the much slower routing network of real SIMD machines. The global "or" of all processors on a real SIMD machine generally requires an "or" tree depth equal to the log of the number of processors. On a typical SIMD machine, the time needed to generate this signal is in the range of 300–500 nanoseconds.

On a vector machine, this global "or" signal may still have to be computed across all processors but in general can be short-stopped once one processor is found to be nonzero. Therefore, the typical time to generate the global "or" on a vector machine is only one scalar memory access, or typically 30–50 nanoseconds. This is a significant performance advantage for vector machines and clearly demonstrates that it may be much better to pipeline instructions than to broadcast them.

As stated earlier, PASSWORK was originally developed as a research tool to explore the semantics of parallel SIMD computation. It now represents a new approach to SIMD computing on conventional machines and even has some specific advantages over real SIMD machines. One of these distinct advantages is the physical mapping of real problems onto real machines. Many times the natural parallelism of a problem does not directly map onto the physical number of SIMD processors. In PASSWORK the natural parallelism of any problem is easily matched to a physical number of simulated processors (to within the next higher power of four processors).

This tradeoff between the number of processors and the speed of the processors is most important when the natural parallelism of the problem is significantly less than the physical number of SIMD processors. In this case, a vector machine, although possibly operating in a short vector mode, can always trade space for time and provide a fairly efficient logical-to-physical mapping for SIMD problems. On a real SIMD machine, there is a significant performance degradation in this case because of the underutilization of physical processors.

In a direct comparison between the CRAY-2 and the CM-2, most SIMD problems run about 10 times slower on the CRAY-2 than on the CM-2. If the CRAY-2 had hardware support for corner-turning and a memory bandwidth equivalent to the CM-2, this performance advantage would completely disappear. Most of this performance loss is due to memory subsystem design, not to basic architectural differences between the two machines; i.e., the CRAY-2 was designed with a bank depth of eight, and the CM-2 was designed with a bank depth of one. As a result, the CRAY-2 can cycle only one-eighth of its memory chips every memory cycle, whereas the CM-2 can cycle all of its memory chips every memory cycle.

As shown in Figure 3, PASSWORK basically models the MPP arithmetic logic unit (ALU), with extensions for indirect addressing and floating point. This MPP model supports both a one- and two-dimensional toroidal mesh of processors. Corner-turning is used extensively for interprocessor routing, floating point, and indirect addressing/table lookup. The PASSWORK library supports a full complement of bit-serial operations that treat bits as full-class objects. Both the massively parallel dimension and the bit-serial dimension are fully exposed to the programmer for algorithmic space/time tradeoff. Other features include

- software support for interactive bit-plane graphics on SUN 3/4 workstations with single-step/animation display at 20 frames per second (512 × 512 images);
- input/output of variable-length integers expressed as decimal or hexadecimal values;
- variable-precision, unsigned-integer arithmetic, including addition, subtraction, multiplication, division, and GCD computations; and
- callable procedures from both C and Fortran.

In summary, the PASSWORK system demonstrates that a vector machine can provide the best of both SIMD and MIMD worlds in one shared-memory machine architecture. The only significant performance limits to SIMD computing on a vector machine are memory bandwidth, the ability to efficiently corner-turn data in a vector register, and the ability to perform multiple logical operations in a single tick.

In contrast to real SIMD machines, a vector machine can more easily trade space for time and provide the exact amount of parallelism needed to solve an actual problem. In addition, global operations like processor tally and global "or" are performed much faster on vector machines than on real SIMD machines.

In my opinion, the SIMD model of computation is much more applicable to general problem solving than is realized today. Causes for this

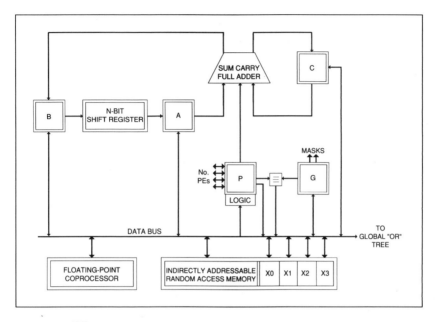

Figure 3. MPP model.

may be more psychological than technical and are possibly due to a Catch 22 between the availability of SIMD research tools and real SIMD machines. Simulators like PASSWORK are keys to breaking this Catch 22 by providing a portable SIMD programming environment for developing new parallel algorithms on conventional machines. Ideally, results of this research will drive the design of even higher-performance SIMD engines.

Related to this last remark, SRC has initiated a research project called PETASYS to investigate the possibility of doing SIMD computing in the memory address space of a general-purpose machine. The basic idea here is to design a new kind of memory chip (a process-in-memory chip) that associates a single-bit processor with each column of a standard RAM. This will break the von Neumann bottleneck between a CPU and its memory and allow a more natural evolution from MIMD to a mixed MIMD/SIMD computing environment.

Applications in this mixed computing environment are just now beginning to be explored at SRC. One of the objectives of the PETASYS Project is to design a small-scale PETASYS system on a SUN 4 platform with 4K SIMD processors and a sustained bit-serial performance of 10 gigabit operations per second. Scaling this performance into the supercomputing arena should eventually provide a sustained SIMD performance of 10^{15} bit operations per second across 64 million SIMD processors. The Greek prefix *peta*, representing 10^{15}, suggested a good name for this SRC research project and potential supercomputer—PETASYS.

Vectors Are Different

Steven J. Wallach

Steven J. Wallach, a founder of CONVEX Computer Corporation, is Senior Vice President of Technology and a member of the CONVEX Board of Directors. Before founding CONVEX, Mr. Wallach served as product manager at ROLM for the 32-bit mill-spec computer system. From 1975 to 1981, he worked at Data General, where he was the principal architect of the 32-bit Eclipse MV superminicomputer series. As an inventor, he holds 33 patents in various areas of computer design. He is featured prominently in Tracy Kidder's Pulitzer Prize-winning book, The Soul of a New Machine.

Mr. Wallach received a B.S. in electrical engineering from Polytechnic University, New York, an M.S. in electrical engineering from the University of Pennsylvania, and an M.B.A. from Boston University. He serves on the advisory council of the School of Engineering at Rice University, Houston, and on the external advisory council of the Center for Research on Parallel Computation, a joint effort of Rice/Caltech/Los Alamos National Laboratory. Mr. Wallach also serves on the Computer Systems Technical Advisory Committee of the U.S. Department of Commerce and is a member of the Board of Directors of Polytechnic University.

In the late 1970s, in the heyday of Digital Equipment Corporation (DEC), Data General, and Prime, people were producing what we called minicomputers, and analysts were asking how minicomputers were different from an IBM mainframe, etc. We used to cavalierly say that if it was made east of the Hudson River, it was a minicomputer, and if was made west of the Hudson River, it was a mainframe.

At the Department of Commerce, the Computer Systems Technical Advisory Committee is trying to define what a supercomputer is for export purposes. For those who know what that entails, you know it can be a "can of worms." In Figure 1, I give my view of what used to be the high-end supercomputer, which had clock cycle X. What was perceived as a microcomputer in the late 1970s was 20X, and perhaps if it came out of Massachusetts, it was 10- to 15X. Over time, we have had three different slopes, as shown in Figure 1. The top slope is a RISC chip, and that is perhaps where CONVEX is going, and maybe this is where Cray Research, Inc., and the Japanese are going. We are all converging on something called the speed of light. In the middle to late 1990s, the clock-cycle difference between all levels of computing will be, at best, four to one, not 20 to one.

The next question to consider is, how fast do things have to be to eliminate vectors and use scalar-only processors? That is why I titled my talk "Vectors are Different."

I think the approach to take is to look at both the hardware and the software. If you only look at the hardware, you will totally miss the point.

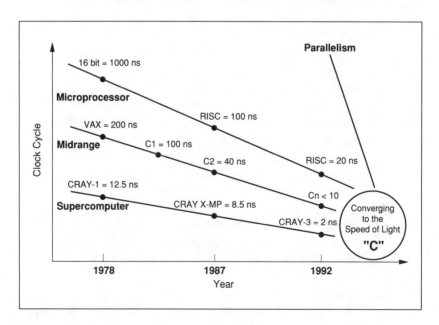

Figure 1. Clock cycle over time since the late 1970s.

I used to design transistor amplifiers. However, today I tend to be more comfortable with compiler algorithms. So what I am going to do is talk about vectors, first with respect to software and then with respect to hardware, and see how that relates to scalar processing.

First, the reason we are even asking the question is because we have very highly pipelined machines. But, as Harvey Cragon pointed out earlier in this session, as have others, we have a history of vectorizing both compilers and coding styles. I would not say that it is "cookie-cutter" technology, but it is certainly based on the pioneering work of such people as Professors David Kuck and Ken Kennedy (presenters in Session 5).

I ran the program depicted in Figure 2 on a diskless SUN with one megabyte of storage. I kept wondering why, after half an hour, I wasn't getting the problem done. So I put a PRINT statement in and realized that about every 15 to 20 minutes I was getting another iteration. The problem was that it was page-faulting across the network. I then ran that benchmark on five different RISC machines or workstations, which are highly pipelined machines that are supposed to put vector processing out of business.

Examine Table 1 row by row, not column by column. What is important is the ratio of row entries for the same processor. The right way to view Table 1 means i is on the inner loop; the wrong way means j is on the inner loop. On the MIPS (millions of instructions per second), it was a

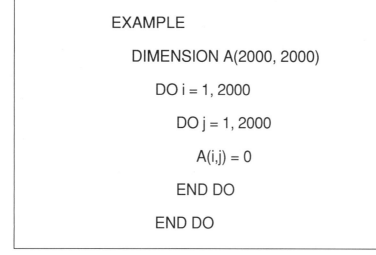

EXAMPLE

DIMENSION A(2000, 2000)

DO i = 1, 2000

DO j = 1, 2000

A(i,j) = 0

END DO

END DO

Figure 2. Vectors—software.

Table 1. Vectors: Software Running Time, in Seconds

	MIPS[a] R6000 60 MHz f77-03	MIPS[b] R3000 33 MHz f77-03	Solbourne[c] 33 MHz SPARC	RIOS 530[d]	RIOS 520[e]
Real *4 Right Way	0.45	0.43	1.7	0.36	0.5
Real *8 Right Way	0.88	2.43	X	0.403	0.64
Real *4 Wrong Way	1.25	4.76	5.4	7.095	8.56
Real *8 Wrong Way	2.0	10.4	X	7.148	26.15

[a] 16 Kb/512 Kb Cache
[b] 64 Kb Cache
[c] 128 Kb Cache
[d] 40 ns, 64 Kb Cache
[e] 40 ns, 32 Kb Cache

difference of three to one. That is, if I programmed DO j, DO i versus DO i, DO j, it was three to one for single precision.

Double precision for the R6000 was approximately the same, which was the best of them because it had the biggest cache—a dual-level cache of R6000 (Table 1). Single precision for the R3000 was about 10 to one, and double precision was five to one. Now, that may sound counter-intuitive. Why should it be a smaller ratio with a bigger data type? The reason is that you start going nonlinear faster because it is bigger than the cache size. The Solbourne, which is the SPARC, is three to one.

We couldn't get the real *8 to run because it had a process size that was too big relative to physical memory, and we didn't know how to fix it. Then we ran it on RIOS (reduced instruction set computing), the hottest thing. The results were interesting—20 to one and 20 to one. Now you get the smaller cache, and in this case, you get almost 30 to one.

What does this all mean? What it really means to me is that to use these highly pipelined machines, especially if they all have caches, we're basically going to vectorize our code anyway. Whether we call it vectors or not, we're going to code DO j, DO i, just to get three or 10 times the performance. It also means that the compilers for these machines are going to have to do dependency analysis and all these vector transformations to overcome this.

In our experience at CONVEX, we have taken code collectively and "vectorized" it, so that 99 times out of 100, you put it back on the scalar machine and it runs faster. The difference is, if it only runs 15 per cent faster, people say, "Who cares? It's only 15 per cent." Yet, if my mainline machine is going to be this class of scalar pipeline machines, 15 per cent actually is a big deal. As the problem size gets bigger, that increase in performance will get bigger also. By programming on any of those workstations, I got anywhere from three to 20 times faster.

No matter what we do, I believe our compiler is going forward, and soon these will be vectorizing compilers. We just won't say it, because it would do vector transformations anyway.

Now let's look at hardware. The goals of hardware design are to maximize operations on operands and use 100 per cent of available memory bandwidth. When I consider clock cycles and other things, I design computers by starting with the memory system, and I build the fastest possible memory system I can and make sure the memory is busy 100 per cent of the time, whether it's scalar or vector. It's very simple—you tell me a memory bandwidth, and I'll certainly tell you your peak performance. As we all know, you very rarely get the peak, but at least I know the peak.

The real issue as we go forward is, if we want to get performance, we have to design high-speed memory systems that effectively approach the point where we get an operand back every cycle. If we get an operand back at every cycle, we're effectively designing a vector memory system, whether we call it that or not. In short, memory bandwidth determines the dominant cost of a system.

Also, as we build machines in the future, we're beginning to find out that the cost of the CPU is becoming lower and lower. All the cost is in the memory system, the crossbar, the amount interleaving, and the mechanics. Consequently, again, we're paying for bandwidth. Figure 3 shows you some hard numbers on these costs, and these numbers are not just off the top of my head.

The other thing that we have to look at is expandability, which adds cost. If you have a machine that can go from baseline X to 4X, whatever that means, you have to design that expandability in from the start. Even if you buy the machine with the baseline configuration, there is overhead, and you're paying for the ability to go to 4X. For example, look at the IBM RS/6000, a workstation that IBM lists for $15,000. But it's not expandable in the scheme of things. It's a uniprocessor.

```
┌─────────────────────────────────────────────────┐
│                                                 │
│   • IBM RS/6000                                 │
│                                                 │
│      Workstation           Servers              │
│      ──────────            ───────              │
│       $15,000             $100,000              │
│                                                 │
│   • Same company, same chip set                 │
│                                                 │
│   • Cost in expandability of I/O, CPU, and memory│
│                                                 │
└─────────────────────────────────────────────────┘
```

Figure 3. Vectors—hardware.

Now let's examine the server version of the RS/6000. We are using price numbers from the same company and using the same chips. So the comparison in cost is apples to apples. When I run a single-user benchmark with little or no I/O (like the LINPACK), I probably get the same performance on the workstation as I do on the server, even though I pay six times more for the server. The difference in price is due to the server having expandability in I/O, CPU, and physical memory.

Workstations generally don't have this expandability. If they do, they start to be $50,000 single-seat workstations. There were two companies that used to make these kinds of workstations, which seems to prove that $50,000 single-user workstations don't sell well.

What is the future, then? For superscalar hardware, I think every one of these machines eventually will have to have vector-type memory systems to sustain CPU performance. If I can't deliver an operand every cycle, I can't operate it on every cycle. You don't need a Ph.D. in computer science to figure that out. If you have a five-inch pipe feeding water to a 10-inch pipe, you can expand the 10-inch pipe to 20 inches; but you're still going to have the same flow through the pipe. You have to have it balanced. Because all these machines so far have caches, you still need loop interchange or other vector transformations.

You're going to see needs for "blocking algorithms," that is, how to take matrices or vectors and make them submatrices to get the performance. I call it "strip mining." In reality, it's the same thing. That is, how do I take a data set and divide it into a smaller data set to fit into a register or memory to get high-speed performance? Fundamentally, it's the same type of algorithm.

So I believe that, again, we're all going to be building the memory systems, and we're going to be building the compilers. Superscalar hardware will become vector processsors in practice, but some people won't acknowledge it.

Now, what will happen to true vector machines? As Les Davis pointed out in the first presentation in this session, they're still going to be around for a long time. I think more and more we're going to have specialized functional units. A functional unit is something like a multiplier or an adder or a logical unit. Very-large-scale integration will permit specialized functional units to be designed, for example, $O(N^2)$ operations for $O(N)$ data. The type of thing that Ken Iobst was talking about with corner-turning (see the preceeding paper, this session) is an example of that. A matrix-multiply type of thing is another example, but in this case it uses N^2 data for N^3 operations. More and more you'll see more emphasis on parallelism (MIMD), which will evolve into massive parallelism.

The debate will continue on the utility of multiple memory paths. I view that debate as whether you should have one memory pipe, two memory pipes, or N memory pipes. I've taken a very straightforward position in the company: there's going to be one memory pipe. Not all agree with me.

I look at benchmarks, and I look at actual code, and I find that the clock cycle (the memory bandwidth of a single pipe) seems to be more of a determining factor. If I'm going to build a memory system to tolerate all the bandwidth of multiple pipes, because of parallelism, I believe, rather than having eight machines with three pipes, I'd rather have a 24-processor machine and utilize the bandwidth. I'd rather utilize the bandwidth that way because I get more scalar processing. Because of parallel compilers, I'll probably get more of that effective bandwidth than someone else will with eight machines using three memory pipes.

One other issue: there shall be no Cobol compiler. I have my lapel button that says, "Cobol can be eliminated in our lifetime if we have to." In reality, ever since we've been programming computers, we've had the following paradigm, although we just didn't realize it: we had that part of the code that was scalar and that part of the code that was vector. We go back and forth. You know, nothing is ever 100 per cent vectorizable, nothing is 100 per cent parallelizable, etc. So I think, going forward, we'll see machines with all these capabilities, and the key is getting the compilers to do the automatic decomposition for this.

If we're approaching the speed of light, then, yes, we're all going to go parallel—there's no argument, no disagreement there. But the real issue is, can we start beginning to do it automatically?

With gallium arsenide technology, maybe I can build an air-cooled, four- or five-nanosecond machine. It may not be as fast as a one-nanosecond machine, but if I can get it on one chip, maybe I can put four of them on one board. That may be a better way to go than a single, one-nanosecond chip because, ultimately, we're going to go to parallelism anyway.

The issue is that, since we have to have software to break this barrier, we have to factor that into how we look at future machines, and we can't just let clock cycle be the determining factor. I think if we do, we're "a goner." At least personally I think, as a designer, that's the wrong way to do it. It's too much of a closed-ended way of doing it.

4

Scalable Parallel Systems

This session focused on the promise and limitations of architectures that feature a large number of homogeneous or heterogeneous processing elements. Panelists discussed the limitations, dependencies, sustained performance potential, processor performance, interconnection topologies, and application domains for such architectures.

Session Chair

Stephen Squires,
Defense Advanced Research Projects Agency

Symbolic Supercomputing

Alvin Despain

Alvin M. Despain is the Powell Professor of Computer Engineering at the University of Southern California, Los Angeles. He is a pioneer in the study of high-performance computer systems for symbolic calculations. To determine design principles for these systems, his research group builds experimental software and hardware systems, including compilers, custom very-large-scale integration processors, and multiprocessor systems. His research interests include computer architecture, multiprocessor and multicomputer systems, logic programming, and design automation. Dr. Despain received his B.S., M.S., and Ph.D. degrees in electrical engineering from the University of Utah, Salt Lake City.

This presentation discusses a topic that may be remote from the fields most of you at this conference deal in—symbolic, as opposed to numeric, supercomputing. I will define terms and discuss parallelism in symbolic computing and architecture and then draw some conclusions.

If supercomputing is using the highest-performance computers available, then symbolic supercomputing is using the highest-performance symbolic processor systems. Let me show you some symbolic problems and how they differ from numeric ones.

If you're doing the usual supercomputer calculations, you use LINPAC, fast Fourier transforms (FFTs), etc., and you do typical, linear-algebra kinds of operations. In symbolic computing, you use programs like MACSYMA, MAPLE, Mathematica, or PRESS. You provide symbols, and you get back not numbers but formulae. For example, you get the solution to a polynomial in terms of a formula.

Suppose we have a problem specification—maybe it is to model global climate. This is a big programming problem. After years of effort programming this fluid-dynamics problem, you get a Fortran program. This is then compiled. It is executed with some data, and some results are obtained (e.g., the temperature predicted for the next hundred years). Then the program is generally tuned to achieve both improved results and improved performance.

In the future you might think that you start with the same problem specification and try to reduce the programming effort by automating some of the more mundane tasks. One of the most important things you know is that the programmer had a very good feel for the data and then wired that into the program. If you're going to automate, you're going to have to bring that data into the process.

Parameters that can be propagated within a program constitute the simplest example of adjusting the program to data, but there are lots of other ways, as well. Trace scheduling is one way that this has been done for some supercomputers. You bring the data in and use it to help do a good job of compiling, vectorizing, and so on. This is called partial evaluation because you have part of the data, and you evaluate the program using the data. And this is a symbolic calculation.

If you're going to solve part of the programming problem that we have with supercomputers, you might look toward formal symbolic calculation. Some other cases are optimizing compilers, formal methods, program analysis, abstract interpretation, intelligent databases, design automation, and very-high-level language compilers.

If you look and see how mathematicians solve problems, they don't do it the way we program Cray Research, Inc., machines, do they? They don't do it by massive additions and subtractions. They integrate together both symbolic manipulations and numeric manipulations. Somehow we have to learn how to do that better, too. It is an important challenge for the future. Some of it is happening today, but there's a lot to be done.

I would like to try to characterize some tough problems in the following ways: there is a set of problems that are numeric—partial differential equations, signal processing, FFTs, etc.; there are also optimization problems in which you search for a solution—linear programming, for example, or numerical optimization of various kinds. At the symbolic level you also have simulation. Abstract interpretation is an example. But you also have theorem proving, design automation, expert systems, and artificial intelligence (AI). Now, these are fundamentally hard problems. In filter calculations (the easy problems), the same execution occurs no matter what data you have. For example, FFT programs will always execute the same way, no matter what data you're using.

With the hard problems, you have to search. Your calculation is a lot more dynamic and a lot more difficult because the calculation does depend upon the data that you happen to have at the time. It is in this area that my work and the work of my group have focused: how you put together symbols and search. And that's what Newell and Simon (1976) called AI, actually. But call it what you like.

I want to talk about two more things: concurrency and parallelism. These are the themes of this particular session. I'd like to talk about the instruction-set architecture, too, because it interacts so strongly with concurrency and parallelism. If you're building a computer, one instruction type is enough, right? You build a Turing machine with one instruction. So that's sufficient, but you don't get performance.

If you want performance, you'd better add more instructions. If you have a numeric processor, you include floating-point add, floating-point multiply, and floating-point divide. If you have a general-purpose processor, you need operations like load, store, jump, add, and subtract. If you want a symbolic processor, you've got to do things like binding two symbols together (binding), dereferencing, unifying, and backtracking. To construct a symbolic processor, you need the general-purpose features and the symbolic operations.

Our latest effort is a single-chip processor called the Berkeley Abstract Machine (BAM). This work has been sponsored by the Defense Advanced Research Projects Agency. For our symbolic language, we have primarily used Prolog, but BAM is not necessarily dependent on it.

Now, I'd like to tell you a little bit about it to illustrate the instruction-set architecture issues involved, especially the features of the BAM chip that boost performance. These are the usual general-purpose instructions—load, store, and so on. There's special support for unification. Unification is the most general pattern match you can do. Backtracking is also supported so that you can do searching and then backtrack if you find it wrong. The architecture features include tags, stack management, special registers, and a microstore—that is, internal opcodes. There is pipeline execution to get performance and multiple I/O ports for address, data, and instructions.

In this processor design we considered what it costs to do symbolic calculations in addition to the general-purpose calculations. We selected a good set of all possible general-purpose instructions to match with the symbolic, and then we added what was needed to get performance.

If you add a feature, you get a corresponding improvement in performance. See Figure 1, which graphs the percentage increase in performance. The cycle count varies between one and two, one being BAM as a benchmark. We took all combinations of features that we could find and with simulation tried to understand what cost-performance tradeoffs can be achieved.

Some cost-performance combinations aren't very good. Others are quite good, and the full combination is quite good. The net result is that an 11 per cent increase in the silicon area of a single-chip microcomputer, BAM, results in a 70 per cent increase in the performance on symbolic calculations. So that's what we chose for BAM. It doesn't cost very much to do the symbolic once you have the general-purpose features.

The BAM chip features 24 internal microinstructions and 62 external ones. It achieves about 1.4 cycles per instruction. However, because of dereferencing, the number of cycles per instruction is indefinitely large. Simulations indicated that the chip would achieve about 24 million instructions per second, or about three million logical inferences per second (i.e., about 3 MLIPS). A logical inference is what you execute for symbolic computing. It's a general pattern match, and if it succeeds, you do a procedure call, execution, and return.

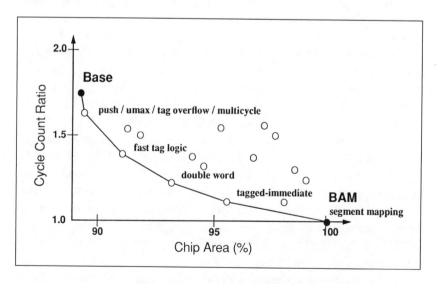

Figure 1. The benefits of adding architectural features: an 11 per cent increase in silicon area yields a 70 per cent increase in performance.

We submitted this chip for fabrication. The chip has now been tested, and it achieved 3.8 MLIPS.

Consider the performance this represents. Compare, for instance, the Japanese Personal Sequential Inference (PSI) machine, built in 1984. It achieved 30,000 LIPS. A few months later at Berkeley, we built something called the Prolog Machine (PLM), which achieved 300,000 LIPS, even then, a 10-fold improvement. The best the Europeans have done so far is the Knowledge Crunch Machine. It now achieves about 600,000 LIPS.

The best the Japanese have done currently in a single processor is an emitter-coupled-logic machine, 64-bit-wide data path, and it achieved a little over an MLIPS, compared with BAM's 3.8 MLIPS. So the net result of all of this is that we've been able to demonstrate in six years' time a 100-fold improvement in performance in this domain.

The PLM was put into a single chip, just as BAM is a single chip. The PSI was not; it was a multiple-chip system. I think what's important is that you really have to go after the technology. You must also optimize the microarchitecture, the instruction-set architecture, and especially the compiler. Architecture design makes a big difference in performance; it's not a dead issue. And architecture, technology, and compilers all have to be developed together to get these performance levels.

Let me say something about scaling and the multiprocessor. What about parallelism? Supercomputers have many levels of parallelism—parallel digital circuits at the bottom level, microexecution, multiple execution per instruction, multiple instruction streams, multiprocessing, shared-memory multiprocessors, and then heterogeneous multiprocessors at the top. And we've investigated how symbolic calculations play across this whole spectrum of parallelism. If you really want performance, you have to play the game at all these different levels of the hierarchy. It turns out that parallelism is more difficult to achieve in symbolic calculations. This is due to the dynamic, unpredictable nature of the calculation. But on the plus side, you get, for instance, something called superlinear speedup during search.

But as in numerics, the symbolic algorithms that are easy to parallelize turn out to be poor in performance. We all know that phenomenon, and it happens here, too. But there are some special cases that sometimes work out extremely well. What we're trying to do with BAM is identify different types of parallel execution so that you can do something special about each type. BAM handles very well the kind of parallelism requiring you to break a problem into pieces and solve all the pieces simultaneously. With BAM, parallelism can spread across networks, so you have all-solution, or-parallelism, where you find a whole set of answers to a problem rather than just one.

However, if you're doing a design, all you want is one good design. You don't want every possible design. There are too many to enumerate. And that's been our interest, and it works pretty well on multiprocessors. Unification parallelism, pattern matching, can be done in parallel, and we do some of that within the processor.

Now, let's say you have a BAM chip and a shared-memory cache with the switch and connections to some external bus memory and I/O. Call that a node. Put that together with busses into what we call multi-multi.

Gordon Bell (1985), a Session 8 presenter, wrote a great paper, called "Multis: A New Class of Multiprocessor Computers," about a shared-memory, single-bus system. It turns out you can do the same trick in multiple dimensions and have yourself a very-large-scale, shared-memory, shared-address-space multiprocessor, and it looks like that's going to work. We'll find out as we do our work.

I think that for a modest cost, you can add powerful symbolic capability to a general-purpose machine. That's one of the things we've learned very recently.

Parallel symbolic execution is still a tough problem, and there is still much to be learned. The ultimate goal is to learn how to couple efficiently, in parallel, both symbolic and numeric calculations.

References

C. G. Bell, "Multis: A New Class of Multiprocessor Computers," *Science* **288**, 462–467 (1985).

A. Newell and H. Simon, "Computer Science as Empirical Inquiry; Symbols and Search," *Communications of the ACM* **19** (3), 113–126 (1976).

Parallel Processing: Moving into the Mainstream

H. T. Kung

H. T. Kung joined the faculty of Carnegie Mellon University in 1974, after receiving his Ph.D. there. Since 1992 he has been serving as the Gordon McKay Professor of Electrical Engineering and Computer Science at Harvard University. During a transition period, he continues his involvement with projects under way at Carnegie Mellon. Dr. Kung's research interests are in high-speed networks, parallel-computer architectures, and the interplay between computer and telecommunications systems. Together with his students, he pioneered the concept of systolic-array processing. This effort recently culminated in the commercial release by Intel Supercomputer Systems Division of the iWarp parallel computer.

In the area of networks, Dr. Kung's team has developed the Nectar System, which uses fiber-optic links and large crossbar switches. A prototype system employing 100 megabits/second links and more than 20 hosts has been operational since early 1989. The team is currently working with industry on the next-generation Nectar, which will employ fibers operating at gigabits/second rates. The gigabit Nectar is one of the five testbeds in a current national effort to develop gigabits/second wide-area networks. His current network research is directed toward gigabit, cell-based local-area networks capable of guaranteeing performance.

I will focus on three key issues in parallel processing: computation models, interprocessor communication, and system integration. These issues are important in moving parallel processing into the mainstream of computing. To illustrate my points, I will draw on examples from the systems we are building at Carnegie Mellon University—iWarp and Nectar. iWarp is a fine-grain parallel machine developed under a joint project with Intel. Nectar is a network backplane that connects different kinds of machines together.

In discussing computation models, we are really trying to address some of the most difficult problems that people are having with parallel computers. Namely, it can be very difficult to write code for these machines, and the applications codes are not portable between parallel machines or between sequential machines and parallel machines. That has been very troublesome.

There have been attempts to solve these problems. For example, some theoretically oriented researchers have come up with the PRAM model, which presents to the user a parallel computation model that hides almost all the properties of a parallel machine. However, because of its high degrees of generality and transparency, the PRAM model does not exploit useful properties such as the locality or regularity that we worked on so diligently for the last 30 years in order to achieve high-performance computing. What we want is more specialized models—in the beginning, at least.

Therefore, I propose to work on those models that users really understand. For example, one thing we have been doing is that, for each specific area for which we have a good understanding of its computation characteristics, we will develop a parallelizing compiler, although we do not call it a compiler because it is not so general purpose. Instead, we call it a parallel program generator (Figure 1). We start with the specifications, without any detailed knowledge about the underlying parallel machines. Then we have a compiler that will generate code for the specific parallel machines that would have SEND and RECEIVE instructions. So the users can work on a high level in a machine-independent manner.

A concrete example is the APPLY compiler, or parallel program generator (Figure 2) used for many image-processing computations defined in terms of localized operations. That is, each output pixel depends on the small neighborhoods of the corresponding input pixel. In APPLY, the software can generate code for each processor and also do the boundary operations, all automatically.

Scalable Parallel Systems 103

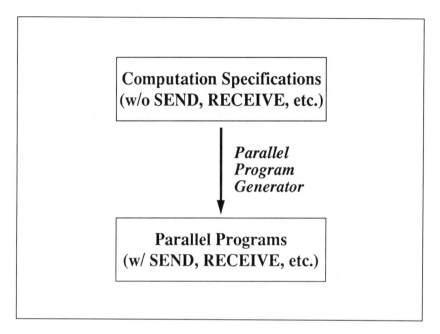

Figure 1. Parallel program generators for special computation models.

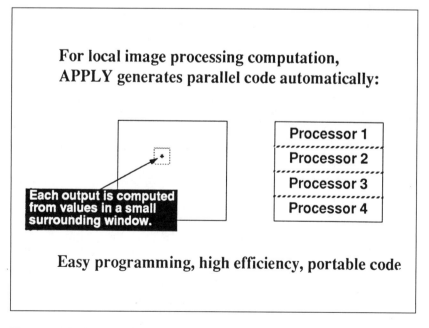

Figure 2. APPLY: a parallel program generator for data parallelism using local operators.

Currently APPLY is operational for Warp, iWarp, Transputers, and a couple of other machines. Actually, it generates code for sequential machines, as well. So you can develop your APPLY programs in a comfortable environment of a sequential machine. Intel is going to support APPLY for the iWarp.

Another program generator developed by Carnegie Mellon is called AL, which will generate matrix operations automatically. Actually, we generate much of the LINPACK code on Warp using this language. The language basically is like Fortran but allows hints to the compiler about roughly how arrays should be partitioned, e.g., in certain directions, onto a parallel machine. Then the code for the specific parallel machine will be generated automatically.

We also have a very high-level parallel program generator, called ASSIGN, for large signal flow graphs with nodes that are signal processing operations like fast Fourier transforms, filters, and so on (Figure 3). You just use the fact that there are so many graph nodes that you have to deal with; as a result, you usually can load balance them by mapping an appropriate number of graph nodes onto each processor. Again, this mapping has been done automatically.

One of the most difficult things, so far, for a parallel machine to handle is branch-and-bound types of operations (Figure 4), which are similar to searching a tree, for example. In this type of operation you do a lot of backtracking, which can depend on the computation you are doing at run time. For example, you might sometimes like to go deep so that you can do a useful operation, but you also want to go breadth so that you can increase concurrency. Usually only the user has this rough idea of what the priority is between the depth-first and the breadth-first search. Yet, today we do not even have languages that can express that idea. Even if the user knows that you should go depth a little before you go breadth, we still do not know how to say it to the parallel machine. So we are pretty far away from having a general computation model.

A strategy that I propose is to gain experience in a few computation models for special application areas. At the same time, we should develop insights for the more general models.

The second key issue in parallel processing is interprocessor communication. After all, parallel machines are about communications between different processors. A lot of the parallel machines that people have today are really built out of processing elements that are not good for communication. As a result, some vendors are telling people that their parallel machines are great—but only if you don't communicate!

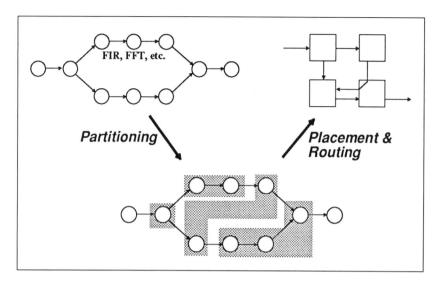

Figure 3. ASSIGN: a parallel program generator for signal flow graphs.

Consider, for example, tree-search processes such as branch-and-bound and α-β pruning:

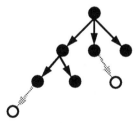

We don't know how to express priority between breadth-first and depth-first search.

Figure 4. General models are not quite there yet.

We are trying to make a processor that is good for computation and good for communication. In the case of iWarp, you can build a single component with communication and computation on the same chip, and then you can use this component to form different parallel processing arrays (Figure 5). Once you have such a processor array, you can program each single cell using C and Fortran in the conventional manner. Then you can use parallelizing compilers, or parallel program generators as described above, to generate parallel code for the array.

The iWarp component itself has both a computation and communication agent (Figure 6). Most significant is that the communication part of the chip can do a total of 320 megabytes/second I/O, whereas other current components can do no more than 10 megabytes/second. In addition, we have 160 megabytes/second I/O for the local memory. If you add them up, that is 480 megabytes/second in the current version.

iWarp has three unique innovations in communication: high-bandwidth I/O, systolic communication, and logical channels.

Obviously, in any parallel machine, you have got to be good in I/O because you have got to be able, at least, to get the input data quickly. For example, if you have a high-performance parallel interface (HIPPI) channel with a 100 megabytes/second I/O bandwidth, you can probably get an interface of such bandwidth into the array (see Figure 7). The challenge, however, will be how to distribute the input data onto multiple

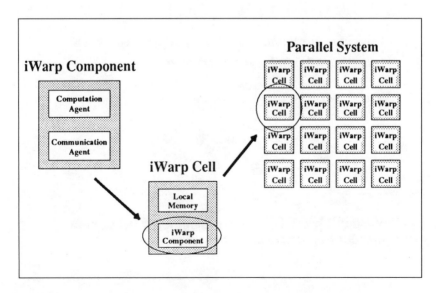

Figure 5. iWarp: a VLSI building block for parallel systems.

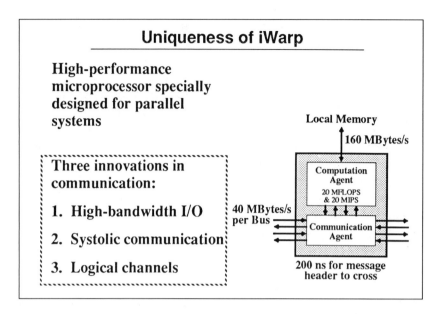

Figure 6. A single-chip component capable of 480 megabytes/second I/O.

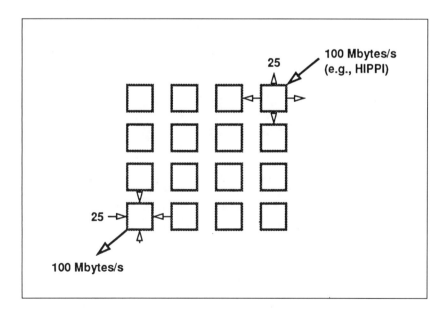

Figure 7. High-bandwidth distribution and collection of data in processor array.

processing elements. For this distribution we need a large I/O bandwidth per processor element.

For iWarp we have eight links, and each of them is 40 megabytes/second. In particular, we can simultaneously use four links each at 25 megabytes/second to distribute a 100 megabyte/second input in parallel and another four links to distribute the 100 megabyte/second output.

Besides the I/O bandwidth, the other important thing about each processor is easy and direct access to I/O. This is done in iWarp through the systolic communication mechanism. Figure 8 compares systolic communication with the traditional memory-based, message-passing communication.

One of the things that any parallel machine should do well is to distribute an array—say, a row of an image into a number of processing elements. In iWarp a processor can send out a row in a single message to many other processing elements and have the first processing element take one end of the row (Figure 9). Then the first processor can change the message header, redirect the rest of the message to go into the second processing element, and so on. Note that the sender only needs to send out one message to distribute the data to many other processors.

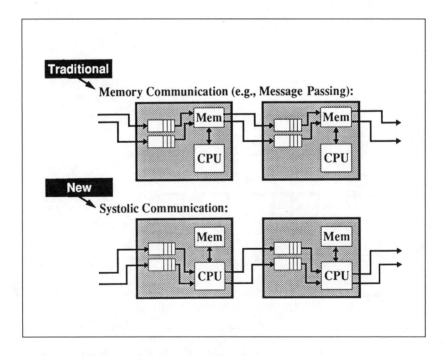

Figure 8. Systolic communications in iWarp.

Scalable Parallel Systems 109

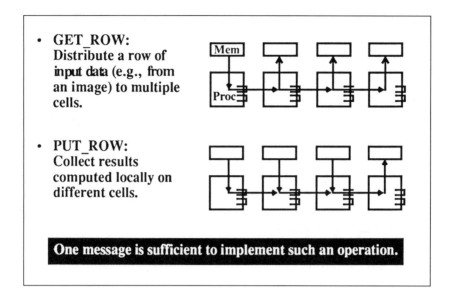

Figure 9. Message redirection in iWarp.

Logical channels are an important concern. No matter how many pins or how many connectors you have, you never have enough to support some applications. People always want more connectivity so that they can map the computation on the parallel machine easily and so that they can do a reconfiguration of the parallel-processor array more easily. Therefore, it is very useful to time-multiplex the wire by using hardware support so that logically you can imagine having many, many wires instead of having one physical wire.

For iWarp we can have 20 logical connections in and out from each chip. For example, we can have both blue and red connections happening at the same time. A programmer can use the blue connection for some computation. In the meantime, the system message can always go on using the red connection without being blocked by the programmer's computation. Therefore, you can reserve some bandwidth logically for system use (Figure 10). Once you have these kinds of logical connections, you will also find it very easy to reconfigure a processor array. For example, to avoid a faulty node, you can just route it around using logical connections. Because you have so many logical connections available, the routing will be easy.

The message here is that although we have already seen some successful parallel machines, we have not yet seen the really good parallel

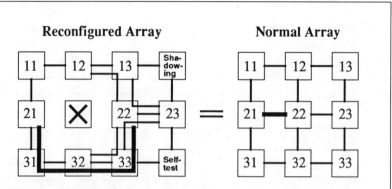

Figure 10. Reconfiguration for fault tolerance: an application of logical channels.

machine that would support many very basic interprocessor communication operations that applications will typically need. In the future, parallel machines built out of processing building blocks that inherently support efficient and flexible interprocessor communication will be much easier to use.

The last key issue in parallel processing I will address is system integration. Parallel machines are not suited for sequential operations, by definition. Thus, parallel machines typically need to be integrated in a general computing environment. The more powerful a machine is, the more accessible it ought to be. Ideally, all of these computers should be connected together. Actually, this kind of configuration is happening in almost all high-performance computing sites (Figure 11).

At Carnegie Mellon, we are building such a network-based high-performance computing system, called Nectar (Figure 12). Nectar has a general network that supports very flexible, yet high-bandwidth and low-latency, communication. The Nectar demo system (Figure 13) connects about 26 hosts. Currently, we can do interhost communication at about 100 megabits/second, with a latency of about 170 microseconds. Nectar supports transmission control protocol/Internet protocol (TCP/IP). If TCP checksum is turned out, the CAB-to-CAB (i.e., without

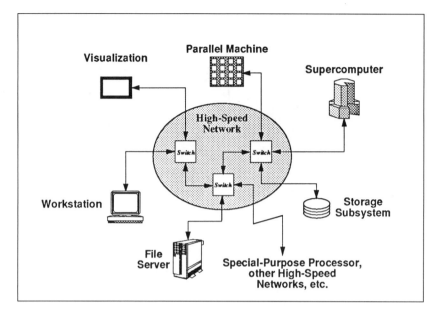

Figure 11. Accessibility in a general-computing environment.

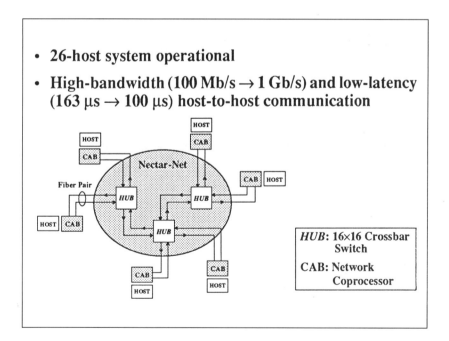

- **26-host system operational**
- **High-bandwidth (100 Mb/s → 1 Gb/s) and low-latency (163 μs → 100 μs) host-to-host communication**

HUB: 16×16 Crossbar Switch

CAB: Network Coprocessor

Figure 12. The Nectar System at Carnegie Mellon University.

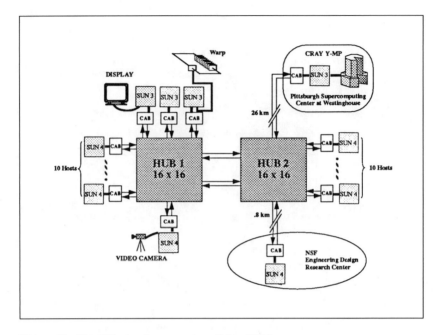

Figure 13. Host Nectar demo system (May 1990).

crossing the host VME bus) TCP/IP bandwidth can be close to the peak link bandwidth of 100 megabits/second for large packets of 32 kilobytes. It's reasonable to consider the case where TCP checksum is turned off because this checksum can be easily implemented in hardware in the future.

We believe that over the next 18 months, we can build a Nectar-like system with much improved performance. In particular, we're building a HIPPI Nectar, which will support the 800 megabits/second HIPPI channels. For TCP/IP, the goal is to achieve at least 300 megabits/second bandwidth. With Bellcore we are also working on an interface between HIPPI networks and telecommunication ATM/SONET networks.

Figure 14 shows a network-based multicomputer configuration that we proposed about six months ago. With this kind of a network, you can literally have a system capable of 10^{11} or even 10^{12} floating-point operations per second, without even building any new machines.

In summary, at Carnegie Mellon we are making progress on the key issues in parallel processing that I have discussed.

> High-Speed = "High-Bandwidth" & "Low-Latency"

- **High-bandwidth** network can sustain high-speed hosts.
- **Low-latency** network enables concurrent processing of small-grain computations by different hosts.

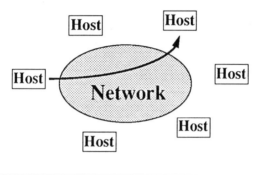

Figure 14. High-speed networks make high-performance computing over network feasible.

It's Time to Face Facts

Joe Brandenburg

Joe Brandenburg is manager of the Computational Sciences Group at Intel Supercomputer Systems Division. He is an expert in parallel programming, having applied his expertise to the fields of artificial intelligence, finance, and general high-performance computing and architectures. He received a Ph.D. in mathematics from the University of Oregon in 1980 and an M.S. in computer science from the University of Maryland in 1983. He has been a member of the technical staff at Intel since 1985.

I am sure that everyone is aware of the efforts of Intel in the last five years in building parallel processors. We started out with what was really a toy in order to learn something about writing parallel programs, and then we developed a machine that was actually competitive with minis and mainframes. We now have a machine that does compete with supercomputers, with the goal to build what we are starting to call ultracomputers. We want to go beyond the supercomputer type of machine.

We see the building of these kinds of computers—these high-performance computers—as having lots of pieces, all of which I will not be able to discuss here. Fundamentally, in building these types of machines, we have to deal with the processor, the memory, and the interconnection hardware. We have to deal with the operating system, the compilers, and the actual software needed for the applications.

Let me begin by discussing building a machine that is capable of 10^{12} floating-point operations per second (TFLOPS) and that is based on traditional supercomputing methods. If I take today's supercomputing capabilities—that is, an MFLOPS processor, approximately a GFLOPS machine—and want to build a TFLOPS machine, I'd have to have 1000 processors put together (Figure 1). Of course, we don't really know how to do that today, but that would be what it would take. If I want to build it in a 10-year time frame, I still am going to have to put together hundreds of these supercomputers—traditional kinds of CPUs. That is not going to be easy, and it is not going to be cheap.

If I want to build a TFLOPS machine with the microprocessors and I want to do that today, I'd have to put together some 10,000 processors (Figure 2). By the year 1995, that would be 3000 or 4000 of these processors, and by the year 2000, I'd need only 1000 processors.

Today we know how to put together hundreds of these things. We believe that within a couple of years, we will be producing machines that have thousands of processors, and therefore in the year 2000, we will definitely be able to put together 1000 of these machines. That will actually be a step backward for us. From a processor point of view, that means we can achieve TFLOPS performance if we concentrate on using stock, off-the-shelf microprocessors.

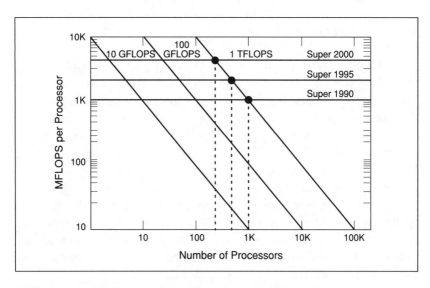

Figure 1. Achieving TFLOPS performance with traditional supercomputers (parallelism versus processor speed).

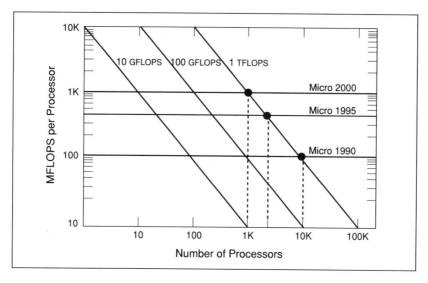

Figure 2. Achieving TFLOPS performance with micro-based multicomputers (parallelism versus processor speed).

We will have to be able to make all of these machines talk with each other, so we have to deal with the interconnectivity, which will mean that we will have to concentrate on building the appropriate interconnectivity. We will need to take the type of the research that has been done over the last few years at places like Caltech and actually put into a higher and higher silicon the necessary connections to move the bytes into the 40-, 50-, and 100-megabyte-per-channel networks and, with them, build the scalable interconnection networks. There are now known technology paths that can get us to the appropriate level of support for building the interconnections that will provide sufficient hardware bandwidth and for moving messages between these machines (Figure 3).

That leaves us, then, with the problem of making sure we can then get the messages out the door fast enough, that is, the latency issue of the time it takes to set a message up, push it out, and bring it back. That problem is solved with a combination of having good interfaces between the processors and building good architecture on the nodes. In addition, you need very lightweight operating systems so that you won't have to pay a large software overhead. If we carefully build these systems, we will be able to support both MIMD and single-processor, multiple-data architectures.

The Touchstone program is a joint project of Intel and the Defense Advanced Research Projects Agency to develop a series of prototypes of

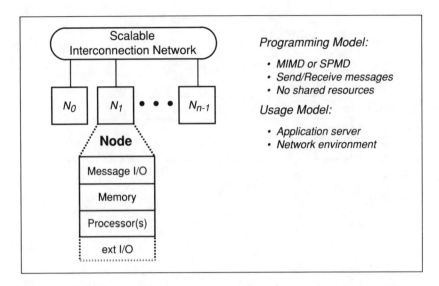

Figure 3. Message-passing multicomputers (scalable parallel computing systems).

these machines (Figure 4). It is Intel's responsibility to commercialize products out of that. The first machine, Iota, was actually the add-I/O capability to our second-generation iPSC. The next step was to add a higher-performance microprocessor to prove the theory that we can continue to leverage the overall technologies we develop by placing a new processor each time. The Gamma project is based, essentially, on the Intel i860.

Delta was scheduled for September 1990 but has been delayed until September 1991. The idea behind Delta is to use the same essential processor but to raise the number of nodes. To match the processing with the communication, we have to go to a new generation of communications. From a hardware point of view, the Gamma step is to go to the i860 processor; the Delta step is to go to these higher-performance networks.

Sigma will be the final prototype in this particular project, which will go to 2000 processors, again based on the same kind of interconnect but with high performance. The latest Intel process technology is to build the writing chips. Perhaps the most significant advance for Sigma is the packaging issues. That is, we will have to handle 2000 processors and find ways of applying packaging technologies so that we can fit them into a package no bigger than that which contains the 512-processor Delta machine. Thus, a series of processor machines, interconnects, and

Scalable Parallel Systems 119

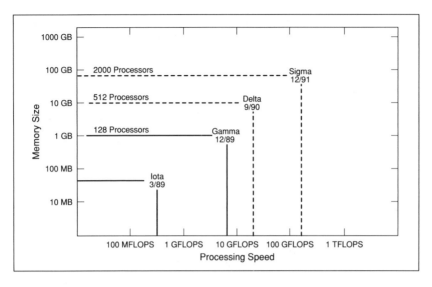

Figure 4. Touchstone prototypes (processing speed versus memory size).

packaging will culminate in what should be a 500-GFLOPS machine by the year 1992.

We now realize we need to put much more effort into the software, itself. From a hardware-design philosophy, that original technology we believed in was using microprocessors to build what we call fat nodes. That is, each node, itself, will be very powerful, with 32 to 64 micros that have large local memories. In fact, in the 1992–1995 time frame, a node will mean multiple processors. How many, I'm not sure, but it will be the latest and greatest processor Intel can produce. The nodes will be very fat and very powerful. The 2000-processor machine will be able to take real advantage of the compiler technology that is being used now. It will still be in the mode where you have to write parallel programs, but you will write it to the node rather than to the processor. Furthermore, the programmer will be free to choose models, algorithms, and tools.

Using this hardware design philosophy, flat interconnects would make it possible for the machine to behave as though it's fully connected, although it is not. Stock processors would be off-the-shelf instead of custom-made microprocessors. Thus, the programmer would get the best chip and software technology. Commodity disks would consist of many inexpensive disks instead of a few expensive ones; they would be cheap and fast enough that the user could ignore disk staging.

If you look at the amount of R&D dollars placed into workstation- and microcomputer-level I/O, it far outweighs the amount of money placed in any other form of I/O research. Thus, we want to leverage that, just as we are leveraging the R&D placed into the microprocessors.

In hardware design, I started with the hardware and built up to the design philosophy. For software, I will start with the philosophy and work backward.

As I mentioned, there is a necessity for dealing with the latency of messages in that you have to deal fundamentally with the operating system—that is, lightweight operating systems—because we can't pay the expense of dealing with heavyweight processes. To make the machine usable, we are going to have to deal with putting onto the compute nodes themselves the appropriate services, such as layered support for specific programming models, applications, or tools.

We will also need distributed services, such as high-level services through service nodes or across local area networks. UNIX compatibility without space or time overhead will be needed.

If we really are serious about delivering these machines as commercial products, then it is clear that we have to deal with standard languages, such as C and Fortran, with minimal augmentation. So we will have to deliver C, Fortran, and C++ and hope that the programming models won't be too convoluted by matching the appropriate programming model. So far, we seem to have a fairly simple addition to the language: you send something, you receive something, you broadcast something out, you accumulate something in some kind of reduction or global operation, etc.

The last items in software philosophy I want to discuss are interactive parallelization tools. These tools help the programmer restructure for parallel execution. They also allow use of standard languages and, therefore, minimize staff retraining.

I now switch from philosophy to multicomputer software needs. One of the needs is in operating systems improvements. Fortunately, there is a network of people, mostly in the university research area, that is working on multiuser resource management and efficient shared-memory emulation.

Another need is in parallel libraries. Parallel libraries involve mathematical routines, solvers, data management, visualization utilities, etc. However, these machines are too hard to program. When I talk to users about how to program these machines, I find out from optimists that the big problems come from modeling natural phenomena, which results in a great deal of natural parallelism. The reasoning is that because there is all of this natural parallelism, it should be very easy to write parallel programs. When I talk to the pessimists, the pessimists ask about solving these equations, which requires a certain amount of stability that is very hard to do.

Interestingly, both the optimistic and pessimistic viewpoints are correct. That is, there is a great deal of parallelism in natural phenomena, and when you deal with that part of it, it really is easy to express. But you come up to the part of the problem where the physicist, the financial analyst, or the chemist usually hands it off and goes and looks up the appropriate numerical algorithm for solving the right set of equations or for generating the right set of random numbers for applying the Monte Carlo method or for doing the appropriate transform. It is exactly at that point where it becomes difficult.

This is exactly the point where we are now as a community of designers. If we are going to make these machines usable, we will have to deal with building the libraries. We used to talk about kernels—building the kernels of applications. However, that was a mistake because it is not the kernels that we need but the solvers.

Therefore, to make these machines more usable, we need better solvers, better random-number generators, better transforms, etc. There is a handful of compute-intensive procedures that cause difficulties when the chemist, the physicist, and the financial analyst run into them in their code, and that is the area where we need to apply computer science.

The other reason why these machines are hard to use is the fact that the compilers, the debuggers, and the tools are very young. The problem is not that it is hard to conceptualize the parallelism but that it is hard to deal with these very young tools. I hope we can deal with the hard-to-program problem by dealing with better tools and by building the solvers.

In summary, it's time to face facts. High-performance computing is at the crossroads. Conventional supercomputers are not improving at a fast enough rate to meet our computational needs. Conventional supercomputers have become too costly to develop or own, given their delivered performance. Microprocessor performance will double every two years for at least the next 10 years. Micro-based, ultraperformance multicomputers are the only viable means to achieve the TFLOPS goal.

Large-Scale Systems and Their Limitations

Dick Clayton

Richard J. Clayton is responsible for the strategy and management of Thinking Machines Corporation's product development, manufacturing, and customer support operations. Since joining Thinking Machines in late 1983—shortly after its founding—he has built both the product and organization to the point where the company's 75 installed Connection Machines represent about 10 per cent of the U.S. supercomputer base.

Before joining Thinking Machines, Mr. Clayton was a Vice President at Digital Equipment Corporation. In his 18 years at Digital, he held numerous positions. As Vice President for Computer System Development, he was responsible for development of products representing 40 per cent of the business. As a Product Line Manager, he was directly responsible for the strategy, marketing, development, and profit and loss for 20 per cent of the company's revenue. As Vice President for Advanced Manufacturing Technology, he was responsible for upgrading the company's manufacturing capabilities.

Mr. Clayton received his bachelor's and master's degrees in electrical engineering from MIT in 1962 and 1964, respectively. During this period he also did independent research on neuroelectric signals in live animals, using the latest computer technology then available.

I'm going to talk about hardware—the applicability of these large-scale systems and their limitations. I have some comments about programming, including some thoughts on the economics, and a few conclusions.

Let me start with the premise that a TFLOPS machines is on the way, i.e., a machine capable of handling 10^{12} floating-point operations per second. It will certainly be built in the 1990s. I might argue sooner rather than later, but that's not the purpose of this discussion.

I want to focus on this idea of data-intensive, or large-scale, computing, where large scale equates to lots of data. This is where the idea of heterogeneous computing fits. This idea has given us a clearly defined context where the large-scale vector machines, on which we've written large amounts of software, are very, very important to computing for as far as I can see into the future. And that's, for me, where the heterogeneous part fits.

As we get up into many millions or billions of data objects, as we go from two to three to four to five dimensions, the amount of computing—the sheer quantity to be done—is enormous. And how we keep that all together is the question.

So for me, the idea of data parallelism has to do with very large amounts of data, and that's where a lot of the action really is in massive parallelism. The generalized space is another one of these all-possible things you might some day want to do with computers.

With apologies to Ken Olson (an ex-boss of mine, who never allowed semi-log or log-log charts) but in deference to Gordon Bell (who is here with us at this conference), I'm going to consider a log-log chart incorporating the number of processors, the speed expressed in millions of instructions per second, and the general space in which to play.

The speed of light gets you somewhere out at around a nanosecond or so. I don't care, move it out faster if you want. Your choice. But somewhere out there, there are problems, one of which is called communication limits. And I know from experience in design work, software accomplishments, and customer accomplishments, it's a long way up there. I would argue that we can build scalable architectures well up into a million elements and I think, beyond. But how far beyond and with what economics are complications we're going to mess with for quite a while. It's one of those science-versus-engineering problems. It's far enough away that it doesn't matter for the near future, like 10 years hence.

Giving some other names to this space, let me use the concepts of serial processing, shared memory, message passing, and data parallel. Those are styles of programming, or styles of computer construction. And they're arbitrarily chosen.

Let me use that same context, and let me talk about this whole idea of computer design and the various styles of computing. If you're starting out fresh, standing back and looking at things objectively, you say, "Gee, the issue probably is interconnectivity and software, so go for it—figure this stuff out, and then pour the technology in it over time." That's an interesting way to go at the question. Slightly biased view, of course.

The problem with this path, as you confront these boundaries (I sort of view them as hedgerows that you have to take your tanks over, like in Europe during World War II), is that you basically have to do everything over again as you change the number of processors while you're finding different algorithms.

But I want to change this whole software discussion from a debate about tools (i.e., a debate about whether it's easy or hard to program) to a more fundamental one—a debate about algorithms. Now we say, "Gee, let's jump right in. If we're lucky, we can skip a few of these hedgerows, and everybody's going to be absolutely sure we're totally crazy, everybody!" I didn't attend the supercomputing conference held here seven years ago, so for me this is the first visit. I hadn't joined the company quite yet. But for the first few years, this looked like total insanity because it made no sense whatsoever.

The important part of this is our users. They're the ones who are helping us really figure this out.

Of course, Gordon wants it to be hard to program (with Mobil winning the Gordon Bell prize, and then a student at Michigan winning the second prize. We didn't even know what the student was up to; he was using the network machine that the Defense Advanced Research Projects Agency had, in fact, helped us sponsor). So I'm sure it's hard to program. It's really tough. But one way or another, people are doing it. You know, there are videotapes in the computer museum in Boston. There's Edward R. Murrow, there's Seymour Cray. And Seymour gives a really beautiful speech about—I think it was the CRAY-1 or something. He was being interviewed, and somebody asks, "Well, Seymour, isn't software compatibility really important?" And Seymour has kind of a twinkle in his eye, and he says, "Yeah, but if I give them a computer that's three or four times faster, it doesn't matter."

Although I may not subscribe to that exact model, I'll admit that if you give the user a machine that's a lot faster, and if there's a promise of cost-performance for these really humongous piles of data, then there's kind of an interesting problem here. And that, to me, is what this idea of massive parallelism is about.

So I think what's interesting is that the really data-intensive part coexists very well with the heterogeneous model and with vector computers that have been around for quite a while.

Let me say one more thing about this software idea. The software problem is one we've seen before. The problem is, we've got this large amount of investment in applications—in real programs—and they're important, and we're going to use them forever. And they don't work on these kinds of machines.

The reasons they don't work are actually fairly simple. The correct algorithms for these large machines are algorithms that essentially have a fair bit of data locality and a machine model that essentially is shared memory, but humongous.

The whole idea in designing machines like Thinking Machines Corporation's Connection Machine or like these massively parallel machines is that you start with an interconnect model. And the interconnect model really supports a programming model. Let me ask, why not make it a shared-memory programming model—start right from the beginning with a shared-memory model and let the interconnect support the software models so that you can develop algorithms to make it happen? You've got to do something else, though, with this interconnect model. You've also got to move data back and forth real fast.

But there's no free lunch. When you build machines with thousands or tens of thousands of processors, it's probably true that getting to a piece of data in another processor ain't nearly so fast as getting to the data locally. And we heard, 10 different ways, the statement that memory bandwidth is where it's at. I completely agree.

So you've got a pile of memory and then you've got processors. But you've got to have a model, an interconnect model, that lets you get at all of that data simply and at relatively low cost. That's what drives us. In fact, I think that's how you design these machines; you start with this interconnect and software model and then memories; in some sense, processors are commodities that you put under this interconnect/software model.

The one thing that's different about these machines is this new idea of locality. Not all memory references are created equal. That was the implicit assumption of software compatibility: all memory references are created equal. Gee, they're not any longer. Some are faster than others if they're local.

Do we have any history of a similar example? Once upon a time, about 15 or 18 years ago, some people in Minnesota led us from serial computing to vector computing. It wasn't, you know, a megabyte of local data that was fast; it was a few hundred words. But there is a model, and it does make the transition.

There are no free lunches. Some real matrix-multiply performance on a 64K Connection Machine is now—this is double-precision stuff—is now up to five GFLOPS. The first serial algorithms we did were 10 MFLOPS, and we've gone through several explorations of the algorithm space to figure out how to get there. More importantly, our users have led us through it by the nose. I was reminded, during another presentation, that we are not fully acknowledging all the people that have beat us over the head out here.

Where do we see Fortran going? Algorithms coded in Fortran 77, twisted around, go fairly well. Start to parallelize them, and then go to Fortran 8X, where you can express the parallelism directly. We see that's where it's headed in the 1990s. We really feel good about the Fortran compilers we've now got and where they're going—very much in line with where we're taking the machines.

New algorithms are required. Standard languages will do fine; the problem is education and the problem is learning. Our users are really helping us get shaped up in the software. We've now got a timesharing system out that's beginning to help speed the program development for multiple users.

There is work being done at Los Alamos National Laboratory on a variable grid-structured problem for global weather modeling. And in fact, this dynamically changes during the calculation. We all *know*, of course, that you can't have irregular grid structures on SIMD or massively parallel machines. And we've, of course, learned that we were wrong.

Hardware conclusions: building TFLOPS machines will be possible fairly early in the 1990s. It's a little expensive. A big socioeconomic problem, but it's going to happen.

Massively parallel computers work well for data-intensive applications. You've got to have a lot of data to make this really worth doing. But where that's the case, it really does make sense. And there's still plenty of room for the software that is already written and for the problems that don't have this massively parallel, data-intensive kind of characteristic.

Now for some ideas about where all this might be going in the mid-1990s. Everybody wins here, everybody's a winner. Have it any way you want. You can pick the year that satisfies you.

By number of programs, this massively parallel stuff is going to be pretty small. By number of cycles, it's going to be pretty big, pretty fast. Gordon's got a bet that it isn't going to happen very soon, but he's already lost.

And finally, by number of dollars it's just so cost effective a way to go that it may be that way longer than we think is smart. But then, you never know.

To conclude, we're having a ball. We think it's a great way to build computers to work with really very large amounts of data. The users—whether at Los Alamos, at Mobil, anyplace—they're all beating the heck out of us. They're teaching us really fast. The machines, the algorithms, and the hardware are getting so much better. And it's a ball. We're really enjoying it.

A Scalable, Shared-Memory, Parallel Computer

Burton Smith

Burton J. Smith is Chief Scientist of Tera Computer Company in Seattle, Washington. He has a bachelor's degree in electrical engineering from the University of New Mexico, Albuquerque, and a doctorate from MIT. He was formerly chief architect of the HEP computer system, manufactured by Denelcor, in Denver, Colorado. His abiding interest since the mid-1970s has been the design and implementation of general-purpose parallel computer systems.

I would like to investigate with you what it means for a parallel computer to be scalable. Because I do not know what a scalable implementation is, I would like to talk about scalable architecture.

An architecture is $\sigma(p)$-scalable with respect to the number of processors, p, if

- the programming model does not change with p and is independent of p,
- the parallelism needed to get $Sp = \theta(p)$, that is, linear speedup, is $O(p \cdot \sigma(p))$, and
- the implementation cost is $O(p \cdot \sigma(p) \cdot \log(p))$.

The meaning of the term "parallelism" depends on the programming model. In the case of a shared-memory multiprocessor, the natural parallelism measure is how many program counters you have. The log term in the last expression is there because we are going from a conventional complexity model into a bit-complexity model, and hence, we need a factor of log to account for the fact that the addresses are getting wider, for example.

Most architectures scale with respect to some programming model or other. Unfortunately, there are some architectures that do not scale with respect to any model at all, although most scale with respect to something that might be called the "nearest-neighbor message-passing" model. Many an architecture is routinely used with a programming model that is stronger than its scaling model. There are no "scaling police" that come around and say, "You can't write that kind of program for that kind of machine because it's only a 'nearest-neighbor' machine."

I would now like to discuss the shared-memory programming model. In this model, data placement in memory does not affect performance, assuming that there is enough parallel slackness. The parallel slackness that Leslie Valiant (1990) refers to is used to tolerate synchronization latency, or in Valiant's case, barrier synchronization latency, as well as memory latency.

In the shared-memory programming model, the memory should be distributed with addresses hashed over what I believe should be a hierarchy or selection of neighborhoods rather than merely two different neighborhoods, as is common practice today. Also, synchronization using short messages is desirable. Message-passing is a good idea because it is the best low-level synchronization and data-communication machinery we have.

Many of us today think a k-ary n-cube network with an adaptive routing algorithm is probably best because adaptive routing avoids certain difficulties that arise with pessimistic permutations and other phenomena.

Tera Computer is developing a high-performance, scalable, shared-memory computer system. Remember, a shared-memory machine has the amusing property that the performance is independent of where the data is placed in memory. That means, for example, there are no data caches.

The Tera system architecture has a scaling factor of $p^{1/2}$. We build a pretty big network to get shared memory to work and to make performance insensitive to data location. The factor $p^{1/2}$ is optimal for scalable, shared-memory systems that use wires or fibers for network interconnections. Using VLSI-complexity arguments (i.e., the implications of very-large-scale integration) in three dimensions instead of two for messages that occupy volume, one can show that scalable, shared-memory machines cannot be built for a lesser exponent of p.

The network has a bisection bandwidth of around 1.6 terabytes per second. Each processor has a sustained network bandwidth of around 3.2

gigabytes per second. The bandwidth of the switch nodes that compose the network is about five times that amount, or 16 gigabytes per second.

However, if free-space optics were employed, one could conceivably use four of the six dimensions available and thereby pack more messages into the computer, thereby decreasing $\sigma(p)$ to $p^{1/3}$.

As far as I know, no other company is developing a scalable, shared-memory system. However, there is a lot of research in scalable, shared-memory systems at Stanford University and MIT, for example. Most architectures that purport to be scalable are less so than Tera's machine, and with respect to a weaker model than shared memory.

Shared memory is better than nonshared memory. One can dynamically schedule and automatically balance processor workloads. One can address irregularly without any difficulties, either in software or hardware. Shared memory is friendlier for explicit parallel programs, although certainly explicit parallelism is perhaps the only salvation of some machine models. Most important, shared memory is needed for machine-independent parallel languages, that is, portable parallel languages and their optimizing compilers. What is surprising about all this is that performance and price/performance need not suffer.

I would like to point out some of the Tera hardware characteristics. The processors are fast, both in millions of instructions per seconds (MIPS) and millions of floating-point operations per second (MFLOPS). There are

- 1.2 scalar GFLOPS per processor (64 bits),
- 1200 equivalent MIPS per processor,
- 16 or 32 megawatts (128 or 256 megabytes) of data memory per processor,
- one gigabyte of I/O memory per processor,
- two 200-megabytes-per-second high-performance parallel interface channels per processor, and
- disk arrays (RAID) for local storage.

The gigabyte of I/O memory per processor is the layer in the storage hierarchy lying between processors and the disk arrays.

These processor characteristics add up to 300 gigaflops and 300,000 MIPS for a 256-processor system, which is interconnected by a 16-ary 3-cube of network routing nodes with one-third of the links missing. Details on the hardware are available in Alverson et al. (1990).

You may be asking why we need to use fast, expensive logic and processors yielding 1.2 GFLOPS. The Tera system clock period will be three nanoseconds or less. Why doesn't Tera use a slower clock and more processors? Although emitter-coupled logic (ECL) and gallium arsenide gates both cost about three times more than complementary metal oxide semiconductor (CMOS) gates do, ECL and gallium arsenide gates are six times as fast as CMOS. BiCMOS, by the way, with bipolar output drivers on some cells, could reduce that number a bit. If most of the logic is pipelined and kept usefully busy, ECL and gallium arsenide are, therefore, twice as cost effective as CMOS.

Our interconnection network achieves a performance of 2X because the network gate count grows faster than p. As wires become more expensive, we must use them better. I think we will see more fast networks because of this. We will also find not-too-fast processors of all sorts being multiplexed to very fast network nodes, maybe even built from Josephson logic.

How massively parallel is a 256-processor Tera machine? Each Tera processor will need to have 64 or so memory references "in the air" to keep it busy. This is comparable to the needs of a fast vector processor. Main memory chip latency is about 20 nanoseconds these days and is not going to improve too quickly.

If one is seeking 100 gigawords per second of memory bandwidth, a latency of 20 nanoseconds per word implies 2000-fold parallelism simply to overcome memory chip latency. Every latency or bandwidth limitation in an architecture will consume still more parallelism in time or space, respectively. One could rightly conclude that all fast computers are massively parallel computers.

Tera's system software characteristics include the following:
- automatic whole-program analysis and parallelization,
- Fortran and C (C++), with parallel extensions,
- parallel extensions that are compatible with automatic analysis and parallelization,
- symbolic debugging of optimized programs,
- workstation-grade UNIX, including network file system, transmission control protocol/Internet protocol, and sockets, and
- parallel I/O to a log-structured file system.

It is the architecture that will make this software feasible.

In the remainder of the decade, supercomputers will continue to creep up in price. A dynamic random-access memory chip is $40 million per terabyte today, and it will halve in cost every three years. Tera will build and deliver a TFLOPS system sometime in 1996, when it becomes affordable. Also by 1996, 64-bit multi-stream microprocessors will appear.

My last predictions are that single-application, "shrink-wrapped" supercomputers will be popular for circuit simulation, structural analysis, and molecular modeling in chemistry and biology. These systems will be highly programmable, but not by the customers.

References

L. Valiant, "A Bridging Model for Parallel Computation," *Communications of the ACM* **33** (8), 103 (1990).

R. Alverson et al., "The Tera Computer System," in *Conference Proceedings. 1990 International Conference on Supercomputing*, ACM, New York, pp. 1–6 (1990).

Looking at All of the Options

Jerry Brost

Gerald M. Brost, Vice President of Engineering for Cray Research, Inc. (CRI), has been with the company since 1973 and has made significant contributions to the development and evolution of CRI supercomputers, from the CRAY-1 through the Y-MP C90. His responsibilities have included overall leadership for projects involving the CRAY X-MP, the CRAY-2, the CRAY Y-MP, CRAY Y-MP follow-on systems, and Cray's integrated-circuit facilities and peripheral products. Today, his responsibilities include overall leadership for the CRAY Y-MP EL, the CRAY Y-MP C90, the MPP Project, and future product development.

Before joining Cray, Mr. Brost worked for Fairchild Industries on a military system project in North Dakota. He graduated from North Dakota State University (NDSU) with a bachelor of science degree in electrical and electronics engineering and has done graduate work in the same field at NDSU.

To remain as the leaders in supercomputing, one of the things that we at Cray Research, Inc., need to do is continue looking at what technology is available. That technology is not just circuits but also architecture. We need to keep looking at all the technological pieces that have to be examined in order to put together a system.

Cray Research looked at technologies like gallium arsenide about eight years ago and chose gallium arsenide because of its great potential.

Today it still has a lot of potential, and I think someday it is going to become the technology of the supercomputer.

We also looked at optical computing and fiber optics, which is an area in which we will see continued growth. However, we are not committed to optical-circuit technology to build the next generation of Cray supercomputers.

Several years ago, we looked at software technology and chose UNIX because we saw that was a technology that could make our systems more powerful and more usable by our customers.

Superconductors look like a technology that has a lot of potential. However, we are unable to build anything with superconductors today.

It may come as a surprise to some that massively parallel architectures have been out for at least 20 years. Some people might say that these architectures have been out longer than that, but they have been out at least 20 years.

Even in light of the available technologies, are we at a point where, to satisfy the customers, we should incorporate the technologies into our systems? Up to now, I think the answer has been no.

We have gone out and talked to our customers and surveyed the customers on a number of things. First of all, when we talked about architectures, we were proposing what all of you know as the C90 Program. What should that architecture look like? We have our own proposal, and we gave that to some of the customers.

We talked about our view of massive parallelism. We asked the customers where they saw massive parallelism fitting into their systems. Is it something that really works? Although you hear all the hype about it, is it running at a million floating-point operations per second—a megaflop? Or is it just a flop? One of the things that we learned in our survey on massive parallelism is that there are a number of codes that do run at two- to five-billion floating-point operations per second (two to five GFLOPS).

If I listen to my colleagues today, I hear that there are large numbers of codes all running at GFLOPS ranges on massively parallel machines. Indeed, there has been significant progress made with massively parallel machines. There has been enough progress to convince us that massively parallel is an element that needs to be part of our systems in the future. Today at Cray we do have a massively parallel program, and it will be growing from now on.

Massively parallel systems do have some limitations. First of all, they are difficult architectures to program. For many of the codes that are running at the GFLOPS or five-GFLOPS performance level, it probably took someone a year's time to get the application developed. But that is because the tools are all young, the architecture is young, and there are a lot of unknowns.

Today there are probably at least 20 different efforts under way to develop massively parallel systems. If we look at progress in massive parallelism, it is much like vector processing was. If we go back in time, basically all machines were scalar processing machines.

We added vector processing to the CRAY-1 back in 1976. At first it was difficult to program because there were not any compilers to help and because people didn't know how to write special algorithms. It took some time before people started seeing the advantage of using vector processing. Next, we went on to parallel processors. Again, it took some time to get the software and to get the applications user to take advantage of the processors.

I see massively parallel as going along the same lines. If I look at the supercomputer system of the future, it is going to have elements of all of those. Scalar processing is not going to go away. Massively parallel is at an infant stage now, where applications are starting to be moved into and people are starting to learn how to make use of them.

Vector processing is not going to go away either. If I look at the number of applications that are being moved to vector processors, I find a great many.

Our goal at Cray is to integrate the massively parallel, the vector processor, and the scalar processor into one total system and make it a tightly coupled system so that we can give our customers the best overall solution. From our view, we think that for the massively parallel element, we will probably have to have at least 10 times the performance over what we can deliver in our general-purpose solution. When you can take an application and move it to the massively parallel, there can be a big payback that can justify the cost of the massively parallel element.

Work that has been done by the Defense Advanced Research Projects Agency and others has pushed the technology along to the point where now it is usable. More work is going to have to be done in optical circuits. Still more work has to be done in gallium arsenide until that is truly usable.

These are all technologies that will be used in our system at Cray. It is just a matter of time before we incorporate them into the system and make the total system operative.

To conclude, Cray does believe that massively parallel systems can work. Much work remains to be done that will be a part of a Cray Research supercomputer solution in the future. We see that as a way to move up to TFLOPS performance. I think when we can deliver TFLOPS performance, the timing will have been determined by somebody's ability to afford it. We could probably do it in 1995, although I don't know if users will have enough money to buy a TFLOPS computer system by 1995.

Massively parallel developments will be driven by technology—software and the architecture. A lot of elements are needed to make the progress, but we are committed to putting a massively parallel element on our system and to being able to deliver TFLOPS performance to our customers by the end of the decade.

5

Systems Software

This session focused on developments and limitations of systems software for supercomputers and parallel processing systems. The panelists discussed operating systems, compilers, debuggers, and utilities, as well as load balancing, multitasking, automatic parallelization, and models of computation.

Session Chair

Paul Schneck, Supercomputing Research Center

Parallel Software

Fran Allen

Frances E. Allen is an IBM Fellow and Senior Manager of Parallel Software at the IBM Thomas J. Watson Research Center in Yorktown Heights, New York. She joined IBM Research in 1957 and has since specialized in compilers, compiler optimization, programming languages, and parallelism. She has been an Institute of Electrical and Electronics Engineers (IEEE) Distinguished Visitor, an Association for Computing Machines (ACM) National Lecturer, a member of the NSF Computer Science Advisory Board, a Visiting Professor at New York University, and a Consulting Professor at Stanford University. Dr. Allen has also served as General, Program, and Tutorial Chair of the ACM SIGPLAN Compiler Construction Conference.

In 1987, Dr. Allen was elected to the National Academy of Engineering, and she was appointed the Chancellor's Distinguished Lecturer and the Mackay Lecturer for 1988–89 at the University of California, Berkeley. In 1990, she became an IEEE Fellow and in 1991 was awarded an Honorary Doctor of Science degree from the University of Alberta.

Parallel hardware is built and bought for performance. While cost/performance is a significant, sometimes overriding factor, performance is the primary motivation for developing and acquiring supercomputers and massively parallel machines. Advances in hardware such as the astounding increase in CPU speeds, increases in communication bandwidth, and reduction in communication latencies are changing both the raw performance of available hardware systems and their structure. An even more profound change is in the diversity of computational models provided by the hardware. These range from the multiple-CPU, shared-memory systems typified by Cray Research, Inc., to the single instruction, multiple data (SIMD) Connection Machine from Thinking Machines Corporation, to a variety of multiple instruction, multiple data (MIMD), distributed-memory machines. However, these technological advances have not resulted in widespread use of parallelism, nor has the promised performance been easily and frequently realized. Parallel software is the key to realizing the potentials of parallelism: performance and pervasive use. The purpose of this discussion is to assess briefly the state of parallel software for numeric-intensive computing on parallel systems and to indicate promising directions.

Fortran, the lingua franca for scientific and engineering programming, was designed for shared-memory uniprocessors. Traditional optimizing Fortran compilers generate high-performance code, sometimes perfect code, for such a machine. Traditionally, however, the user is responsible for managing storage so that memory and caches are used effectively. With the advent of vector hardware on the uniprocessors, Fortran compilers needed to compose vector instructions out of the multiple statements used to iterate through the elements of arrays. Dependence analysis to determine the relative order of references to the same storage location and loop transformations to reorder the references have been developed. When applied to Fortran codes, these transformations are now powerful enough so that vectorization is usually left to the compiler, with little or no intervention by the user.

Parallelism in any form involving more than one processor presents a very different challenge to the user and to the parallel software. There are two fundamental problems: forming parallel work and reducing overheads. Solutions to these problems involve the user and all aspects of the parallel software: the compiler, the run-time environment, the operating system, and the user environment. However, the user is often unwilling to rework the code unless there is a significant performance gain; currently, at least a factor of 10 speedup is expected for rework to be worth the effort. Even when the user does rework the algorithm and the code, what target machine model should be targeted? Given the plethora of supercomputers and massively parallel machines and, more importantly, the variety of computational models supported, users are reluctant to choose one. Thus, the challenge of parallel software is both performance and portability.

Two recent developments are very promising for numeric-intensive computing: Fortran 90 and a user-level single computational model. Fortran 90 array language allows the user to succinctly express computations involving arrays. An operand in a statement can be an entire array or a subsection of an array, and because all operands in a statement must be conformable, all the implicit, element-level operations such as add or multiply can be executed in parallel. While the designers of Fortran 90 primarily had vectors in mind, the array language is very appropriate as a computational model potentially applicable to several hardware computational models.

Fortran 90 array statements can also be compiled easily to shared-memory machines, though to do this, compilers must eliminate the transformational functions used to provide conformable operands, replacing them with appropriate index reference patterns to the operand elements. Fortran 90 array statements can be compiled very effectively to SIMD architectures. The distributed-memory MIMD architectures pose a bigger problem.

Computation can only be performed on data accessible to the computation. Data must be in registers, in cache, or in addressable memory. The delay or latency in accessing the data is critical to the performance of the machine because delays due to data accesses only reduce the effective speed of the computational units. As stated earlier, one of the two challenges for achieving parallel performance is the reduction of overheads. Delays in data movement in the system are often the most significant contributors to overhead. For shared-memory machines, this is alleviated by effective use of registers and caches, with appropriately placed synchronization and refreshing of shared memory so that shared values are correct. For distributed-memory MIMD machines, the cost of assuring that correct data is available to the computation can be very significant, hundreds or thousands of cycles in many cases, with explicit sends and receives being required at the hardware level to move information from one machine to another. It becomes important, therefore, to preplan the placement and movement of information through the system. An effort is currently under way to augment the Fortran 90 language with user directives indicating how data can be partitioned.

Another factor contributing to overhead is synchronization. While the SIMD architectures do not incur explicit synchronization overhead because the machine is synchronized at every instruction, all other forms of parallel hardware assume software-controlled synchronization to coordinate work and data availability. Programs running on massively parallel MIMD machines often have a data-driven, intermachine model. As soon as data is computed, it is sent to the appropriate machine(s), which can then start computing with it. This model of computation, a direct extension of the basic hardware-level, send-and-receive model, generally requires a total recasting of the solution and rewrite of the code.

An emerging computational model that has great promise is SPMD: single program, multiple data. A program written in Fortran 90, augmented with data-partitioning directives, can be executed concurrently on multiple processors but on different data partitions. It has some of the characteristics of SIMD but is much more flexible in that different control paths can be taken on different parallel executions of the same code. But the most important aspect of SPMD is that it may not require extensive problem rework, yet the programs may be portable across very diverse parallel systems. It should be emphasized that parallel software systems do not currently support the approach, so the hoped-for performance and portability goals may not be achievable.

Advances in parallel software toward the goals of performance, portability, and ease of use requires enlarging the scope of software in two directions. Whole programs or applications should be analyzed, transformed, and managed by the system, and the whole parallel software system should participate as an integrated unit.

The traditional functional boundaries among operating systems, runtime, compilers, and user environments must be revised. For example, scheduling of work to processors can be done by the compiler and the run-time system and even by the user, not just the operating system. Fast protocols avoiding operating system calls are needed to reduce overheads. The user and compiler need information about the performance of the running program so that adjustments can be made in the source code or the compilation strategy. Debugging of parallel code is a major problem, and integrated tools to facilitate that solution are needed. For the foreseeable future, the user must be an active participant in enabling the parallelism of the application and should be able to participate interactively with the entire system.

Whole-program analysis and transformation is essential when compiling programs to target machines such as distributed-memory MIMD systems because of the relatively large granularity of parallel work and the need to partition data effectively. This analysis is both broad and deep, involving interprocedural analysis and analyses for data and control dependence and for aliases. The transformations involve transforming loops, regions, and multiple procedures to create appropriate parallelism for the target machine. They also involve modifying the size, shape, location, and lifetimes of variables to enable parallelism.

It is widely recognized that the major challenge in parallelism is in developing the parallel software. While there are products and active research on whole-program analysis and transformation, there is less effort on whole-system integration, including an integrated user environment. A totally unanswered question is how effective these parallel software directions will be on massively parallel MIMD machines. What does seem certain is that without advances in parallel software and in problem-solving languages, the true potential of parallel systems will not be realized.

Supercomputer Systems-Software Challenges

David L. Black

David L. Black is a Research Fellow at the Cambridge office of the Open Software Foundation (OSF) Research Institute, where he participates in research on the evolution of operating systems. Before joining OSF in 1990, he worked on the Mach operating system at Carnegie Mellon University (CMU), from which he received a Ph.D. in computer science. Dr. Black also holds an M.S. in computer science from CMU and an M.A. in mathematics from the University of Pennsylvania. His current research is on microkernel-based operating system environments, incorporating his interests in parallel, distributed, and real-time computation.

Abstract

This paper describes important systems-software challenges to the effective use of supercomputers and outlines the efforts needed to resolve them. These challenges include distributed computing, the availability and influence of high-speed networks, interactions between the hardware architecture and the operating system, and support for parallel programming. Technology that addresses these challenges is crucial to ensure the continued utility of supercomputers in the heterogeneous, functionally specialized, distributed computing environments of the 1990s.

Introduction

Supercomputers face important systems-software challenges that must be addressed to ensure their continued productive use. To explore these issues and possible solutions, Lawrence Livermore National Laboratory and the Supercomputing Research Center sponsored a workshop on Supercomputer Operating Systems and related issues in July 1990. This paper is based on the results of the workshop* and covers four major challenges: distributed computing, high-speed networks, architectural interactions with operating systems (including virtual memory support), and parallel programming.

Distributed Computing

Distributed computing is an important challenge because supercomputers are no longer isolated systems. The typical supercomputer installation contains dozens of systems, including front ends, fileservers, workstations, and other supercomputers. Distributed computing encompasses all of the problems encountered in convincing these systems to work together in a cooperative fashion. This is a long-standing research area in computer science but is of increasing importance because of greater functional specialization in supercomputing environments.

Functional specialization is a key driving force in the evolution of supercomputing environments. The defining characteristic of such environments is that the specialization of hardware is reflected in the structure of applications. Ideally, applications are divided into components that execute on the most appropriate hardware. This reserves the supercomputer for the components of the application that truly need its high performance and allows other components to execute elsewhere (e.g., a researcher's workstation, a graphics display unit, etc.). Cooperation and coordination among these components is of paramount importance to the successful use of such environments. A related challenge is that of partitioning problems into appropriate components. Communication costs are an important consideration in this regard, as higher costs require a coarser degree of interaction among the components.

Transparency and interoperability are key system characteristics that are required in such environments. Transparent communication mechanisms work in the same fashion, independent of the location of the communicating

---

* The views and conclusions in this document are those of the author and should not be interpreted as representing the workshop as a whole, its sponsors, other participants, or the official policies, expressed or implied, of the Open Software Foundation.

components, including whether the communication is local to a single machine. Interoperability ensures that communication mechanisms function correctly among different types of hardware from different manufacturers, which is exactly the situation in current supercomputing environments. Achieving these goals is not easy but is a basic requirement for systems software that supports functionally specialized supercomputing environments.

High-Speed Networks

High-speed networks (gigabit-per-second and higher bandwidth) cause fundamental changes in software at both the application and systems levels. The good news is that these networks can absorb data at supercomputer rates, but this moves the problem of coping with the high data rate to the recipient. To illustrate the scope of this challenge, consider a Cray Research, Inc., machine with a four-nanosecond cycle time. At one gigabit per second, this Cray can handle the network in software because it can execute 16 instructions per 64-bit word transmitted or received. This example illustrates two problem areas. The first is that a Cray is a rather expensive network controller; productive use of networks requires that more cost-effective interface hardware be employed. The second problem is that one gigabit per second is slow for high-speed networks; at least another order of magnitude in bandwidth will become available in the near future, leaving the Cray with less than two instructions per word.

Existing local area networking practice does not extend to high-speed networks because local area networks (LANs) are fundamentally different from their high-speed counterparts. At the hardware level, high-speed networks are based on point-to-point links with active switching hardware rather than the common media access often used in LANs (e.g., Ethernet). This is motivated both by the needs of the telecommunications industry (which is at the forefront of development of these networks) and the fact that LAN media access techniques do not scale to the gigabit-per-second range. On a 10-megabit-per-second Ethernet, a bit is approximately 30 meters long (about 100 feet); since this is the same order of magnitude as the physical size of a typical LAN, there can only be a few bits in flight at any time. Thus, if the entire network is idled by a low-level media-management event (e.g., collision detection), only a few bits are lost. At a gigabit per second, a bit is 30 centimeters long (about one foot), so the number of bits lost to a corresponding media-management event on the same-size network is a few hundred; this can be a significant source of lost bandwidth and is avoided in high-speed network protocols. Using point-to-point links can reduce these management events to the individual link level (where they are less costly) at the cost of active switching and routing hardware.

The bandwidth of high-speed networks also raises issues in the areas of protocols and hardware interface design. The computational overhead of existing protocols is much more costly in high-speed networks because the bandwidth losses for a given amount of computation are orders of magnitude larger. In addition, the reduced likelihood of dropped packets may obviate protocol logic that recovers from such events. Bandwidth-related issues also occur in the design of hardware interfaces. The bandwidth from the network has to go somewhere; local buffering in the interface is a minimum requirement. In addition, the high bandwidth available from these networks has motivated a number of researchers to consider memory-mapped interface architectures in place of the traditional communication orientation. At the speeds of these networks, the overhead of transmitting a page of memory is relatively small, making this approach feasible.

The importance of network management is increased by high-speed networks because they complement rather than replace existing, slower networks. Ethernet is still very useful, and the availability of more expensive, higher-bandwidth networks will not make it obsolete. Supercomputing facilities are likely to have overlapping Ethernet, fiber-distributed data interface, and high-speed networks connected to many machines. Techniques for managing such heterogeneous collections of networks and subdividing traffic appropriately (e.g., controlling traffic via Ethernet, transferring data via something faster) are extremely important. Managing a single network is challenging enough with existing technology; new technology is needed for multinetwork environments.

Virtual Memory

Virtual memory originated as a technique to extend the apparent size of physical memory. By moving pages of memory to and from backing storage (disk or drum) and adjusting virtual to physical memory mappings, a system could allow applications to make use of more memory than existed in the hardware. As applications executed, page-in and page-out traffic would change the portion of virtual memory that was actually resident in physical memory. The ability to change the mapping of virtual to physical addresses insulated applications from the effects of not having all of their data in memory all the time and allowed their data to occupy different physical pages as needed.

Current operating systems emphasize the use of virtual memory for flexible mapping and sharing of data. Among the facilities that depend on this are mapped files, shared memory, and shared libraries. These features provide enhanced functionality and increased performance to applications. Paging is also provided by these operating systems, but it is less important than the advanced mapping and sharing features supported by virtual memory. Among the operating systems that provide such features are Mach, OSF/1,* System V Release 4,** and SunOS.*** These features are an important part of the systems environment into which supercomputers must fit, now and in the future. The use of standard operating systems is important for interoperability and commonality of application development with other hardware (both supercomputers and other systems).

This shift in the use of virtual memory changes the design tradeoffs surrounding its use in supercomputers. For the original paging-oriented use, it was hard to justify incorporating virtual memory mapping hardware. This was because the cycle time of a supercomputer was so short compared with disk access time that paging made little sense. This is still largely the case, as advances in processor speed have not been matched by corresponding advances in disk bandwidth. The need for virtual memory to support common operating systems features changes this tradeoff. Systems without hardware support for virtual memory are unable to support operating systems features that depend on virtual memory. In turn, loss of these features removes support for applications that depend on them and deprives both applications and the system as a whole of the performance improvements gained from these features. This makes it more difficult for such systems to operate smoothly with other systems in the distributed supercomputing environment of the future. The next generation of operating systems assumes the existence of virtual memory; as a result, hardware that does not support it will be at a disadvantage.

---

*OSF/1 is a trademark of the Open Software Foundation.
**System V is a trademark of UNIX Systems Laboratories, Inc.
***SunOS is a trademark of Sun Microsystems, Inc.

Resource Management

The increasing size and scale of supercomputer systems pose new resource-management problems. Enormous memories (in the gigabyte range) require management techniques beyond the LRU-like paging that is used to manage megabyte-scale memories. New scheduling techniques are required to handle large numbers of processors, nonuniform memory access architectures, and processor heterogeneity (different instruction sets). A common requirement for these and related areas is the need for more sophisticated resource management, including the ability to explicitly manage resources (e.g., dedicate processors and memory to specific applications). This allows the sophistication to be moved outside the operating system to an environment- or application-specific resource manager. Such a manager can implement appropriate policies to ensure effective resource usage for particular applications or specialized environments.

Parallel Processing

Supercomputer applications are characterized by the need for the fastest possible execution; the use of multiple processors in parallel is an important technique for achieving this performance. Parallel processing requires support from multiple system components, including architecture, operating systems, and programming language. At the architectural level, the cost of operations used to communicate among or synchronize processors (e.g., shared-memory access, message passing) places lower bounds on the granularity of parallelism (the amount of computation between successive interactions) that can be supported. The operating system must provide fast access to these features (i.e., low-overhead communication mechanisms and shared-memory support), and provide explicit resource allocation, as indicated in the previous section. Applications will only use parallelism if they can reliably obtain performance improvements from it; this requires that multiple processors be readily available to such applications. Much work has been done in the areas of languages, libraries, and tools, but more remains to be done; the goal should be to make parallel programming as easy as sequential programming. A common need across all levels of the system is effective support for performance analysis and debugging. This reflects the need for speed in all supercomputer applications, especially in those that have been structured to take advantage of parallelism.

Progress

Progress has been made and continues to be made in addressing these challenges. The importance of interconnecting heterogeneous hardware and software systems in a distributed environment has been recognized, and technology to address this area is becoming available (e.g., in the forthcoming Distributed Computing Environment offering from the Open Software Foundation). A number of research projects have built and are gaining experience with high-speed networks, including experience in the design and use of efficient protocols (e.g., ATM). Operating systems such as Mach and OSF/1 contain support for explicit resource management and parallel programming. These systems have been ported to a variety of computer architectures ranging from PCs to supercomputers, providing an important base for the use of common software.

Future Supercomputing Elements

Bob Ewald

Robert H. Ewald is Executive Vice President for Development at Cray Research, Inc., in Chippewa Falls, Wisconsin, having joined the company in 1984. From 1977 to 1984, he worked at Los Alamos National Laboratory, serving during the last five years of his tenure as Division Leader in the Computing and Communications Division.

This discussion will focus on the high-performance computing environment of the future and will describe the software challenges we see at Cray Research, Inc., as we prepare our products for these future environments.

Figure 1 depicts the environment that Cray Research envisions for the future and is actively addressing in our strategic and product plans. This environment consists of machines that have traditional, or general-purpose, features, as well as special architectural features that provide accelerated performance for specific applications. It's currently unclear how these elements must be connected to achieve optimum performance—whether by some processor, memory, or network interconnect. What is clear, however, is that the successful supercomputer architecture of the future will be an optimal blend of each of these computing elements. Hiding the architectural implementation and delivering peak performance, given such a heterogeneous architecture, will certainly be a challenge for software. We currently have a number of architectural study teams looking at the best way to accomplish this.

Another key consideration in software development is what the future workstation architecture will look like. The view that we have today is something like that depicted in Figure 2. These will be very fast scalar machines with multiple superscalar and specialized processors combined to deliver enhanced real-time, three-dimensional graphics. Not only will tomorrow's software be required to optimize application performance on a heterogeneous-element supercomputer, as described earlier, but it will also be required to provide distributed functionality and integration with these workstations of tomorrow.

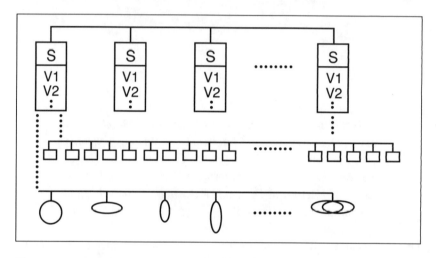

Figure 1. Future supercomputer elements.

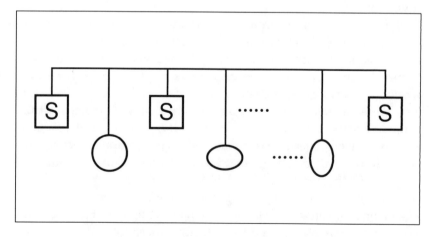

Figure 2. Future workstation elements.

Figure 3 is Cray Research's view of what our customer networks look like or will look like. It is a heterogeneous network with systems and network components from a broad array of vendors. The four key elements of the network are workstations, networks of networks, compute servers (either general purpose or dedicated), and file servers. The networks are of varying speed, depending on the criticality and bandwidth of the resources that are attached. The key point of this scenario is that every resource is "available" to any other resource. Security restrictions may apply and the network may be segmented for specialized purposes, but basically the direction of networking technology is toward more open architectures and network-managed resources.

Figure 4 portrays the conceptual model that we are employing in our product development strategy. The idea of the client-server model is that the workstation is the primary user interface, and it transparently draws upon other resources in the network to provide specialized services. These resources are assigned to optimize the delivery of a particular service to the user. Cray Research's primary interest is in providing the highest performance compute and file servers. This hardware must be complemented with the software to make these systems accessible to the user in a transparent manner.

Currently, Cray Research provides a rich production environment that incorporates the client-server model. Our UNICOS software is based on AT&T UNIX System V with Berkeley Standard Distribution extensions and is designed for POSIX compliance. This enables application portability and a common cross-system application development environment. Through the use of X Windows and the network file system, CRAY-2s, X-MPs, and Y-MPs have connected to a variety of workstations from Sun Microsystems, Inc., Silicon Graphics IRIS, IBM, Digital Equipment Corporation, Apollo (Hewlett-Packard), and other vendors. Coupled with our high-speed interconnects, distributing an application across multiple Cray systems is now a practical possibility. In fact, during the summer of 1990, we achieved 3.3×10^9 floating-point operations per second (GFLOPS) on a matrix multiply that was distributed between a CRAY-2 and a Y-MP. Later that summer, a customer prospect was able to distribute a three-dimensional elastic FEM code between three Cray systems at the Cray Research Computing Center in Minnesota and achieved 1.7 GFLOPS sustained performance. Aside from the performance, what was incredible about this is that this scientist was running a real application, had only three hours of time to make the coding changes, and did this all without leaving his desk in Tokyo. The technology is here today to do this kind of work as a matter of course. I think

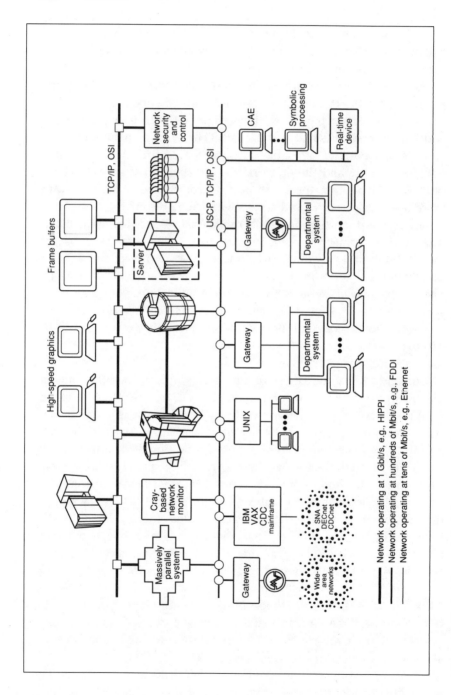

Figure 3. Network supercomputing.

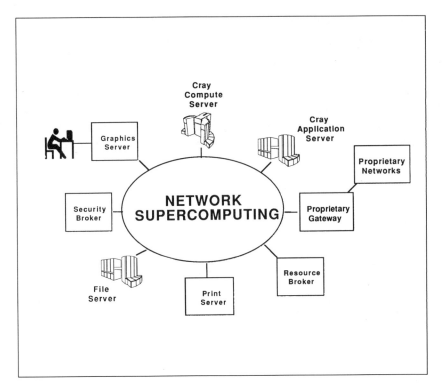

Figure 4. Client-server model.

you'll see significant progress in the next year or two in demonstrating sustained high performance on real-world distributed applications.

These technologies present a number of software challenges to overcome. First, there is a growing need to improve automatic recognition capabilities. We must continue to make progress to improve our ability to recognize and optimize scalar, vector, and parallel constructs. Data-element recognition must utilize long vectors or massively parallel architectures as appropriate. Algorithm recognition must be able to examine software constructs and identify the optimal hardware element to achieve maximum performance. As an example, an automatic recognition of fast Fourier transforms might invoke a special processor optimized for that function, similar to our current arithmetic functional units.

Second, because this environment is highly distributed, we must develop tools that automatically partition codes between heterogeneous systems in the network. Of course, a heterogeneous environment also implies a mix of standard and proprietary software that must be taken into account. Security concerns will be a big challenge, as well, and must extend from the supercomputer to the fileserver and out to the workstation.

Not only must we make it easy to distribute applications, but we must also optimize them to take advantage of the strengths of the various network elements. The size of the optimization problem is potentially immense because of different performance characteristics of the workstation, network, compute, and fileserver elements. The problem is exacerbated by discontinuities in such performance characteristics as a slow network gateway or by such functional discontinuities as an unavailable network element. Ultimately, we will have to develop expert systems to help distribute applications and run them.

Finally, to operate in the computing environment of the future, the most critical software components will be the compilers and the languages that we use. If we cannot express parallelism through the languages we use, we will have limited success in simulation because the very things we are trying to model are parallel in nature. What we have been doing for the last 30 years is serializing the parallel world because we lack the tools to represent it in parallel form. We need to develop a new, non-von Neumann way of thinking so that we do not go through this parallel-to-serial-back-to-parallel computational gyration. A language based on physics or some higher-level abstraction is needed.

In terms of existing languages, Fortran continues to be the most important language for supercomputer users, and we expect that to continue. Unfortunately, its current evolution may have some problems. Because of the directions of the various standards organizations, we may see three different Fortran standards emerge. The American National Standards Institute (ANSI), itself, will probably have two standards: the current Fortran 77 standard and the new Fortran 9X standard. The International Standardization Organization may become so frustrated at the ANSI developments that they will develop yet another forward-looking Fortran standard. So what was once a standard language will become an unstandard set of standards. Nevertheless, we still envision that Fortran will be the most heavily used language for science and engineering through the end of this decade.

The C language is gaining importance and credibility in the scientific and engineering community. The widespread adoption of UNIX is partly responsible for this. We are seeing many new application areas utilizing C, and we expect this to continue. We also can expect to see additions to C for parallel processing and numeric processing. In fact,

Cray Research is quite active with the ANSI numerical-extensions-to-C group that is looking at improving its numeric processing capabilities.

Ada is an important language for a segment of our customer base. It will continue to be required in high-performance computing, in part because of Department of Defense mandates placed on some software developed for the U.S. government.

Lisp and Prolog are considered important because of their association with expert systems and artificial intelligence. In order to achieve the distributed-network optimization that was previously discussed, expert systems might be employed on the workstation acting as a resource controller on behalf of the user. We need to determine how to integrate symbolic and numeric computing to achieve optimal network resource performance with minimal cost. A few years ago we thought that most expert systems would be written in Lisp. We are seeing a trend, however, that suggests that C might become the dominant implementation language for expert systems.

In summary, today's languages will continue to have an important role in scientific and engineering computing. There is also, however, a need for a higher-level abstraction that enables us to express the parallelism found in nature in a more "natural" way. Nevertheless, because of the huge investment in existing codes, we must develop more effective techniques to prolong the useful life of these applications on tomorrow's architectures.

On the systems side, we need operating systems that are scalable and interoperable. The implementation of the system may change "under the hood" and, indeed, must change to take advantage of the new hybrid architectures. What must not change, however, is the user's awareness of the network architecture. The user interface must be consistent, transparent, and graphically oriented, with the necessary tools to automatically optimize the application to take advantage of the network resources. This is the high-performance computing environment of the future.

Compiler Issues for TFLOPS Computing

Ken Kennedy

Ken Kennedy received a B.A. in mathematics from Rice University, Houston, in 1967, an M.S. in mathematics from New York University in 1969, and a Ph.D. in computer science from New York University in 1971. He has been a faculty member at Rice since 1971, achieving the rank of Professor of Mathematical Sciences in 1980. He served as Chairman of the Department of Computer Science from its founding in 1984 until 1989 and was appointed Noah Harding Professor in 1985. He is currently Director of the Computer and Information Technology Institute at Rice and heads the Center for Research on Parallel Computing, an NSF Science and Technology Center at Rice, Caltech, and Los Alamos National Laboratory.

From 1985 to 1987, Professor Kennedy chaired the Advisory Committee to the NSF Division of Computer Research and has been a member of the Board of the Computing Research Association since 1986. In 1990, he was elected to the National Academy of Engineering.

Professor Kennedy has published over 60 technical articles on programming support software for high-performance computer systems and has supervised the construction of two significant software systems for programming parallel machines: a vectorizer for Fortran and an integrated scientific software development environment.

> *Professor Kennedy's current research focuses on extending techniques developed for automatic vectorization to programming tools for parallel computer systems and high-performance microprocessors. Through the Center for Research on Parallel Computation, he is seeking to develop new strategies for supporting architecture-independent parallel programming, especially in science and engineering.*

I would like to ponder the notion of TFLOPS computing, which is a sort of subgoal of our high-performance computing initiative. I opposed this subgoal when I was first on the subcommittee of the Federal Coordinating Committee on Science, Engineering, and Technology that wrote the report on high-performance computing, but now I've warmed up to it a lot, primarily because I've seen the kind of scientific advances that we can achieve if we're able to get to the level of a TFLOPS and the corresponding memory sizes.

I want to talk about what I see as the compiler issues if we build such a machine. There are many ways one might consider building such a machine. I'll pick the way Session 13 presenter Justin Rattner would build it, which is to take Intel's latest microprocessors. Although I don't know exactly what the peaks are going to be, I'm sure that we won't be able to achieve more than about half of peak out of those things. Intel's study indicates they're going to achieve all of these in the middle-to-late 1990s and that they're going to have all sorts of performance out of single-chip processors. I'm counting on about 250 in the middle-to-late 1990s. That means that we'll have to have at least 8000 processors. If we are able to get Intel to give them to us for $10,000 a processor, which is less than they're currently offering for their iPSC/860s, that will be an $80 million machine, which is reasonable in cost. This means that we have to think about how to use 4000 to 8000 processors to do science. So I'd like to reflect on how we're using parallelism today.

The good news is that we have machines, and we're doing science with them. Some people in the audience can talk about the real achievements in which science and engineering calculations have been done at very low cost relative to conventional supercomputers, with high degrees of parallelism. Unfortunately, the other side of the coin is that there's some bad news. The bad news is that we have a diverse set of architectures, and the programming systems for those machines are primitive in the sense that they reflect a great deal of the architecture of the machine in the programming system. Thus, the user is, in fact, programming a specific architecture quite frequently rather than writing general parallel programs.

In addition, there are the ugly new kinds of bugs that we have to deal with and all sorts of performance anomalies that, when you have a thousand processors and you get a speedup of two, people wonder what's gone wrong. You have to have some mechanism to help people to deal with these things.

So all sorts of problems exist in the programming. I think that the really critical issue is that we have not achieved the state where commercial firms with those zillions of dollars of investment will leap into the massive-parallelism world; they're afraid of losing that investment, and they're afraid of a major reprogramming effort for a particular machine that will be lost the next time a new machine has a different architecture.

I think people in the commercial world are standing back and looking at parallelism with a bit of a skeptical eye. Some of the scientists aren't, but I think a lot of scientists are. That means if we want parallelism to be successful in the way that previous generations of supercomputers have been successful, we have to provide some form of machine-independent parallel programming, at least enough so that people feel comfortable in protecting their investments.

It is useful for us to look at the legacy of automatic vectorization, which I think really made it possible for vector supercomputers to be well programmed. I want to dispel what may be a misconception on the part of some people. Vectorization technology did not make it possible for people to take dusty decks and run them on their local supercomputer and expect high performance. What it did was provide us with a subdialect of Fortran 77, a sort of vectorizable Fortran 77, which the whole community learned. Once they had this model of how to write vectorizable programs in Fortran 77, people could write them and run them with high performance on a variety of different vector machines. In that sense, a great deal of the architectural specificity was factored out. I think that's one of the main reasons for the success of this technology.

Now, the question is, can we achieve the similar success with automatic parallelization? As Fran Allen pointed out earlier in this session, there are a lot of people on this panel and elsewhere who are doing very hard work on this problem of automatic parallelization. I've got to say that our system, like that of Fran and of Dave Black (also a Session 5 presenter)—all of those systems—can do many impressive things. I think each of us can give you some examples in which our system does not just do impressive things but amazing things.

Unfortunately, in general we can't bet on that, which has to do with all sorts of technical problems. Mainly it has to do with the fact that if we're dealing with parallelism, at least of the MIMD asynchronous variety, we have to deal with the overhead, and that means we have to find larger and larger regions of parallelism. The imprecision of dependence analysis and all the interprocedural effects that Fran talked about are really causing problems.

Thus, I think this technology is going to make a contribution, but it's not going to be the answer in the same way vectorization technology was. That means we have to support explicit parallel programming. However, can we support that in a machine-independent way? The goal of our efforts, I think, should be to let the programmer specify parallelism at a fairly high level of abstraction, which may be contradictory if we stay within Fortran. But even within Fortran, I think we should be able to do it at a high level of abstraction, in the sense that it shouldn't depend on the machine. The programmer should specify the strategy, and the compiler should take care of the details.

If we're going to address this problem, what are the issues we have to deal with? The first issue is that we don't even know what the right programming paradigm is for these machines. If I pick anyone in the audience, I can probably get that person to tell me which paradigm he or she "knows" to be the right one. But the fact of the matter is that we can't bet on any of these because they're not proven.

There are a lot of candidates for parallel languages that we have to consider. We have to consider that the shared-memory community is centering on the Parallel Computing Forum (PCF) effort, which is, I think, going to provide at least some machine independence across all shared-memory machines. Unfortunately, not many of the massively parallel machines, except for that of the Cedar Project and IBM's RP3-Mach project, are still trying to implement shared-memory hardware. So we have to concern ourselves with the generality of PCF.

Then there's Fortran 90, which is the language of choice for the SIMD community led by Thinking Machines Corporation; also there's Linda and other languages. So one can have Fortran extensions.

There are also all sorts of efforts in compositional specifications of programming, where one takes Fortran and combines it with some specification of how to combine various Fortran modules and runs them in parallel. Then there are people who are using STRAND[88] and its successors to specify parallel compositions, with Fortran as the components. Also there is the functional community, which makes the very good point that in functional languages, the dependence analysis is precise. Unfortunately, we have a problem with mapping that down to a finite amount of memory. So the efficiencies in this area are still a major question, although there have been some strides in that direction in getting better efficiencies.

I think the language and abstract-programming model we provide is not yet clear. If we're going to have machine-independent programming, we have to have some model, and there's no resolution on what that should be like.

Another question is, how much of the tradeoff of machine independence versus efficiency can we tolerate? One of the reasons we can't do machine-independent programming right now is because the ways that we could do it are unacceptably inefficient on the machines available. A fraction of the user community agrees that they could do it in a particular language available today, but the programs that come out aren't efficient enough and have to be reprogrammed. Therefore, I think we have to understand how far we can go in abstracting the parallel decomposition process. I think a major issue is how far we can go in combining the compiler and operating system and managing the memory hierarchy, which is a major dependency that we have to deal with.

Yet another issue is, of course, that although compilers exist within an environment, I think we've seen the end of the day when a compiler sits by itself and is an independent component of the programming system. Every compiler will exist in an environment because all the tools in the environment will be intimately intertwined in their design. The debugger has to understand what the compiler does in order to present in an understandable way the execution of the program as it happened to the user. Performance tuners have to be able to explain in a language understandable to the user why the compiler chose to do the strange things that it did, not just spit out lines and lines of assembly code that are hard to trace to anything. The performance tuner has to explain how the program ran and why in somewhat high-level terms that the average user is able to grasp.

Fran talked about interprocedural compilation. Environmental support will need to be provided in order to manage that in a way that programmers will find acceptable. We have to have tools that help people prepare well-formed parallel programs in whatever language we have.

Finally, I think there's an issue of what we do if we're going to manage a memory hierarchy in a way that's transparent to the user so that we can have true machine-independent programming. To do so, I think we have to have some help from the architecture. What the form of that help is, I'm not certain. I think that if I knew, we would be farther along in the research line than we are right now.

I think the ideas of virtual shared memory, when combined with compiler ideas, may be very promising. The Cedar Project, led by the next presenter in this session, Dave Kuck, incorporates some very interesting ideas about what things you can do with memory to support a more programmable interface. The architecture and operating system have to do more and more to support debugging and performance modeling, since those are going to be so critical in the environment that is part of the compiler system. We need to have some support to make sure we're not paying an enormous amount to achieve that.

To summarize, I think we have to have machine-independent parallel programming or we are going to waste a generation of physicists and chemists as low-level programmers. I think we need to provide a level of programming support that is consistent with what we did in vectorization. We have a long research agenda if we're going to achieve that by the end of this decade, when TFLOPS computing is to arrive.

Performance Studies and Problem-Solving Environments

David Kuck

David J. Kuck is Director of the Center for Supercomputing Research and Development, which he organized in 1984, and is a Professor of Computer Science and Electrical and Computer Engineering at the University of Illinois-Urbana/Champaign, where he has been a faculty member since 1965. He is currently engaged in the development and refinement of the Cedar parallel processing system and, in general, in theoretical and empirical studies of various machine and software organization problems, including parallel processor computation methods, interconnection networks, memory hierarchies, and the compilation of ordinary programs for such machines.

I would like to address performance issues in terms of a historical perspective. Instead of worrying about architecture or software, I think we really need to know why the architectures, compilers, and applications are not working as well as they could. At the University of Illinois, we have done some work in this area, but there is a tremendous amount of work remaining. Beyond programming languages, there are some efforts ongoing to make it possible to use machines in the "shrink-wrapped" problem-solving environment that we use in some PCs today.

Figure 1 shows a century-long view of computing, broken into a sequential (or von Neumann) era and a parallel era. The time scales show how long it takes to get the applications, architectures, compilers, etc.,

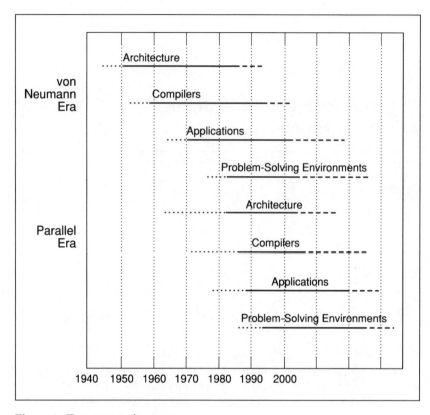

Figure 1. Two computing eras.

that we have now and will have in the future. Because architectures, compilers, and applications are three components of performance, if you get any one of them slightly wrong, performance may be poor.

With respect to sequential machines, at about the time the RISC processors came along, you could for $5000 buy a box that you really did not program. It was a turnkey machine that you could use for lots of different purposes without having to think about it too much; you could deal with it in your own terms. I regard that kind of use of machines as an area that is going to be very important in the future.

At about the same time, in the early 1980s, companies started producing parallel architectures of all kinds. Then, with about a 30-year delay, everything in the sequential era is repeating itself. Now, for parallel machines we have compilers, and you will be able to start to buy application software one of these days.

In the von Neumann era, the performance issues were instructions, operations, or floating-point operations, reckoned in millions per second, and the question was whether to use high-level or assembly language to achieve a 10–200 per cent performance increase. In the parallel era, the performance issues have become speedup, efficiency, stability, and tunability to vectorize, parallelize, and minimize synchronization, with performance potentials of 10X, 100X, and even 1000X. In both eras, memory-hierarchy management has been crucial, and now it is much more complex.

Some of these performance issues are illustrated in the Perfect data of Figure 2 (Berry et al. 1989), which plots 13 supercomputers from companies like Hitachi, Cray Research, Inc., Engineering Technology Associates Systems, and Fujitsu. The triangles are the peak numbers that you get from the standard arguments. The dots are the 13 codes run on all those different machines. The variation between the best and worst is about a factor of 100, all the way across. It doesn't matter which machine you run on. That is bad news, it seems to me, if you're coming from a Digital Equipment Corporation VAX. The worst news is that if I label those points, you'll see that it's not the case that some fantastic code is right up there at the top all the way across (Corcoran 1991). Things bounce around.

When you move from machine to machine, there's another kind of instability. If you decided to benchmark one machine, and you decided for price reasons to buy another, architecturally similar, one that you didn't have access to, it may not give similar performance.

The bars in Figure 2 are harmonic means. For these 13 codes on all the supercomputers shown, you're getting a little bit more that 10 million floating-point operations per second. That is one per cent of peak being delivered, and so there is a gap of two orders of magnitude there, and a gap of two orders of magnitude in the best-to-worst envelope.

If you think it's just me who is confused about all this, no less an authority than the *New York Times* in a two-month period during the Spring of 1990 announced, "Work Station Outperforms Supercomputer," "Cray Is Still Fastest," "Nobody Will Survive the Killer Micros," and "Japanese Computer Rated Fastest by One Measure." So there's tremendous confusion among experts and the public.

How do we get around this? Let me suggest an old remedy in a new form. Why don't we just forget about all the kinds of programming languages we have right now? If you can just specify problems and have them solved on workstations or on PCs, why can't we do that on these machines? I think we will have to eventually if parallel processing is really going to be a general success. I don't really know how to do this, but we have been thinking about it quite a bit at the Center for

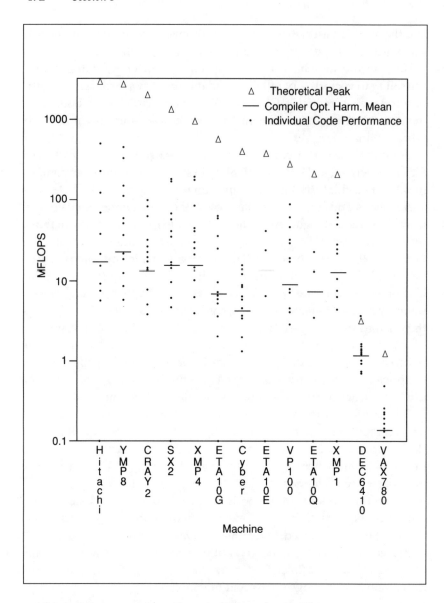

Figure 2. Thirteen Perfect codes (instability scatter plot).

Supercomputing Research and Development (CSRD) for the last year or so on the basis of our existing software.

If you could write mathematical equations or if you could describe circuits by pictures or whatever, there would be some tremendous benefits. The parallelism that exists in nature should somehow come straight through, so you'd be able to get at that a lot easier. You would be able to adapt old programs to new uses in various ways. As examples, I should say that at Lawrence Livermore National Laboratory, there's the ALPAL system, at Purdue University there's ELLPACK, and Hitachi has something called DEQSOL. These are several examples into which parallelism is starting to creep now.

Figure 3 gives a model of what I think an adaptive problem-solving environment is. This is just your old Fortran program augmented a bit and structured in a new way. There's some logic there that represents "all" methods of solution on "all" machines. The data structures and library boxes contain lots of solution methods for lots of machines and come right out of the Perfect approach. Imagine a three-dimensional volume labeled with architectures and then for each architecture, a set of applications and each application broken down into basic algorithms. That's the way we've been attacking the Perfect database that we're building, and it flows right into this (Kuck and Sameh 1987). So you can have a big library with a lot of structure to it.

The key is that the data structures have to represent all algorithms on all machines, and I've got some ideas about how you translate from one of those to the other if you wanted to adapt a code (Kuck 1991). There are two kinds of adaptation. Take an old program and run it in some new way, e.g., on a different machine with good performance, which is the simplest case. But then, more interestingly, take two programs that should work together. You've got one that simulates big waves on the ocean, and you've got one that simulates a little airplane flying over the ocean, and you want to crash that airplane into that ocean. How would you imagine doing that? Well, I'm thinking of 20 or 30 years from now, but I think we can take steps now that might capture some of these things and lead us in that direction.

I'd like to talk a little bit about spending money. How should we spend high-performance computing money in the 1990s? I looked back at the book and the talks from the Frontiers of Supercomputing meeting in 1983, and I recalled a dozen or so companies, such as Ardent-Stellar,

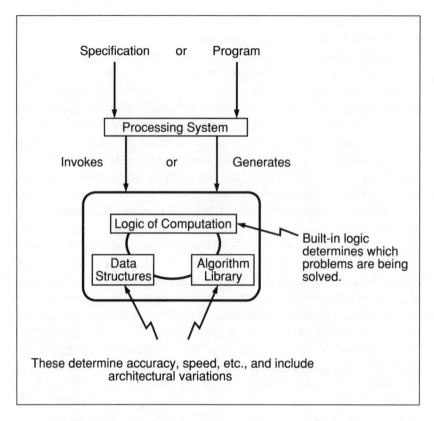

Figure 3. Problem-solving environment model.

Masscomp-Concurrent, and Gould-Encore, that either went out of business or merged together of necessity. There were probably a billion dollars spent on aborted efforts. There are some other companies that are ongoing and have spent much more money.

On the other hand, I looked at the five universities that were represented at the 1983 meeting, and two of us, the New York University people with IBM and the Cedar group at CSRD, have developed systems (as did the Caltech group, which was not at the 1983 Frontiers meeting). Cedar has had 16 processors in four Alliant clusters running for the last year (Kuck et al. 1986). Although the machine is down right now so we can put in the last 16 processors, which will bring us up to 32, we are getting very interesting performances. We did Level 3 of the basic linear algebra subprograms and popularized parallel libraries. We've done parallelization of lots of languages, including Lisp, which I think is a big accomplishment because of the ideas that are useful in many languages. My point here is that all the money spent on university development projects is perhaps only five per cent of the industrial expenditures, so it seems to be a bargain and should be expanded.

In the 1990s the field needs money for
- research and teaching of computational science and engineering,
- development of joint projects with industry, and
- focusing on performance understanding and problem-solving environments.

For the short term, I think there is no question that we all have to look at performance. For instance, I have a piece of software now that instruments the source code, and I'm able to drive a wedge between two machines that look similar. I take the same code on machines from CONVEX Computer Corporation and Alliant Computer Systems, and frequently a program deviates from what the ratio between those machines should be, and then we are able to go down, down, down and actually locate causes of performance differences, and then we see some commonality of causes. For example, all of the loops with the middle- and inner-loop indices reversed in some subscript position are causing one machine to go down in relative performance. So those kinds of tools are absolutely necessary, and they have to be based on having real codes and real data.

I feel that we're not going to stop working on compilers; but I think we need a longer-term vision and goal, and for me that's obviously derived from the PC, after all, because it is a model that works.

References

M. Berry, D. Chen, P. Koss, D. J. Kuck, et al., "The Perfect Club Benchmarks: Effective Performance Evaluation of Supercomputers," *International Journal of Supercomputer Applications* **3** (3), 5–40 (1989).

E. Corcoran, "Calculating Reality," *Scientific American* **264** (1), 100–109 (1991).

D. J. Kuck, *A User's View of High-Performance Scientific and Engineering Software Systems in the Mid-21st Century, Intelligent Software Systems I*, North-Holland Press, Amsterdam (1991).

D. J. Kuck, E. S. Davidson, D. H. Lawrie, and A. H. Sameh, "Parallel Supercomputing Today and the Cedar Approach," *Science* **231**, 967–974 (1986).

D. J. Kuck and A. H. Sameh, "Supercomputing Performance Evaluation Plan," in *Lecture Notes in Computer Science, No. 297: Proceedings of the First International Conference on Supercomputing, Athens, Greece*, T. S. Papatheodorou, E. N. Houstis, C. D. Polychronopoulos, Eds., Springer-Verlag, New York, pp. 1–17 (1987).

Systems and Software

George Spix

George A. Spix has been Director, Software Development, at Supercomputer Systems, Inc., since 1987. From 1983 to 1987, Mr. Spix was Chief Engineer, Software Development, MP Project, at Cray Research, Inc., responsible for the design and development of the user environment and operating system for the MP. From 1980 to 1983, he served in a variety of positions at Cray Research, at Los Alamos National Laboratory, and at Lawrence Livermore National Laboratory. Mr. Spix holds a B.S. in electrical engineering from Purdue University.

At Supercomputer Systems, Inc. (SSI), my main responsibility is software and systems. Our teams are responsible for the user environments; operating systems; peripherals and networks; design verification, diagnostics, testing, and quality assurance; and documentation, publications, and technical operations. The operating-systems activity, of course, has to support what we are doing in the compilation arena in terms of exposing the power of the machine to the user.

I have responsibility for peripherals and networks. This is a large area, especially because we change the balance and the resources in the machines. Every time we do that, we basically are asking the user to reprogram the application. Again, as the definition of the supercomputer changes in terms of those resources, the codes also change. Every time the codes have to be changed, that results in a loss of almost a year in terms of work hours and resources spent.

Design verification and quality assurance are currently my most difficult areas because software is typically the first victim of a hardware problem. As a result, 1990 is the year that we finished testing the machine before we built it. Also, the system does not do you much good unless you have written up how to use it, and that is another part of the challenge.

Probably our main objective at SSI is what we call minimum-time solution. This means that the users of the instrument decide they have a problem that they want to solve at the time they understand the solution. Indeed, from the first days of the company, we have been focused on that objective, which starts at the user arena in terms of how you set up your applications to the operating-system level to the I/O level. We are not just trying to build a throughput machine; we are really trying to solve the general problem and trying to lower the response time for an individual application.

Our approach has been to build architecture and hardware that have the highest performance applications and that are parallel at every level, even by default. I think, as we review hardware issues, we will see that we have succeeded in a fairly rational fashion at exploiting parallelism at almost every layer of the architecture on the machine.

Another objective at SSI is that we are focused on a visual laboratory paradigm. As you know, the early work of von Neumann at Los Alamos National Laboratory focused on the bandwidth match between the eyeball and the machine. A last, but not least, objective is to make sure that reliable software is delivered on time.

We believe we have a comprehensive parallel-processing strategy that does not leave too many stones unturned, although the massively parallel developers might look at the machine and say that we have not focused enough on massive parallelism. Our approach is to prepare for the customer an application base in terms of the relationships that SSI President and CEO Steve Chen and the company have set up with various important industrial customers or prospects, plus the relationships we have with the national laboratories. Unlike systems that were started 15 to 20 years ago, we are starting from quite a different point. We can focus our energies less on operating-system issues per se, in terms of building another type of operating system, and more on compiler problems.

There are a lot of penalties for parallel processing in the traditional world, some of which are perceptual, some architectural, and some very real. There is certainly a perception of high parallel-processing overhead because oftentimes you bring in a $20 or $30 million machine, and the administrative scheduling priorities are such that if you have 20 users and you freeze 19 of them out to support one user well, that is not acceptable. Accounting issues relative to parallel processing come back and bite you, especially if you are doing something "automagically." If you take a 10-hour job, and it runs in two hours because the system decided that was the best way to run it, and you charge the user for 20 hours, your customer will probably be quite upset.

Compilation is viewed as expensive and complex, and it is not foreseen to change. In other words, the expertise around compilation and parallel processing tends to be kind of interesting. If you are willing to factor the user into the equation, like the massively parallel people do, and let them write the code in a form that maps well to the machine, then you have avoided some of that issue. That is a pretty elegant thing to do, but it's pretty painful in terms of user time.

Another point is that the expertise required to optimize spans a lot of disciplines. If you take the massively parallel approach, you have a physicist who not only has to understand the problem very well but also has to understand the topology of the machine that the problem is being mapped to. In the case of classical parallel processing, you have to teach a practicing scientist or engineer what it means to deal with operating-systems-type asynchronous issues and all of the problems those issues cause. We have not done a lot in terms of the Fortran development or standards development to alleviate those problems. On top of that, you get nonreproducible results. If you give a programmer asynchronous behavior in a problem, then you have to provide the tools to help with that.

I think that although we talk about parallel processing, we are actually working it in traditional fashion, somewhat against the user's interest in the sense that we are placing additional algorithmic constraints on the user's ability to get the work done. We have not provided standard languages and systems to the point where you can implement something and move it across systems. A whole different set of disciplines is required in terms of understanding and mapping a problem. In that, I think we are creating a full-employment role for system programmers, although I suspect that is against our interest as a country. We need to basically deliver tools to get the job done for the end user who is being productive and the end user who is not a programmer.

In 1976, Seymour Cray said that one of the problems with being a pioneer is that you always make mistakes. In view of that remark, I never, never want to be a pioneer; it is always best to come second, when you can look at the mistakes the pioneers made. Put another way, the fast drives out the slow, even if the fast is wrong, which kind of goes with the idea that good software is never noticed. As a software developer, I would not like to be in a position of saying that our software will never be noticed.

Lagging I/O is certainly a focus at SSI because as the machines change balances, you add compute power and memory power and do not necessarily deal with the I/O problem. You end up again bending the user's view of the machine and bending the user's ability to get an application to run.

I think Los Alamos National Laboratory is to be commended for Don Tolmie's efforts in the arena of high-performance parallel interface (HIPPI). I think because of the 200-megabyte HIPPI, there is some chance that we will be able to bring I/O back into balance in the next generation of machines.

We can start looking at data-parallel applications and start talking about what it means to really deliver 10^{12} floating-point operations per second (TFLOPS), or even hundreds of TFLOPS. And having done so, we will realize what it means when memory is the processor. The challenge of the future architectures is less about the processor as an entity unto itself with some memory access and more about the memory actually being a processor.

My bottom line is that the massively parallel work that is going on—in which people actually look at algorithms in terms of data-parallel applications and look at vector machines and multiple-processor vector machines—does us nothing but good in the traditional world. The memory-as-processor is probably what will develop in the next five years as the real solution to get a TFLOPS at a reasonable rate, perhaps bypassing the people that have focused on the floating-point, 32- or 64-bit domain.

6

User-Interface Software

This session focused on developments and limitations of systems software for supercomputers and for parallel processing systems. Panelists discussed operating systems, compilers, debuggers, and utilities, as well as load balancing, multitasking, automatic parallelization, and models of computation.

Session Chair

Jack T. Schwartz, New York University

Parallel Architecture and the User Interface

Gary Montry

Gary R. Montry is a freelance consultant for Southwest Software in Albuquerque, New Mexico. Until 1990, he was a Consulting Systems Engineer at Myrias Computer Corporation in Albuquerque. Gary has worked in the area of computer simulations for over 20 years. Since 1985, he has concentrated on simulation codes for massively parallel computers. In 1988, he was a recipient of the first Gordon Bell Award for achievements in parallel processing. He also received the Karp Challenge Award for his work as part of the first team ever to demonstrate parallel speedup of 200 on a parallel MIMD computer. In 1989, he received an R&D 100 Award for Contributions in Parallel Computing Software for Scientific Problems. He was the first person to successfully run a computer program using over 1000 MIMD processors. He also has the unique experience of having run parallel codes on both distributed-memory and shared-memory computers with more than 1000 processing elements.

I want to focus on parallel architectures and the user interface. My background is in parallel processing, and I have been working for vendors specializing in parallel software and shared-memory computers since 1989.

Figure 1 is a tree showing what I consider to be a parallel-architecture hierarchy. If you look at it very closely, you will see that none of the branches have any relationship to the other branches. The reason I drew it this way is because it helps me to enforce the points of view that I want to discuss.

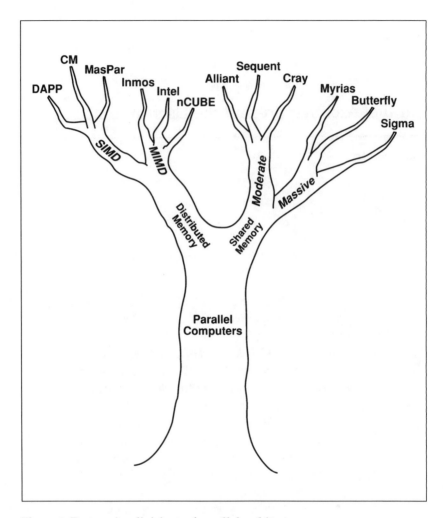

Figure 1. Four major divisions of parallel architectures.

On the left side of the tree, I have the distributed-memory architectures—the SIMD branch and the MIMD branch. On the right-hand side of the tree, I have the shared-memory machines grouped into two branches, which I call the moderately parallel and the massively parallel branches. Massively parallel would be 200 or more processors, but the number is kind of nebulous. The reason I did the tree this way is because the interface difficulties in providing usability for the user of parallel machines are directly dependent on which side of the tree you happen to be.

On the left side of the tree, the MIMD machines provide us with Fortran and C and a couple of message-passing primitives. The same holds true for SIMD. Thus, the interface is pretty much fixed. We are given the language, a few instructions, and a few directives that allow us to pass messages. From there it is possible to build more complex interfaces that make it easier for the user to interact with the machine. A good example of that, I think, is the way Thinking Machines Corporation has brought out a new Fortran language to support their SIMD architecture, which helps to do automatic layout of the variables across the memory.

If you are on the right-hand side of the tree, it is a lot more difficult because you have a lot more options. On the massively parallel side, I have identified three particular architectures or companies that are working on massively parallel, shared-memory machines. One company, Myrias Computer Corporation, has tried to build and market a machine that has distributed memory physically but supports shared memory globally, with a global address base across disparate processors that are connected in some kind of hierarchical fashion. (The figure reflects developments through mid-1990; a Kendall Square Research machine based on similar architecture had already been produced at that time, but the work had not yet been made public.)

Butterfly architecture—represented by the TC2000 from Bolt Beranek & Newman (BBN), the RP3 from IBM, and the Ultra Project at New York University—uses shuffle-exchange networks in order to support shared memory. One difference between these three machines and the Myrias is that while for the former there is only one copy of the operating system in the memory, for the Myrias there is a copy of the operating system on every processor.

I have included the Sigma machine, which is actually the first machine of a family of data-flow machines being built at Electro-Technical Laboratories in Japan. It has already been superseded by their new M machine. Their goal is to have a 1000-processor machine by 1994. Although they are currently lacking floating-point hardware and a little software, that does not mean they won't succeed—their gains have been very impressive.

The typical view of looking at the software and hardware together is a layered approach, where you have the high-level language at the top of the stack, followed by the compilers, the debuggers, and the linkers, some kind of a supervisor that is used to do I/O or to communicate with the user, and the kernel, which interacts with the hardware at the bottom level. This structure is considered to be layered because it has been viewed historically as the way to develop software for workstations and single-processor machines. For those single-processor machines, software development can usually take place separately in each level, which contributes to the notion of independent layers of software.

On top of that, everybody wants another level of abstraction, some kind of user interface or metalanguage—maybe a graphic user interface (GUI)—to help them interact with the actual language that interacts with the machine (Figure 2). My assertion is that for parallel processors, it is too soon to have to worry about this. There are several problems in the lower layers that we have to address for massively parallel machines before we can even think about the user interface.

If we correctly design the actual interface at the language level for shared-memory parallel processors, then we will be able to leverage off work that is done for GUIs with metalanguages and user interfaces from serial-like machines in the future. So really, the programming language is the user interface. It is what you have now, and what you are going to have for the next five, six, or seven years, and it is not really going to change much. The difficulty with parallel processors is that the user interface reflects hardware and architecture dependency through the use of language extensions and/or compiler directives.

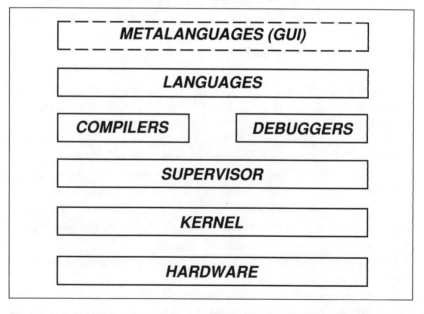

Figure 2. A layered approach to developing software for workstations and single-processor machines, with metalanguages as another level on top to help the user interact with the languages that in turn interact with the machine.

For distributed-memory machines we have message-passing subprogram calls. For a shared-memory machine we can have message-passing or locks and semaphores. For parallel processors like those from Alliant Computer Systems and the Cedar Project, in which you have lots of compiler assist, there are compiler directives, or pragmas, to indicate to the compiler and to the computer exactly what we are trying to do and how to decompose the problem ideally.

For shared-memory machines, the design of the human-machine interface at the language level is intimately tied to the particular underlying hardware and software. That means that you cannot think of the whole stack of software and hardware in the classical sense, as you would for a workstation or a serial processor. You have to think of it as a seamless environment in which the software and the hardware all have to work together at all different levels in order to cooperate and provide the parallelism.

From a functional point of view, what you really want is for all the software and hardware infrastructure to sit below a level at which the user has to view it. At the top, you only want the high-level interface of the languages to show, like the tip of an iceberg (Figure 3). To do that, you have to develop an interface that is very complex underneath but looks very simple to the user. The complexity that the user sees is inversely proportional to the amount of work that went into the interface to support it.

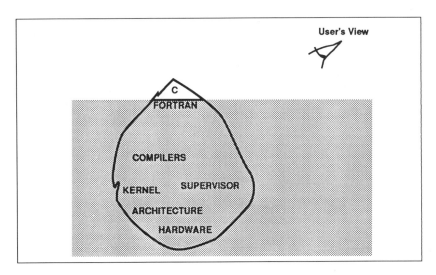

Figure 3. Functionality of the software and hardware of a simple interface based on a complex system.

The corollary to this concept is that complex interfaces are cheap, and simple interfaces are expensive, which is the reason you have complex interfaces on distributed-memory machines, or on the first generations thereof, because they were simple to implement and they were not very costly. An iceberg model of this interface would result in an apparent violation of the laws of physics in which the iceberg would float 90 per cent above the water and 10 per cent below the water (Figure 4).

There is another reason for wanting to have a serial look and feel to a shared-memory machine: you would like to get the feeling that when a program is run, you get the right result. You would like to have the feeling that you have an interface that is actually reproducible and trustworthy.

I can tell you some stories about the early days on the ELXSI—about a code of mine that ran for six months. Bob Benner (a Session 7 panelist) and I put it together. It was a finite-element code, and we used it to test the machine's integrity. We ran it for six months, and it always gave the same answers. I went in one day and I ran it, and it didn't give the same answer anymore. It gave different answers. So I went out and had a drink and came back later and ran it, and there was no one on the machine, and it gave the right answers. I studied the problem for a week, and I finally called up the ELXSI company and said, "You have a problem with your machine." They said, "Oh, no, you're a user; you have a problem with your code." We went around and around, and a week later I finally

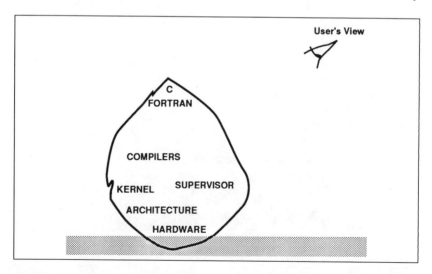

Figure 4. Functionality of the software and hardware of a complex interface based on a simple system.

convinced them to look at the microcode. It turned out that they had changed the microcode in their caching scheme and had failed to fully flush all caches during context switch.

So there are problems with shared-memory parallel processors in the sense of really trying to give people the feeling that both the vendor and the user have something solid. We really do want to have the look and feel of serial interfaces. To accomplish these goals, there are several requirements:
- language and compiler extensions that have the look and feel people are used to when working with computer codes;
- automatic generation of independent tasks for concurrent execution, whereby the generation is initiated by the compiler at compile time and then instantiated by the operating system during the run;
- dynamic scheduling and load balancing of independent tasks;
- automatic distribution and collection of data from concurrently executed tasks; and
- the ability to relieve the user of the burden of data locality, which is the toughest constraint of all (the easiest way to meet this last requirement is to build a common-memory machine, although such a machine is very expensive).

The machine discussed by Burton Smith in Session 4 is really a common-memory machine. It is a single-level machine, and it is very expensive. The hope is that we can build machines with memory that is distributed with the processors and still relieve the user of the burden of data locality. We can explicitly relieve the user of the burden if we have an interconnect that is fast enough, but no one can afford to build it. Actually, it is not even technologically possible to do that right now. The alternative way of eliminating the data-locality problem is with latency hiding, and asynchronous machines can do quite a bit of that.

When you look at all of these requirements and you say you want to have a number of independent tasks that you would like to be able to schedule across some kind of interconnect, you have to decide as a computer designer and builder what kind of tasks you are going to have. Are you going to have lightweight threads with very little context that could move very easily, although you might have to generate many of them? If you do, that puts particular constraints on the interconnect system that you have. Or are you going to have fewer heavyweight processes, which are larger-grained, which don't have to move quite as often, but which also don't put quite the strain on the interconnect? The interconnect is very costly in shared-memory machines, so you need to make the decision about what you want to do at the top level, which will affect your compiler and perhaps the constructs that you show to the user, which in turn will affect the hardware at the bottom of the software/hardware hierarchy.

Now that I have outlined the technical requirements to build the machine, I will discuss the programmatic requirements.

There are five requirements I have identified that are important and that every project has to meet to build a massively parallel, shared-memory machine:

- You have to have a lot of good parallel software expertise to write the operating system and the kernel and the compilers, and they all have to work together.
- You have to have parallel hardware expertise. For that, you have to have hardware resources and people who understand how parallel machines work and what the costs and tradeoffs are for doing certain things in certain ways.
- You have to have teamwork. This may sound corny, but this is one of the most important points. There has got to be teamwork that is driven by end-users, people who are actually going to use the machine at the end so you have something that is usable. A good example of a lack of teamwork here would be the Floating Point Systems T-Series machine, for which there was plenty of hardware built, but the software was left out.
- You have to have commitment. These are long projects that last many years. If you don't think commitment is important, ask the people at Evans & Sutherland about commitment, for example.
- Finally, you need on the order of $50 to $100 million to solve the problems.

Many companies and various entities in the U.S. meet some of those five requirements, but they don't meet them all. Unless the situation changes substantially in the next few years, you probably won't see a massively parallel, shared-machine from a new entity in the U.S. It will either have to come from BBN or Myrias or the companies that are already in business.

In summary, we need to develop a somewhat standardized language implementation at the top level for shared-memory parallel processors so that we can start designing hardware and building the machines to execute that code. These hardware experiments are expensive, and in the current political and economic climate, it is not too likely that private industry is going to take on this particular challenge.

Object-Oriented Programming, Visualization, and User-Interface Issues*

David Forslund

David W. Forslund has served since 1989 as Deputy Director of the Advanced Computing Laboratory, Los Alamos National Laboratory. Dr. Forslund, a specialist in theoretical plasma physics, is credited with such accomplishments as the discovery of heat conduction instabilities in the solar wind and development of the first model to account for intense visible harmonics of laser light produced in CO_2 laser plasmas. Much of Dr. Forslund's most original work has focused on simulation codes for applications in plasma physics. For example, he developed and has maintained the international laser fusion plasma simulation code WAVE on a multitude of operating systems and machine architectures. Further, he codeveloped the first implicit electromagnetic plasma simulation code, VENUS, with which he discovered the surface magnetic fields that have since explained many of the interactions between intense CO_2 light with plasmas. Currently, he is investigating advances in human-computer interfaces and has recently demonstrated successful coupling of a scientific workstation and a supercomputer via the Network extensible Window System^{TM}.

---

* Rendering of the computer graphics reproduced in this paper was carried out at the Los Alamos National Laboratory Advanced Computing Laboratory using the Advanced Visualization System on a Stardent Inc. computer.

Dr. Forslund holds an M.A. (1967) and a Ph.D. (1969) from Princeton University. During his tenure at Los Alamos, which began in 1969, he has served as a Staff Member at the Experimental Physics Division, the Associate Group Leader of the Laser Division, and the Alternate Group Leader of the Applied Theoretical Physics Division. In 1981, he was named a Laboratory Fellow. He has published widely on topics relating to plasma physics and plasma simulations and referees numerous professional journals, including the Journal of Geophysical Research, *the* Journal of Applied Physics, *and the* Journal of Computational Physics.

There are two important but disparate elements of the user interface that we will discuss in this presentation. The first is the object-oriented paradigm, which provides a useful framework for writing parallel applications, and the second is the use of visualization tools, which can provide an intuitive interface to complex applications.

Object-Oriented Parallel Programming

Object-oriented programming has become fairly common and popular and has been used in numerous computer-science projects. However, it has not yet been utilized to any degree in large-scale scientific computing. Nevertheless, we believe it is well suited to scientific computing, which frequently deals with well-defined, loosely interacting, physical objects. In particular, the paradigm is particularly useful in distributed, parallel computing because the objects help to encapsulate and clearly define the movement of data. The message interface maps well to the distributed memory model of parallel computing by constraining the mapping of data into memory. The additional features of inheritance and data abstraction also promise to reduce significantly the cost of software maintenance for large-scale scientific programs.

Distributed Computing

An important tool in defining the objects to be used in scientific computing comes from a mapping of the physical model into the computational environment. If the objects are made to correspond to physical elements in the underlying model, the paradigm fits very well. Since all of the information an object needs is stored internally, an object can also provide a good representation for a thread of control. This can greatly assist in the development of parallel applications in a way that is independent of the hardware on which it is being run. One can logically define the necessary number of threads (or processes) required for a given problem without worrying about the number of physical processors. This is analogous to not worrying about the size of vector registers in a supercomputer or the actual number of processors in a Thinking Machines Corporation Connection Machine.

A number of these ideas have been implemented in a distributed particle simulation code, which is reported on in the 1990 USENEX C++ conference proceedings (Forslund et al. 1990). The distributed environment for this code is the ISIS programming environment developed at Cornell University by Ken Birman (1990) and his colleagues.

Data Parallel Programming

Another area in which object-oriented programming has had success is data parallel programming. Rob Collins (personal communication), from the University of California, Los Angeles, has built an efficient C++ library for the Thinking Machines Corporation CM-2, called CM++. It gives full access to the C/Paris functionality without any loss in performance. This allows writing in a higher-level abstraction without sacrificing speed. Collins and Steve Pope of the Advanced Computing Laboratory at Los Alamos National Laboratory (personal communication) have been working on porting this library to a more general environment, called DPAR. Thus, the data-parallel paradigm has been abstracted and now runs on a workstation with good optimization. We are optimistic that this library could be made to work on a Cray Research, Inc., supercomputer, with comparable efficiency to that provided on the CM-2.

We are also trying to combine this data-parallel programming paradigm with the distributed environment we mentioned before, following some of the ideas of Guy Steele. In a paper entitled "Making Asynchronous Parallelism Safe for the World," Steele (1990) describes a programming style that tries to unify SIMD and MIMD computing. The essence of the proposal is to allow asynchronous threads to have only a restricted shared-memory access providing only commutative operations. This removes the dependence of the order of sibling parallel threads.

Threads then communicate (synchronize) only on their death. Complex operations are provided by a hierarchy of communications. This is the style we have used in our distributed particle code mentioned earlier. In one sense, this style is the simplest extension of the SIMD programming model.

Visualization Requirements

As one solves large-scale problems on massively parallel machines, the data generated become very difficult to handle and to analyze. In order for the scientist to comprehend the large volume of data, the resulting complex data sets need to be explored interactively with intuitive tools that yield realistic displays of the information. The form of display usually involves polygons and lines, image processing, and volume rendering. The desired interface is a simple, flexible, visual programming environment for which one does not have to spend hours writing code. This might involve a dynamic linking environment much like that provided by the Advanced Visualization System (AVS) from Stardent Computer or apE from the Ohio State University Supercomputer Center in Columbus.

The output need not always be precisely of physical variables but should match what we expect from our physical intuition and our visual senses. It also should not be just a collection of pretty artwork but should have physical meaning to the researcher. In this sense, we don't try to precisely match a physical system but rather try to abstract the physical system in some cases.

To handle the enormous computational requirements involved in visualization, we must also be able to do distributed processing of the data and the graphics. Besides being useful in the interpretation of significant physical and computational results, this visualization environment should be usable both in algorithmic development and debugging of the code that generates the data. The viewing should be available in both "real time" and in a postprocessing fashion, depending on the requirements and network bandwidth. To optimize the traversal of complex data sets, advanced database techniques such as object-oriented databases need to be used.

As mentioned above, there are two graphical environments available today (and possibly others) that attempt to provide the sort of capability described above. They are AVS and apE. The idea is to provide small, strongly typed, modular building blocks out of which one builds the graphical application. These are illustrated in Figure 1, which displays a schematic of the user's workspace with AVS. The data flows through the graphical "network" from the input side all the way through the graphical display. In AVS, there are four basic types of components out of which one builds the application: data that is input, filters that modify the data, mappers that change the data from one format to another, and renderers that display them on the screen. Figure 2 illustrates AVS's ability to interactively analyze data.

User-Interface Software   195

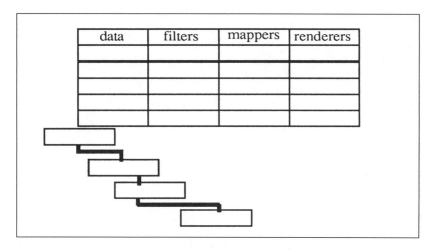

Figure 1. A schematic of an AVS-network user's workspace.

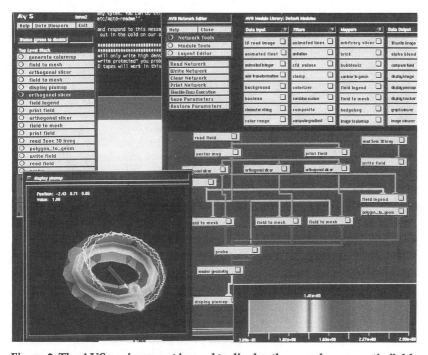

Figure 2. The AVS environment is used to display the complex magnetic field in a numerical model of a Tokamak Fusion Reactor system. AVS provides a simple visual environment, which is useful to interactively analyze the data. Several different magnetic field surfaces are shown, as well as the trajectory of a particle in the system.

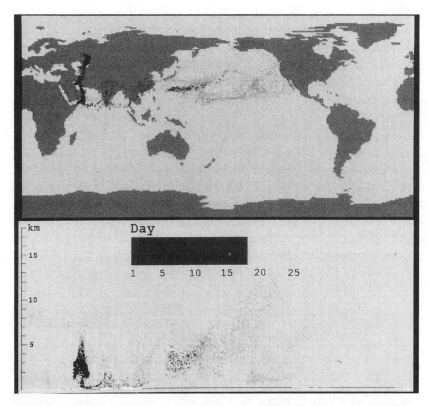

Figure 3. This still from a computer animation illustrates the modeling of the propagation of smoke in elevation, latitude, and longitude generated by the oil fires in Kuwait. A full global climate model was used, including rain washing out the smoke. The picture shows that the smoke does not loft into the stratosphere to cause a global climate modification.

The limit to the network complexity is only the memory and display limits of the workstation. However, this limitation can frequently be a major problem, as the size of the data set produced on current supercomputers can far exceed the capabilities of this software, even on the most powerful graphics workstations.

Because this data-flow style is, in fact, object-oriented, this model can be readily distributed or parallelized, with each module being a thread or distributed process. By placing the nodes on different machines or processors, this data-flow model can, at least in principle, be distributed or parallelized. In fact, the apE environment provides for this kind of functionality. For high performance in a graphics environment, these nodes need to be connected with a very-high-speed (e.g., gigabit/second) network if they are not running out of shared memory on the same machine. The next generation of graphics environments of this type will hopefully operate in this manner.

A number of real physical applications are using this graphics environment, including problems that run on a CRAY Y-MP and the CM-2. For example, at the Institute of Geophysics and Planetary Physics, Los Alamos National Laboratory, a three-dimensional climate model has been run on the Cray (see Figure 3). A layer of the resulting temperature data has been taken and mapped onto a globe and displayed in an animated manner as a function of time. Using AVS, one can rotate the spherical globe while the data is being displayed, allowing one to investigate the polar regions, for example, in more detail. This is one simple example of how the data can be explored in a manner that is hard to anticipate ahead of time.

Figures 4 and 5 further illustrate the capability of high-performance graphics environments as applied to physical processes. Realistic displays like these, which can be explored interactively, are powerful tools for understanding complex data sets.

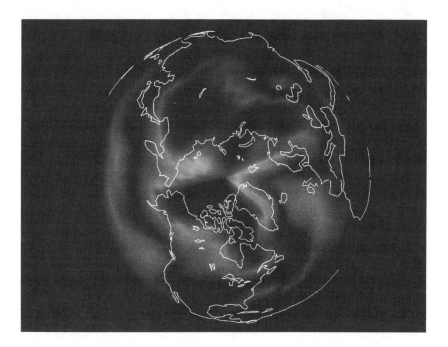

Figure 4. Temperature distribution over the Arctic, generated by the global climate simulation code developed at the Earth and Environmental Sciences Division, Los Alamos National Laboratory.

Figure 5. Model of the penetration of a high-speed projectile through a metal plate. Note the deformation of the projectile and the splashing effect in the metal plate. This calculation was done on a Thinking Machines Corporation CM-2.

References

K. Birman, *ISIS Reference Manual, Version 2.0*, ISIS Distributed Systems, Inc., Ithaca, New York (1990).

D. Forslund, C. Wingate, P. Ford, J. Junkins, J. Jackson, S. Pope, "Experience in Writing a Distributed Particle Simulation Code in C++," in *Proceedings of the USENEX C++ Conference*, USENEX Association, Berkeley, California, pp. 177–190 (1990).

G. Steele, Jr., "Making Asynchronous Parallelism Safe for the World," in *The Conference Record of the Seventeenth Annual ACM Symposium on the Principles of Programming Languages, San Francisco, California, 17–19 January 1990*, Association for Computing Machinery, New York, pp. 218–231 (1990).

Software Issues at the User Interface*

Oliver McBryan

Oliver A. McBryan studied theoretical physics at Harvard University, where he received a Ph.D. in 1973. After postdoctoral appointments at the University of Toronto and the Rockefeller University, New York, he joined the Mathematics Department faculty at Cornell University, Ithaca, New York, as an assistant professor. He moved to New York University's Courant Institute of Mathematical Sciences in 1979, taking up a post as an associate professor of mathematics and, later, as a full professor. In 1987, Dr. McBryan moved to the University of Colorado at Boulder, where he is currently Professor of Computer Science and Director of the Center for Applied Parallel Processing. His interests include parallel computation, graphics and visualization, computational fluid dynamics, statistical mechanics, and quantum field theory. He has published over 120 technical papers and reports on these subjects.

Abstract

In this presentation we review software issues that are critical to the successful integration of parallel computers into mainstream scientific computing. Clearly, on most systems, a compiler is the most important

---

* This research was supported by the Air Force Office of Scientific Research, under grant AFOSR-89-0422.

software tool available to a user. We discuss compilers from the point of view of communication compilation—their ability to generate efficient communication code automatically. We illustrate with two examples of distributed-memory computers, on which almost all communication is handled by the compiler rather than by explicit calls to communication libraries.

Closely related to compilation is the need for high-quality debuggers. While single-node debuggers are important, parallel machines have their own specialized debugging needs related to the complexity of interprocess communication and synchronization. We describe a powerful simulation tool we have developed for such systems, which has proved essential in porting large applications to distributed memory systems.

Other important software tools include high-level languages, libraries, and visualization software. We discuss various aspects of these systems briefly. Ultimately, however, general-purpose supercomputing environments are likely to include more than a single computer system. Parallel computers are often highly specialized and rarely provide all of the facilities required by a complete application. Over the coming decade we will see the development of heterogeneous environments that will connect diverse supercomputers (scalar, vector, and parallel), along with high-end graphics, disk farms, and networking hubs. The real user-interface challenge will then be to provide a unified picture of such systems to potential users.

Introduction

This paper will survey a selection of issues dealing with software and the user interface, always in the context of parallel computing. The most obvious feature in looking back over the last five to 10 years of parallel computing is that it has been clearly demonstrated that parallel machines can be built. This was a significant issue eight or 10 years ago, particularly in the light of some early experiences. There was a feeling that the systems would be too unreliable, with mean time to failure measured in minutes and with intractable cooling and packaging issues. There are now several examples of real, highly parallel systems, such as Thinking Machine Corporation's Connection Machine CM-2 and the Intel iPSC/860, which are certainly recognized as serious computers and are among the leading supercomputers at present.

It has also been shown that model problems can be solved efficiently on these systems. For example, linear algebra problems, partial differential equation (PDE) solvers, and simple applications such as quantum chromodynamics have been modeled effectively. In a few cases, more complex models such as oil reservoir simulations and weather models have been parallelized.

Although hardware progress has been dramatic, system software progress has been painfully slow, with a few isolated exceptions that I will highlight in the following section. Another area in which there has been almost no progress is in demonstrating that parallel computers can support general-purpose application environments. Most of what we will present here will be motivated to some extent by these failures.

We could begin by asking what a user would like from a general-purpose supercomputer application environment. In simplest terms, he would like to see a computer consisting of a processor that is as powerful as he wants, a data memory as large as his needs, and massive connectivity with as much bandwidth (internal and external) as desired. Basically, he would like the supercomputer to look like his desktop workstation. This also corresponds to the way he would like to program the system.

It is well known that there are various physical limitations, such as the finite speed of light and cooling issues, that prevent us from designing such a system. As a way around, one has been led to parallelism that replicates processors, memories, and data paths to achieve comparable power and throughput. To complicate the situation further, there are many different ways to actually effect these different connections, as well as several distinctly different choices for control of such systems. Thus, we end up with the current complexity of literally dozens of different topologies and connection strategies.

Yet the user wants to think of this whole system as a monolithic processor if possible. We will focus on several software areas where the user obviously interacts with the system and discuss ways in which the software can help with this issue of focusing on the machine as a single entity.

At the lowest level we will discuss compilers, not from the point of view of producing good machine code for the individual processors but rather from the higher-level aspect of how the compiler can help with the unifying aspect for parallel machines. Then there are the all-important debugging, trace, and simulation phases. People actually spend most of their time developing programs rather than running them, and if this phase is not efficient for the user, the system is likely ultimately to be a failure.

We will briefly discuss several related developments: higher-level languages, language extensions, and libraries. Portable parallel libraries provide a key way in which the user's interaction with systems can be simplified. Graphics and the visualization pipeline are, of course, critical areas about which we will make several comments. For each of these topics, we will refer to other papers in these proceedings for more coverage.

Finally, we will discuss the development of software for heterogeneous environments, which is in many ways the most important software issue. No one parallel machine is going to turn out to provide the general-purpose computer that the user would really like to have. There will be classes of parallel machines, each well suited to a range of problems but unable to provide a solution environment for the full range of applications. Heterogeneous systems will allow complex algorithms to use appropriate and optimal resources as needed. So ultimately we will be building heterogeneous environments, and the software for those systems is perhaps the greatest challenge in user-interface design in the near future.

Compilers and Communication

There are three roles that compilers play in the context of parallel machines. First of all, they provide a mechanism for generating good scalar and vector node code. Since that topic is covered adequately in other papers in this volume, we will not focus on it here. Rather, we will focus on the fact that the compiler can help the user by taking advantage of opportunities for automatic parallelization, and particularly important in the context of distributed machines, there is the possibility for compilers to help the user with some of the communication activities.

The current compilers do a very good job in the area of scalar/vector node code generation, although some node architectures (e.g., i860) are quite a challenge to compiler writers. Some of the compilers also make a reasonable effort in the area of parallelization, at least in cases where data dependencies are obvious. However, there is very little to point to in the third area of compilers helping on distributed machines. The picture here is not completely bleak, so we will refer to two examples that really stand out, namely, the CM-2 and Myrias Research Corporation's SPS-2 computers. In both of these systems, the compilers and the associated runtime system really help enormously with instantiation and optimization of communication.

Myrias SPS-2: Virtual Memory on a Distributed System

The Myrias SPS-2 system was introduced in Gary Montry's presentation earlier in this session. It is a typical distributed-memory machine, based on local nodes (Motorola MC68020) with some memory associated and connected by busses organized in a three-level hierarchy. The SPS-2 has the remarkable feature that it supports a virtual shared memory, and that feature is what we want to focus on here. For further details on the SPS-2, see McBryan and Pozo (1990).

On the system side, virtual shared memory is implemented by the Fortran compiler and by the run-time system. The result is to present a uniform 32-bit address space to any program, independent of the number of processors being used. From the user's point of view, he can write a standard Fortran F77 program, compile it on the machine, and run it as is, without any code modification. The program will execute instructions on only one processor (assuming it is written in standard Fortran), but it may use the memory from many processors. Thus, even without any parallelization, programs automatically use the multiple memories of the system through the virtual memory. For example, a user could take a large Cray application with a data requirement of gigabytes and have it running immediately on the SPS-2, despite the fact that each node processor has only eight megabytes of memory.

With the sequential program now running on the SPS-2, the next step is to enhance performance by exploiting parallelism at the loop level. To parallelize the program, one seeks out loops where the internal code in the loop involves no data dependencies between iterations. Replacing DO with PARDO in such loops parallelizes them. This provides the mechanism to use not only the multiple memories but also the multiple processors. Developing parallel programs then becomes a two-step refinement: first, use multiple memories by just compiling the program, and second, add PARDOs to achieve instruction parallelism.

As discussed in the following section, the virtual-memory support appears to reduce SPS-2 performance by about 40 to 50 per cent. A lot of people would regard a 50 per cent efficiency loss as too severe. But we would argue that if one looks at the software advantages over long-term projects as being able to implement shared-memory code on a distributed-memory system, those 50 per cent losses are certainly affordable. However, one should note that the SPS-2 is not a very powerful supercomputer, as the individual nodes are MC68020 processors with a capacity of 150,000 floating-point operations per second (150 KFLOPS). It remains to be demonstrated that virtual memory can run on more powerful distributed systems with reasonable efficiency.

One other point that should be made is that we are not talking about virtual shared memory on a shared-memory system. The SPS-2 computer is a true distributed-memory system. Consequently, one cannot expect that just any shared-memory program will run efficiently. To run efficiently, a program should be well suited to distributed systems to begin with. For example, grid-based programs that do local access of data will run well on such a system. Thus, while you can run any program on these systems without modification, you can only expect good performance from programs that access data in the right way.

The real benefit of the shared memory to the user is that there is no need to consider the layout of data. Data flows naturally to wherever it is needed, and that is really the key advantage to the user of such systems. In fact, for dynamic algorithms, extremely complex load-balancing schemes have to be devised to accomplish what the SPS-2 system does routinely. Clearly, such software belongs in the operating system and not explicitly in every user's programs.

Myrias SPS-2: A Concrete Example

In this section we study simple relaxation processes for two-dimensional Poisson equations to illustrate the nature of a Myrias program. These are typical of processes occurring in many applications codes involving either elliptic PDE solutions or time-evolution equations. The most direct applicability of these measurements is to the performance of standard "fast solvers" for the Poisson equation. The code kernels we will describe are essentially those used in relaxation, multigrid, and conjugate gradient solutions of the Poisson equation. Because the Poisson equation has constant coefficients, the ratio of computational work per grid point to memory traffic is severe, and it is fair to say that while typical, these are very hard PDEs to solve efficiently on a distributed-memory system.

The relaxation process has the form

$$v(j,i) = s\, u(j,i) + r(u(j+1,i) + u(j-1,i) + u(j,i+1) + u(j,i-1))\ .$$

Here, the arrays are of dimensions $n_1 \times n_2$, and s,r are specified scalars, often 4 and 1, respectively. The equation above is to be applied at each point of the *interior* of a two-dimensional rectangular grid, which we will denote generically as G. If the equations were applied at the boundary of G, then they would index undefined points on the right-hand side. This choice of relaxation scheme corresponds to imposition of Dirichlet boundary conditions in a PDE solver. The process generates a new solution v from a previous solution u. The process is typified by the need to access a number of nearby points. At the point i,j it requires the values of u at the four nearest neighbors.

We implement the above algorithm serially by enclosing the expression in a nested set of DO loops, one for each grid direction:

```
      do 10 j = 2,n1-1
      do 10 i = 2,n2-1
      v(j,i) = s*u(j,i) + r(u(j,i-1) + u(j,i+1)
                        + u(j-1,i) + u(j+1,i))
10    continue
```

To parallelize this code using T parallel tasks, we would like to replace each DO with a PARDO, but this in general generates too many tasks—a number equal to the grid size. Instead, we will decompose the grid G into T contiguous rectangular subgrids, and each of T tasks will be assigned to process a different subgrid.

The partitioning scheme used is simple. Let $T = T_1 T_2$ be a factorization of T. Then we divide the index interval $[2, n_1 - 1]$ into T_1 essentially equal pieces, and similarly we divide $[2, n_2 - 1]$ into T_2 pieces. The tensor product of the interval decompositions defines the two-dimensional subgrid decomposition.

In case T_1 does not divide $n_1 - 2$ evenly, we can write

$$n_1 - 2 = h_1 T_1 + r_1, 0 \quad r_1 < T_1 \ .$$

We then make the first r_1 intervals of length $h_1 + 1$ and the remaining $T_1 - r_1$ intervals of length h_1, and similarly in the other dimension(s). This is conveniently done with a procedure

decompose (a,b,t,istart,iend) ,

which decomposes an interval $[a,b]$ into t near-equal-length subintervals as above and which initializes arrays *istart* (t), *iend* (t) with the start and end indices of each subinterval.

Thus, the complete code to parallelize the above loop takes the form

```
      decompose(2,n1-1,t1,istart1,iend1)
      decompose(2,n2-1,t2,istart2,iend2)
      pardo 10 q1=1,t1
      pardo 10 q2=1,t2
      do 10 i= istart1(q1),iend1(q1)
      do 10 j= istart2(q2),iend2(q2)
      v(j,i) = s*u(j,i) + r(u(j,i-1) + u(j,i+1)
                        + u(j-1,i) + u(j+1,i))
10    continue
```

The work involved in getting the serial code to run on the Myrias using multiple processors involved just one very simple code modification. The DO loop over the grid points is replaced by, first of all, a DO loop over processors, or more correctly, tasks. Each task computes the limits within the large array that it has to work on by some trivial computation. Then the task goes ahead and works on that particular limit. However, the data arrays for the problem were never explicitly decomposed by the user, as would be needed on any other distributed-memory MIMD machine.

This looks exactly like the kind of code you would write on a shared-memory system. Yet the SPS-2 is truly a distributed-memory system. It really is similar to an Intel Hypercube, from the logical point of view. It is a fully distributed system, and yet you can write code that has no communication primitives. That is a key advance in the user interface of distributed-memory machines, and we will certainly see more of this approach in the future.

MYRIAS SPS-2: Efficiency of Virtual Memory

We have made numerous measurements on the SPS-2 that attempt to quantify the cost of using the virtual shared memory in a sensible way (McBryan and Pozo 1990). One of the simplest tests is a SAXPY operation (adding a scalar times a vector to a vector):

$$y_i = y_i + a x_i .$$

We look at the change in performance as the vector is distributed over multiple processors, while performing all computations using only one processor. Thus, we take the same vector but allow the system to spread it over varying numbers of processors and then compute the SAXPY using just one processor. We define the performance with one processor in the domain as efficiency 1. As soon as one goes to two or more processors, there is a dramatic drop in efficiency to about 60 per cent, and performance stays at that level more or less independent of the numbers of processors in the domain. That then measures the overhead for the virtual shared memory.

Another aspect of efficiency related to data access patterns may be seen in the relaxation example presented in the previous section. The above procedure provides many different parallelizations of a given problem, one for each possible factorization of the number of tasks T. At one extreme are decompositions by rows (case $T_1 = 1$), and at the other extreme are decompositions by columns ($T_2 = 1$), with intermediate values representing decompositions by subrectangles. Performance is strongly influenced by which of these choices is made. We have in all cases found that decomposition by columns gives maximum performance. This is not, a priori, obvious; in fact, area-perimeter considerations suggest that virtual-memory communication would be minimized with a decomposition in which $T_1 = T_2$. Two competing effects are at work: the communication bandwidth requirements are determined by the perimeter of subgrids, whereas communication overhead costs (including memory merging on task completion) are determined additionally by a factor proportional to the total number of data requests. The latter quantity is minimized by a column division. Row division is unfavorable because of the Fortran rules for data storage.

It is instructive to study the variation in performance for a given task number T as the task decomposition varies. We refer to this as "varying the subgrid aspect ratio," although in fact it is the task subgrid aspect ratio. We present sample results for two-dimensional relaxations in Table 1. The efficiency measures the deviation from the optimal case. Not all aspect ratios would in fact run. For heavily row-oriented ratios (e.g., $T_1 = 1$, $T_2 = T$), the system runs out of virtual memory and kills the program unless the grid size is quite small.

The Connection Machine CM-2: Overlapping Communication with Computation

The Connection Machine CM-2 affords another good example of how a powerful compiler can provide a highly effective user interface and free the user from most communication issues. The Connection Machine is a distributed-memory (hypercube) SIMD computer, which in principle might have been programmed using standard message-passing procedures. For a more detailed description of the CM-2, see McBryan (1990).

Table 1. Two-Dimensional Effect of Subgrid Aspect Ratio

Grid	D	T1	T2	MFLOPS	Efficiency
512 × 512	64	1	64	0.036	0.022
512 × 512	64	2	32	0.076	0.047
512 × 512	64	4	16	0.217	0.134
512 × 512	64	8	8	0.502	0.310
512 × 512	64	16	4	0.946	0.584
512 × 512	64	32	2	1.336	0.825
512 × 512	64	64	1	1.619	1.000

In fact, the assembly language of the system supports such point-to-point communication and broadcasting. However, Connection Machine high-level software environments provide basically a virtual shared-memory view of the system. Each of the three high-level supported languages, CM Fortran, C*, and *Lisp, makes the system look to the user as if he is using an extremely powerful uniprocessor with an enormous extended memory. These languages support parallel extensions of the usual arithmetic operations found in the base language, which allows SIMD parallelism to be specified in a very natural and simple fashion. Indeed, CM-2 programs in Fortran or C* are typically substantially shorter than their serial equivalents from workstations or Cray Research, Inc., machines because DO loops are replaced by parallel expressions.

However, in this discussion I would like to emphasize that very significant communication optimization is handled by the software. This is best illustrated by showing the nature of the optimizations involved in code generation for the same generic relaxation-type operation discussed in the previous section. We will see that without communication optimization the algorithm runs at around 800 MFLOPS, which increases to 3.8 GFLOPS when compiler optimizations are used to overlap computation and communication.

For the simple case of a Poisson-type equation, the fundamental operation $v = Au$ takes the form (with r and s scalars)

$$v_{i,j} = su_{i,j} + r(u_{i+1,j} + u_{i-1,j} + u_{i,j+1} + u_{i,j-1}) \ .$$

The corresponding CM-2 parallel Fortran takes the form

```
v = s*u + r*(cshift(u,1,1) + cshift(u,1,-1) + cshift(u,2,1)
    + cshift(u,2,-1))   .
```

Here, *cshift (u,d,l)* is a standard Fortran 90 array operator that returns the values of a multidimensional array u at points a distance l away in dimension direction d.

The equivalent *Lisp version of a function *applya* for $v = Au$ is

```
(defun *applya (u v)

    (*set v    (-!! (*!! (!! s) u)
                    (*!! (!! r) (+!! (news!! u -1 0)
                                     (news!! u 1 0)
                                (news!! u 0 -1) (news!! u 0 1)
)))))
```

*Lisp uses !! to denote parallel objects or operations, and as a special case, !! s is a parallel replication of a scalar s. Here (news!! u dx dy) returns in each processor the value of parallel variable u at the processor dx processors away horizontally and dy away vertically. Thus, cshift (i + 1,j) in Fortran would be replaced by (news!! u 1 1) in *Lisp.

The *Lisp source shown was essentially the code used on the CM-1 and CM-2 implementation described in McBryan (1988). When first implemented on the CM-2, it yielded a solution rate of only 0.5 GFLOPS. Many different optimization steps were required to raise this performance to 3.8 GFLOPS over a one-year period. Probably the most important series of optimizations turned out to be those involving the overlap of communication with computation. Working with compiler and microcode developers at Thinking Machines Corporation, we determined the importance of such operations, added them to the microcode, and finally improved the compiler to the point where it automatically generated such microcode calls when presented with the source above.

We will illustrate the nature of the optimizations by discussing the assembly language code generated by the optimized compiler for the above code fragment. The language is called PARIS, for PARallel Instruction Set. The PARIS code generated by the optimizing *Lisp compiler under version 4.3 of the CM-2 system is shown in the code displayed below. Here, the code has expanded to generate various low-level instructions, with fairly recognizable functionality, including several that overlap computation and communication, such as

cmi:get–from–east–with–f–add–always ,

which combines a communication (getting data from the east) with a floating-point operation (addition). Here is the optimized PARIS code for relaxation:

```
(defun *applya (u v)
    (let* ((slc::stack-index *stack-index*)
           (-!!-index-2 (+ slc::stack-index 32))
           (pvar-location-u-11 (pvar-location u))
           (pvar-location-v-12 (pvar-location v))) ,

       (cm:get-from-west-always -!!-index-2
                                pvar-location-u-11 32)
       (cm:get-from-east-always *!!-constant-index4
                                pvar-location-u-11 32)
       (cmi::f+always -!!-index-2 *!!-constant-index4 23 8)
```

```
            (cmi::get-from-east-with-f-add-always -!!-index-2
                                         pvar-location-u-11 23 8)

            (cmi::f-multiply-constant-3-always pvar-location-v-12
                                         pvar-location-u-11 s 23 8)
            (cmi::f-subtract-multiply-constant-3-always
                        pvar-location-v-12 pvar-location-v-12
                                         -!!-index-2 r 23 8)

            (cmi:get-from-north-always -!!-index-2
                                         pvar-location-u-11 32)
            (cmi::f-always slc::stack-index -!!-index-2 23 8)
            (cmi::get-from-north-with-f-subtract-always
                    pvar-location-v-12 pvar-location-u-11 23 8)

            (cmi:get-from-south-always -!!-index-2
                                         pvar-location-u-11 32)
            (cmi::float-subtract pvar-location-v-12 slc::
                                    stack-index -!!-index-2 23 8)
            (cmi::get-from-south-with-f-subtract-always
                        pvar-location-v-12 -!!-index-2 23 8)
            )
    )
```

Obviously, the generated assembly code is horrendously complex. If the user had to write this code, the Connection Machine would not be selling today—even if the performance were higher than 3.8 GFLOPS! The key to the success of Thinking Machines in the last two years has been to produce a compiler that generates such code automatically, and this is where the user interface is most enhanced by the compiler. The development of an optimizing compiler of this quality, addressing communication instructions, as well as computational instructions, is a major achievement of the CM-2 software system. Because of its power, the compiler is essentially the complete user interface to the machine.

Debugging Tools

The debugging of code is a fundamental user-interface issue. On parallel machines and especially on distributed memory systems, program debugging can be extremely frustrating. Basically, one is debugging not one program, but possibly 128 programs. Even if they are all executing the same code, they are not executing the same instructions if the system is MIMD. Furthermore, there are synchronization and communications bugs that can make it extremely difficult to debug anything. For example, one problem that can occur on distributed systems is that intermediate nodes that are required for passing data back for debugging from the node where a suspected bug has developed may themselves be sick in some form or another. Debugging messages sometimes arrive in a different order than they were sent and in any event may well be coming in multiples of 128 (or more). Finally, the overall complexity of the systems can be extremely confusing, particularly when communication data structures involve complex numbering schemes such as Grey codes.

We would like to give an example of a debugging tool that we have developed and worked with for some time, with good experiences. The tool is a parallel distributed-system simulator called PARSIM. One goal of PARSIM was to develop an extremely simple and portable simulator that could be easily instrumented and coupled with visualization.

Portability is achieved by developing a UNIX-based tool, where the lowest-level communication is implemented through a very simple data-transfer capability. The data transfer may be handled using either IP facilities or even just by using the UNIX file system. PARSIM provides library support for Intel Hypercube functionality and also library support for similar communication capabilities. All of the standard communication protocols are supported, including typed messages, broadcasts, and global operations. Finally, PARSIM is usable from Fortran or C. In fact, a user simply links the host and node programs of the application to the PARSIM library.

PARSIM maintains a full trace history of all communication activity. A portable X-11 interface provides a graphical view of all the communication activities so that as the simulation is running, one can monitor all communication traffic between nodes. The graphical display represents nodes by colored, numbered circles and represents messages by arms reaching out from the nodes. A dispatched message is represented by an arm reaching toward the destination, whereas a receive request is represented by an arm reaching out from the receiver. When a requested message is actually received, the corresponding send-and-receive arms are linked to form a single path that indicates a completed transaction. Nodes awaiting message receipt are highlighted, and the types of all messages are displayed. In addition to the main display, separate text windows are used to display the output of all node and host processes. Thus, the user can watch the communication activity on a global scale while maintaining the ability to follow details on individual processors. The display works effectively on up to 32 nodes, although typically a smaller number suffices to debug most programs. Finally, PARSIM provides a history file that records the correct time sequence in which events are occurring. The history file may be viewed later to recheck aspects of a run, without the need to rerun the whole program.

PARSIM has turned out to be a key to porting large programs to a whole range of parallel machines, including the Intel iPSC/860. It is much easier to get the programs running in this environment than it is on the Intel. Once applications are running on the simulator, they port to the machine very quickly. As a recent example, with Charbel Farhat of the University of Colorado, we have ported a large (60,000) line finite-element program to both the iPSC/860 and the SUPRENUM-1 computers in just several weeks. Thus, user-interface tools of this type can be extremely helpful.

High-Level Languages, Extensions, Libraries, and Graphics

There has been substantial progress recently in the area of high-level languages for parallel systems. One class of developments has occurred in the area of object-oriented programming models to facilitate parallel programming. An example that I've been involved with is a C++ library for vector and matrix operations, which was implemented on the Intel Hypercube, the FPS Computing T-Series, the Ametek 2010, the Connection Machine CM-2, and several other systems. Another example is the language DINO developed by R. Schnabel and coworkers at the University of Colorado (Rosing and Schnabel 1989).

There are also some language extensions of several standard languages that are extremely important because they have a better chance of becoming standards. An example here would be the Fortran 90 flavors—for example, Connection Machine CM Fortran. A user can write pure Fortran 90 programs and compile them unchanged on the CM-2, although for best performance it is advisable to insert some compiler directives. This provides the possibility of writing a program for the Connection Machine that might also run on other parallel machines—for example, on a CRAY Y-MP. Indeed, now that many manufacturers appear to be moving (slowly) toward Fortran 90, there are real possibilities in this direction. Several similar extensions of other languages are now available on multiple systems. One good example would be the Connection Machine language extensions C* of the C language. The C* language is now available on various systems, including products from Sequent Computer and MasPar.

There are some problems with the high-level language approach. One is the lack of portability. There is increased learning time for users if they have to learn not only the lower-level aspects of systems but also how to deal with new language constructs. Finally, there is the danger of the software systems simply becoming too specialized.

A few words are in order about libraries. While there is a tremendous amount of very interesting work going on in the library area, we will not attempt to review it. A very good example is the LAPAK work. Most of that work is going on for shared-memory systems, although there are some developments also for distributed-memory machines. It is difficult to develop efficient libraries in a way that includes both shared and distributed systems.

For distributed-memory systems there is now substantial effort to develop communication libraries that run on multiple systems. One example worth noting is the SUPRENUM system, for which they have developed high-level libraries for two- and three-dimensional grid applications (Solchenbach and Trottenberg 1988). That library really helps the user who has a grid problem to deal with. In fact, it allows the user to completely dispense with explicit communication calls. One specifies a few communication parameters (topology, grid size, etc.) and then handles all interprocess communication through high-level, geometrically intuitive operations. The system partitions data sets effectively and calls low-level communication operations to exchange or broadcast data as needed.

One other point to note is that most codes in the real world don't use libraries very heavily, and so one has to be aware that not only is it important to port libraries, but also to make the technology used to design and implement algorithms in the libraries available to the scientific community in a way such that other users can adapt those same techniques into their codes.

The graphics area is certainly one of the weakest features of parallel systems. The Connection Machine is really the only system that has tightly coupled graphics capabilities—it supports up to eight hardware frame buffers, each running at 40 megabytes per second. One disadvantage with the Connection Machine solution is that graphics applications using the frame buffer are not portable. However, the ability to at least do high-speed graphics outweighs this disadvantage. In most systems, even if there is a hardware I/O capability that is fast enough, there is a lack of software to support graphics over that I/O channel. Furthermore, many systems actually force all graphics to pass through a time-shared front-end processor and an Ethernet connection, ensuring poor performance under almost any conditions.

The best solution is certainly to tightly couple parallel systems to conventional graphics systems. This is a way to avoid developing graphics systems that are specialized to specific pieces of parallel hardware. Much effort is now under way at several laboratories to develop such high-speed connections between parallel systems and graphics stations.

We will mention an experiment we've done at the Center for Applied Parallel Processing (CAPP) in Boulder, Colorado, where we have connected a Stardent Titan and a Silicon Graphics IRIS directly to the back end of a Connection Machine CM-2, allowing data to be communicated between the CM-2 hardware and the graphics system at a very high speed, limited, in fact, only by the back-end speed of the CM-2 (Compagnoni et al. 1991). The data are postprocessed on the graphics processor using the very high polygon processing rates that are available on those two systems. This was first implemented as a low-level capability based on IP-type protocols. However, once it was available we realized that it could be used to extend the domain of powerful object-oriented graphics systems, such as the Stardent AVS system, to include objects resident on the CM-2. From this point of view, a Stardent user programs the graphics device as if the CM-2 is part of the system.

One of the advantages here is that the Connection Machine can continue with its own numeric processing without waiting for the user to complete visualization. For example, the CM-2 can go on to the next time step of a fluid code while you graphically contemplate what has been computed to date.

Another point about this approach (based on a low-level standard protocol) is that it is easy to run the same software over slow connections. In fact, we have designed this software so that if the direct connection to the back end is not available, it automatically switches to using Ethernet links. This means you don't have to be in the next room to the CM-2 hardware to use the visualization system. Of course, you won't get the same performance from the graphics, but at least the functionality is there.

Future Supercomputing Environments: Heterogeneous Systems

Over the last 20 years we have seen a gradual evolution from scalar sequential hardware to vector processing and more recently to parallel processing. No clear consensus has emerged on an ideal architecture. The trend to vector and parallel processing has been driven by the

computational needs of certain problems, but the resulting systems are then inappropriate for other classes of problems. It is unlikely that in the near term this situation will be resolved, and indeed one can anticipate further generations of even more specialized processor systems. There is a general consensus that a good computing environment would at least provide access to the following resources:
- scalar processors (e.g., workstations),
- vector processors,
- parallel machines (SIMD and/or MIMD),
- visualization systems,
- mass-storage systems, and
- interfaces to networks.

This leads us to the last topic and what in the long run is probably the most important: heterogeneous systems, heterogeneous environments, and the importance of combining a spectrum of computing resources.

There is a simple way of avoiding the specialization problem described above. The key is to develop seamless, integrated heterogeneous computing environments. What are the requirements for such systems? Obviously, high-speed communication is paramount—that means both high bandwidth and low latency. Because different types of machines are present, a seamless environment therefore requires support for data transformations between the different kinds of hardware. Equally important, as I've argued from previous experience with single machines, is to try to support wherever possible shared-memory concepts. Ease of use will require load balancing. If there are three Connection Machines on that system, one should be able to load-balance them between the demands of different users. Shared file systems should be supported, and so on. And all of this should be done in the context of portability of the user's code because the user may not always design his codes on these systems initially. Obviously, adopting standards is critical.

Such an environment will present to the user all of the resources needed for any application: fast scalar, vector, and parallel processors; graphics supercomputers; disk farms; and interfaces to networks. All of these units would be interconnected by a high-bandwidth, low-latency switch, which would provide transparent access between the systems. System software would present a uniform, global view of the integrated resource, provide a global name space or a shared memory, and control load balancing and resource allocation.

The hardware technology now allows such systems to be built. Two key ingredients are the recent development of fast switching systems and the development of high-speed connection protocols and hardware implementing these protocols standardized across a wide range of vendors. We illustrate with two examples.

Recently, Carnegie Mellon University researchers designed and built a 100-megabit-per-second switch called Nectar, which supports point-to-point connections between 32 processors (Arnould et al. 1989). A gigabit-per-second version of Nectar is in the design stage. Simultaneously, various supercomputer and graphics workstation vendors have begun to develop high-speed (800-megabit-per-second) interfaces for their systems. Combining these approaches, we see that at the hardware level it is already possible to begin assembling powerful heterogeneous systems. As usual, the really tough problems will be at the software level.

Several groups are working on aspects of the software problem. In our own group at CAPP, we have developed, as discussed in the previous section, a simple, heterogeneous environment consisting of a Connection Machine CM-2 and a Stardent Titan graphics superworkstation (Compagnoni et al. 1991). The Titan is connected with the CM-2 through the CM-2 back end, rather than through the much slower front-end interface that is usually used for such connectivity. The object-oriented, high-level Stardent AVS visualization system is then made directly available to the CM-2 user, allowing access to graphical objects computed on the CM-2 in real time, while the CM-2 is freed to pursue the next phase of its computation. Essentially, this means that to the user, AVS is available on the CM-2. Porting AVS directly to the CM-2 would have been a formidable and pointless task. Furthermore, the CM-2 is freed to perform the computations that it is best suited for, rather than wasting time performing hidden surface algorithms or polygon rendering. These are precisely the sorts of advantages that can be realized in a heterogeneous system.

Looking to the future, we believe that most of the research issues of heterogeneous computing will have been solved by the late 1990s, and in that time frame, we would expect to see evolving heterogeneous systems coming into widespread use wherever a variety of distinct computational resources is present. In the meantime, one can expect to see more limited experiments in heterogeneous environments at major research computation laboratories, such as Los Alamos National Laboratory and the NSF supercomputer centers.

An Application for a Heterogeneous System

We will conclude by asking if there are problems that can use such systems. We will answer the question by giving an example typical of many real-world problems.

The scientific application is the aeroelastic analysis of an airframe. The computational problem is complicated because there are two distinct but coupled subproblems to be solved. The real aerodynamic design problem is not just to compute airflow over a plane in the context of a static frame but to design planes the way they are flown, which is dynamically. As soon as you do that, you get into the coupled problem of the fluid flow and the airframe structure. There is a structural engineering problem, which is represented by the finite-element analysis of the fuselage and the wing. There is a fluid-dynamics problem, which is the flow of air over the wing surface. Finally, there is a real interaction between these two problems. Obviously, the lift depends on the wing surface, but correspondingly, the fluid flow can cause vibrations in the wing.

One can represent the computation schematically as a two-dimensional (surface) structural problem with $O(N^2)$ degrees of freedom or as a three-dimensional fluid problem with, typically, $O(N^3)$ degrees of freedom. Here N measures the spatial resolution of each problem.

Typically, the fluid computation can be solved by explicit methods, but the finite-element problem requires implicit solution techniques. The fluid models require solution time proportional to the number of degrees of freedom. The finite-element problems are typically sparse matrix problems, and it is hard to do better than $O(N^3)$ on solution times. Thus, we have a computation that is $O(N^3)$ for both the fluid and structure components.

Thus, the two computational phases are much the same in overall computational complexity. However, the communication of data, the interaction between the two phases, is related to the surface, only, and is therefore an $O(N^2)$ data-transfer problem. Therefore, provided one is solving a large enough problem so that the $O(N^2)$ communication cost is negligible, one can imagine solving the fluid part on a computer that is most effective for fluid computation and the structural part on a machine that is most effective for structural problems.

We have been studying this problem quite actively at the University of Colorado (Farhat et al. 1991, Saati et al. 1990, and Farhat et al. 1990) and have found that the fluid problem is best done on a machine like the

Connection Machine, on which one can take advantage of the SIMD architecture and work with grids that are logically rectangular. The structural problem is best done on machines that can handle irregular grids effectively, for example, a CRAY Y-MP. Thus, ideally we would like to solve the whole problem on a heterogeneous system that includes both Cray and CM-2 machines. One should also remember that both phases of this computation have heavy visualization requirements. Thus, both systems need to be tightly coupled to high-end graphics systems. Furthermore, if a record is to be saved of a simulation, then high-speed access to a disk farm is mandatory because of the huge volumes of data generated per run. A fully configured heterogeneous environment is therefore essential.

Conclusions

We conclude by remarking that the winning strategy in supercomputer design for the coming decade is certainly going to be acquisition of a software advantage in parallel systems. Essentially all of the MIMD manufacturers are likely to be using competitive hardware. It is clear from recent experiences—for example, Evans & Sutherland—that using anything other than off-the-shelf components is too expensive. It follows immediately that different manufacturers can follow more or less the same upgrade strategies. They will end up with machines with basically similar total bandwidths, reliability, and so on. Thus, the key to success in this market is going to be to develop systems that have the best software. Key points there will be virtual shared-memory environments and general-purpose computing capabilities.

References

E. A. Arnould, F. J. Bitz, E. C. Cooper, H. T. Kung, et al., "The Design of Nectar: A Network Backplane for Heterogeneous Multicomputers," in *Proceedings of the Third International Conference on Architecture Support for Programming Languages and Operating Systems (ASPLOS III)*, Association for Computing Machinery, pp. 205–216 (1989).

L. Compagnoni, S. Crivelli, S. Goldhaber, R. Loft, O. A. McBryan, et al., "A Simple Heterogeneous Computing Environment: Interfacing Graphics Workstations to a Connection Machine," Computer Science Department technical report, University of Colorado, Boulder (1991).

C. Farhat, S. Lanteri, and L. Fezoui, "Mixed Finite Volume/Finite Element Massively Parallel Computations: Euler Flows, Unstructured Grids, and Upwind Approximations," in *Unstructured Massively Parallel Computations*, MIT Press, Cambridge, Massachusetts (1991).

C. Farhat, N. Sobh, and K. C. Park, "Transient Finite Element Computations on 65,536 Processors: The Connection Machine," *International Journal for Numerical Methods in Engineering* **30**, 27–55 (1990).

O. A. McBryan, "The Connection Machine: PDE Solution on 65,536 Processors," *Parallel Computing* **9**, 1–24 (1988).

O. A. McBryan, "Optimization of Connection Machine Performance," *International Journal of High Speed Computing* **2**, 23–48 (1990).

O. A. McBryan and R. Pozo, "Performance of the Myrias SPS-2 Computer," Computer Science Department technical report CU-CS-505-90, University of Colorado, Boulder (1990). To appear in *Concurrency: Practice and Experience*.

M. Rosing and R. B. Schnabel, "An Overview of Dino—A New Language for Numerical Computation on Distributed Memory Multiprocessors," in *Parallel Processing for Scientific Computation*, Society for Industrial and Applied Mechanics, Philadelphia, pp. 312–316 (1989).

A. Saati, S. Biringen, and C. Farhat, "Solving Navier-Stokes Equations on a Massively Parallel Processor: Beyond the One Gigaflop Performance," *International Journal of Supercomputer Applications* **4** (1), 72–80 (1990).

K. Solchenbach and U. Trottenberg, "SUPRENUM—System Essentials and Grid Applications," *Parallel Computing* **7**, 265–281 (1988).

What Can We Learn from Our Experience with Parallel Computation up to Now?

Jack T. Schwartz

Jack T. Schwartz is Professor of Mathematics and Computer Science at New York University's Courant Institute of Mathematical Sciences. Trained as a mathematician at Yale University, he has worked in a variety of subjects, including functional analysis, mathematical economics, computer design, parallel architectures, programming-language design, and robotics. Currently, he is interested in the design of interactive computer systems and interfaces.

In this talk I wish to raise a few questions rather than try to answer any. My first question is, what does the way in which parallel computers are currently being used tell us about how they will be used? In this connection, I would like to distinguish among a number of computational paradigms into which particular problems might fit.

The most successful paradigm so far has been the SIMD paradigm, which exists in a number of versions, and all, of course, are evolving. It is worth distinguishing between two types of SIMD computation: the "lockstep" kind of computation (which is what the hardware forces you to if it is vector hardware or some other form of hardware that is centrally driven) and the "relaxed SIMD" paradigm (which you would have on a machine that is basically SIMD, at least in its software structure, but that permits independent branching within a DOALL loop). The latter sort of machine will handle complex loops with subroutine calls in a much more comfortable way than a lockstep machine, but in both cases SIMD software organization will leave one still thinking of problems in a SIMD mode.

Along the spectrum that runs from "relaxed SIMD" to a true MIMD paradigm, there arise other distinctions between a number of types of MIMD calculations. The first of these paradigms is the Monte Carlo class of MIMD calculations, which represents a particularly advantageous pole within the MIMD world. In Monte Carlo, one uses what are essentially independent complex processes, which only need to interact when their results are combined at the end of what may be a very long calculation. One generates almost independent parallel computations at an arbitrary scale—the ideal case for a MIMD machine.

A second class of MIMD computations, which requires more communication, is typified by chaotic relaxation schemes. Here, communication is required, but the numerical method used does not always have to have absolutely up-to-date information about what's going on at the neighbors of a good data point, as long as the available information is updated often. The broad applicability of this general idea, i.e., the fact that it can be expected to lead to stable and convergent calculations, is reflected in the familiar fact that I need read a newspaper only once a day instead of every minute and nevertheless can feel this keeps me reasonably current about what's happening in the world. This remark on communication within an algorithmic process reflects a very fundamental point: communication latencies within parallel systems are likely to increase because it may simply not be feasible as machines get larger and larger to provide absolutely current remote information to all their processors or to keep caches fully current. Indeed, this consideration may define a whole class of computations that theorists haven't considered in depth but that deserve attention—computations that simply ignore the noncurrency of data but work out well, anyhow.

Another class of algorithms that has stood very much in the forefront of the thinking of people who begin from the MIMD end of the parallel computer design spectrum is the "workpile" kind of computation. Let me distinguish between two types of such algorithms. The first is the class of "compute-driven" workpile computations, as typified by Lisp theorem-prover searches. Such searches find more and more branches to explore and simply throw those branches onto a pile; the branches are subsequently picked up by some processor whose prior exploration has ended. The second class is the "data-driven" workpile algorithms, typified by a large commercial database system in which only one processor can do a particular piece of work, that processor being the one on whose disk the data required for the work in question is resident. Parallel applications of this sort suggest the use of highly parallel collections of "servers."

I hope that presentations at this conference can help to answer the following questions, which seem to me to be strategic for the development of parallel computers and computation over the next few years:

- What are the basic paradigms of parallel computation? Will both parallel architectures and the languages that constitute the user-hardware interface evolve around those paradigms? Are there other paradigms that can be defined as broadly as those I have listed and that are equally important?
- Are all the paradigms outlined above really important? Are they all of roughly equal importance, or does our experience to date suggest that some are much more important than others? For example, there seems to have been much more success in the "relaxed SIMD" mode use of parallel computations than in the Monte Carlo mode. Is this true? If so, what does it tell us about the future?
- Are all of the cases that I have listed really populated? For example, are there really any important compute-driven workpile algorithms, or is their existence a myth propagated by people who have thought about these algorithms from a theoretical point of view?

I raise these questions because I believe that they can help give some shape to the question of where parallel computer hardware and software design is going.

7

Algorithms for High-Performance Computing

This session focused on ways in which architectural features of high-performance computing systems can be used to design new algorithms to solve difficult problems. The panelists discussed the effect on algorithm design of large memories, the number of processors (both large and small), and special topologies. They also discussed scalable algorithms, algorithm selection, models of computation, and the relationship between algorithms and architectures.

Session Chair

Gian-Carlo Rota, Massachusetts Institute of Technology

Parallel Algorithms and Implementation Strategies on Massively Parallel Supercomputers*

R. E. Benner

Robert E. Benner is a senior member of the Parallel Computing Technical Staff, Sandia National Laboratories. He has a bachelor's degree in chemical engineering (1978) from Purdue University and a doctorate in chemical engineering (1983) from the University of Minnesota. Since 1984, he has been pursuing research in parallel algorithms and applications on massively parallel hypercubes and various shared-memory machines. He was a member of a Sandia team that won the first Gordon Bell Award in 1987 and the Karp Challenge Award in 1988 and was cited in R&D Magazine's 1989 R&D 100 List *for demonstrating parallel speedups of over 1000 for three applications on a 1024-processor nCUBE/ten Hypercube. Dr. Benner specializes in massively parallel supercomputing, with particular emphasis on parallel algorithms and parallel libraries for linear algebra, nonlinear problems, finite elements, dynamic load balancing, graphics, I/O, and the implications of parallelism for a wide range of science and engineering.*

---

* Special thanks go to my colleagues, whose work has been briefly summarized here. This paper was prepared at Sandia National Laboratories, which is operated for the U.S. Department of Energy under Contract Number DE-AC04-76DP00789.

Introduction

This presentation is on parallel algorithms and implementation strategies for applications on massively parallel computers. We will consider examples of new parallel algorithms that have emerged since the 1983 Frontiers of Supercomputing conference and some developments in MIMD parallel algorithms and applications on first- and second-generation hypercubes. Finally, building upon what other presenters at this conference have said concerning supercomputing developments—or lack thereof—since 1983, I offer some thoughts on recent changes in the field.

We will draw primarily on our experience with a subset of the parallel architectures that are available as of 1990, those being nCUBE Corporation's nCUBE 2 and nCUBE/ten Hypercubes and Thinking Machines Corporation's CM-2 (one of the Connection Machine series). The nCUBE 2 at Sandia National Laboratories has 1024 processors with four megabytes of local memory per processor, whereas the nCUBE/ten has the same number of processors but only 0.5 megabytes of memory per processor. The CM-2 is presently configured with 16K single-bit processors, 128 kilobytes of memory per processor, and 512 64-bit floating-point coprocessors. This conference has already given considerable attention to the virtues and pitfalls of SIMD architecture, so I think it will be most profitable to focus this short presentation on the state of affairs in MIMD architectures.

An interdisciplinary research group of about 50 staff is active in parallel computing at Sandia on the systems described above. The group includes computational scientists and engineers in addition to applied mathematicians and computer scientists. Interdisciplinary teams that bring together parallel-algorithm and applications researchers are an essential element to advancing the state of the art in supercomputing.

Some Developments in Parallel Algorithms

In the area of parallel algorithms and methods, there have been several interesting developments in the last seven years. Some of the earliest excitement, particularly in the area of SIMD computing, was the emergence of cellular automata methods. In addition, some very interesting work has been done on adaptive-precision numerical methods, for which the CM-2 provides unique hardware and software support. In addition, there has been much theoretical and experimental research on various asynchronous methods, including proofs of convergence for some of the most interesting ones.

Algorithms for High-Performance Computing   229

A more recent development, the work of Fredericksen and McBryan (1988) on the parallel superconvergent multigrid method, prompted a surge in research activity in parallel multigrid methods. For example, parallel implementations of classic multigrid have been demonstrated with parallel efficiencies of 85 per cent for two-dimensional problems on 1000 processors and about 70 per cent for three-dimensional problems on 1000 processors (Womble and Young 1990)—well in excess of our expectations for these methods, given their partially serial nature.

A new class of methods that emerged is parallel time stepping. C. William Gear (now with the Nippon Electric Corporation Research Institute, Princeton, New Jersey), in a presentation at the July 1989 SIAM meeting in San Diego, California, speculated on the possibility of developing such methods. Womble (1990) discusses a class of methods that typically extract a factor of 4- to 16-fold increase in parallelism over and above the spatial parallelism in a computation. This is not the dramatic increase in parallelism that Gear speculated might be achievable, but it's certainly a step in the right direction in that the time parallelism is multiplicative with spatial parallelism and therefore attractive for problems with limited spatial parallelism.

At the computer-science end of the algorithm spectrum, there have been notable developments in areas such as parallel load balance, mapping methods, parallel graphics and I/O, and so on. Rather than considering each of these areas in detail, the focus will now shift to the impact of parallel algorithms and programming strategies on applications.

Some Developments in Parallel Applications

The prospects for high-performance, massively parallel applications were raised in 1987 with the demonstration, using a first-generation nCUBE system, of 1000-fold speedups for some two-dimensional simulations based on partial differential equations (Gustafson et al. 1988). The Fortran codes involved consisted of a few thousand to less than 10,000 lines of code. Let's consider what might happen when a number of parallel algorithms are applied to a large-scale scientific computing problem; i.e., one consisting of tens of thousands to a million or more lines of code.

A case study is provided by parallel radar simulation (Gustafson et al. 1989). This is, in some sense, the inverse problem to the radar problem that immediately comes to mind—the real-time processing problem. In radar simulation one takes a target, such as a tank or aircraft, produces a geometry description, and then simulates the interaction of radar with the geometry on a supercomputer (Figure 1). These simulations are generally based on multibounce radar ray tracing and do not vectorize well. On machines like the CRAY X-MP, a radar image simulator such as the Simulated Radar IMage (SRIM) code from ERIM Inc. typically achieves five or six million floating-point operations per second per processor. Codes such as SRIM have the potential for high performance on massively parallel supercomputers relative to vector supercomputers. However, although the novice might consider ray tracing to be embarrassingly parallel, in practice radar ray tracing is subject to severe load imbalances.

The SRIM code consists of about 30,000 lines of Fortran, an amalgamation of many different algorithms that have been collected over a period of 30 years and do not fit naturally into a half megabyte of memory. The implementation strategy was *heterogeneous*. That is, rather than combining all of the serial codes in the application package, the structure of the software package is preserved by executing the various codes simultaneously on different portions of the hypercube (Figure 2), with data pipelined from one code in the application package to the next.

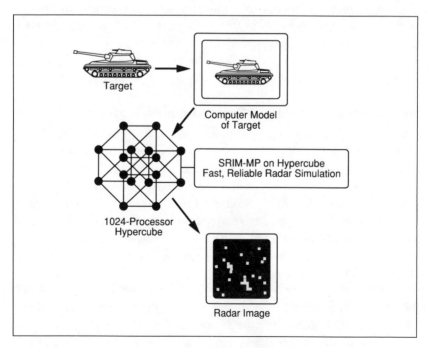

Figure 1. A parallel radar simulation as generated by the SRIM code.

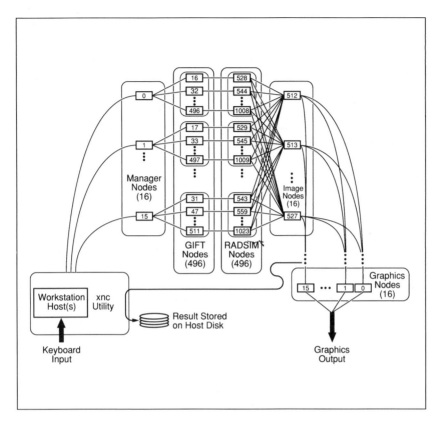

Figure 2. Heterogeneous image simulation using a MIMD computer.

The heterogeneous implementation uses a MIMD computer in a very general MIMD fashion for the first time. An observation made by one of our colleagues concerning the inaugural Gordon Bell Prize was that the parallel applications developed in that research effort were very SIMD-like and would have performed very well on the Connection Machine. In contrast, the parallel radar simulation features, at least at the high level, a true MIMD strategy with several cooperating processes: a load-balance process, a multibounce radar ray-tracing process, an imaging process, and a process that performs a global collection of the radar image, as well as a user-supplied graphic process and a host interface utility to handle the I/O.

The nCUBE/ten version of the radar simulation, for which one had to develop a host process, had six different cooperating programs. There are utilities on second-generation hypercubes that handle direct node I/O, so a host process is no longer needed. By keeping the application code on the nodes rather than splitting it into host and node code, the resulting parallel code is much closer to the original workstation codes of Cray Research, Inc., and Sun Microsystems, Inc. Over the long haul, I think that's going to be critical if we're going to get more people involved with using massively parallel computers. Furthermore, three of the remaining five processes—the load-balance, image-collection, and graphics processes—are essentially library software that has subsequently been used in other parallel applications.

Consider the religious issue of what qualifies as a massively parallel architecture and what doesn't. Massively parallel is used rather loosely here to refer to systems of 1000 or more floating-point processors or their equivalent in single-bit processors. However, Duncan Buell (also a presenter in this session) visited Sandia recently, and an interesting discussion ensued in which he asserted that massively parallel means that a collection of processors can be treated as an ensemble and that one is most concerned about the efficiency of the ensemble as opposed to the efficiencies of individual processors.

Heterogeneous MIMD simulations provide a nice fit to the above definition of massive parallelism. The various collections of processors are loosely synchronous, and to a large extent, the efficiencies of individual processors do not matter. In particular, one does not want the parallel efficiency of the dynamic load-balance process to be high, because that means the load-balance nodes are saturated and not keeping up with the work requests of the other processors. Processor efficiencies of 20, 30, or 50 per cent are perfectly acceptable for the load balancer, as long as only a few processors are used for load balancing and the bulk of the processors are keeping busy.

Some Developments in Parallel Applications II

Some additional observations concerning the development of massively parallel applications can be drawn from our research into Strategic Defense Initiative (SDI) tracking and correlation problems. This is a classic problem of tracking tens or hundreds of thousands of objects, which Sandia is investigating jointly with Los Alamos National Laboratory, the Naval Research Laboratory, and their contractors. Each member of the team is pursuing a different approach to parallelism, with Sandia's charter being to investigate massive parallelism on MIMD hypercubes.

One of the central issues in parallel computing is that a major effort may be expended in overhauling the fundamental algorithms involved as part of the porting process, irrespective of the architecture involved. The course of the initial parallelization effort for the tracker-correlator was interesting: parallel code development began on the hypercube, followed by a quick retreat to restructuring the serial code on the Cray. The original tracking algorithms were extremely memory intensive, and data structures required extensive modification to improve the memory use of the code (Figure 3), but first-generation hypercube nodes with a mere half megabyte of memory were not well suited for the task.

Halbgewachs et al. (1990) developed a scheme for portability between the X-MP and the nCUBE/ten in which they used accesses to the solid-state device (SSD) to swap "hypercube processes" and to simulate the handling of the message-passing. In this scheme one process executes, sends its "messages" to the SSD, and is then swapped out for another "process." This strategy was a boon to the code-development effort. Key algorithms were quickly restructured to reduce the memory use from second order to first order. Further incremental improvements have been made to reduce the slope of the memory-use line. The original code could not track more than 160 objects. The first version with linear memory requirements was able to track 2000 objects on the Cray and the nCUBE/ten.

On the nCUBE/ten a heterogeneous implementation was created with a host process. (A standard host utility may be used in lieu of a host program on the second-generation hypercubes.) The tracking code that runs on the nodes has a natural division into two phases. A dynamic load balancer was implemented with the first phase of the tracking algorithm, which correlates new tracks to known clusters of objects. The second phase, in which tracks are correlated to known objects, is performed on the rest of the processors.

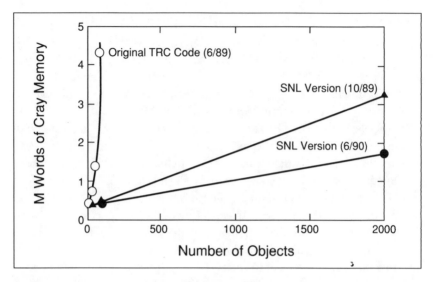

Figure 3. Improved memory utilization: Phase 1, reduced quadratic dependence to linear; Phase 2, reduced coefficient of linear dependence.

Given the demonstration of an effective heterogeneous implementation and the ability to solve large problems, suddenly there is user interest in real-time and disk I/O capabilities, visualization capabilities, etc. When only small problems could be solved, I/O and visualization were not serious issues.

Performance evaluation that leads to algorithm modifications is critical when large application codes are parallelized. For example, the focus of SDI tracking is on simulating much larger scenarios, i.e., 10,000 objects, as of September 1990, and interesting things happen when you break new ground in terms of problem size. The first 10,000-object run pinpointed serious bottlenecks in one of the tracking phases. The bottleneck was immediately removed, and run time for the 10,000-object problem was reduced from three hours to one hour. Such algorithmic improvements are just as important as improving the scalability of the parallel simulation because a simulation that required 3000 or 4000 processors to run in an hour now requires only 1000.

A heterogeneous software implementation, such as the one described above for the SDI tracking problem, suggests ways of producing heterogeneous hardware implementations for the application. For example, we can quantify the communication bandwidth needed between the different algorithms; i.e., the throughput rate needed in a distributed-computing approach where one portion of the computation is done on a space platform, other portions on the ground, etc. In addition, MIMD processors might not be needed for all phases of the simulation. Heterogeneous implementations are MIMD at their highest level, but one might be able to take advantage of the power of SIMD computing in some of the phases. Furthermore, heterogeneous nodes are of interest in these massively parallel applications because, for example, using a few nodes with large memories in a critical phase of a heterogeneous application might reduce the total number of processors needed to run that phase in a cost-effective manner.

Given our experience with a heterogeneous implementation on homogeneous hardware, one can propose a heterogeneous hardware system to carry out the computation efficiently. We note, however, that it would be risky to build a system in advance of having the homogeneous hardware implementation. If you're going to do this sort of thing—and the vendors are gearing up and are certainly talking seriously about providing this capability—then I think we really want to start with an implementation on a homogeneous machine and do some very careful performance analyses.

Closing Remarks

What, besides those developments mentioned by other presenters, has or has not changed since 1983? In light of the discussion at this conference about risk taking, I think it is important to remember that there has been significant risk taking involved in supercomputing the last several years. The vendors have taken considerable risks in bringing massively parallel and related products to market. From the perspective of someone who buys 1000-processor systems before one is even built by the vendor, customers have also taken significant risks. Those who funded massively parallel acquisitions in the 1980s have taken risks.

I've seen a great improvement in terms of vendor interest in user input, including input into the design of future systems. This doesn't mean that vendor-user interaction is ideal, but both sides realize that interaction is essential to the viability of the supercomputing industry.

A more recent, encouraging development is the emerging commercial activity in portability. Commercial products like STRAND[88] (from Strand Software Technologies Inc.) and Express (from Parasoft Corporation) have appeared. These provide a starting point for code portability, at least between different distributed-memory MIMD machines and perhaps also between distributed-memory and shared-memory machines.

We are much further from achieving portability of Fortran between MIMD and SIMD systems, in part due to the unavailability of Fortran 90 on the former.

Another philosophical point concerns which current systems are supercomputers and which are not. We believe that the era of a single, dominant supercomputer has ended, at least for the 1990s if not permanently. Without naming vendors, I believe that at least four of them have products that qualify as supercomputers in my book—that is, their current system provides the fastest available performance on some portion of the spectrum of computational science and engineering applications. Even given the inevitable industry shakeouts, that is likely to be the situation for the near future.

What hasn't happened in supercomputing since 1983? First, language standards are lagging. Fortran 8X has now become Fortran 90. There are no parallel constructs in it, although we at least get array syntax, which may make for better nonstandard parallel extensions. To some extent, the lack of a parallel standard is bad because it certainly hinders portability. In another sense the lack of a parallel standard is not bad because it's not clear that the Fortran community knows what all the parallel extensions should be and, therefore, what all the new standards should be. I would hate to see a new standard emerge that was focused primarily on SIMD, distributed-memory MIMD, or shared-memory MIMD computing, etc., to the detriment of the other programming models.

A major concern is that massive parallelism does not have buy-in from most computational scientists and engineers. There are at least three good reasons for this. First, recall the concern expressed by many presenters at this conference for *education* to get more people involved in supercomputing. This is especially true of parallel computing. A second issue is *opportunity*, i.e., having systems, documentation, experienced users, etc., available to newcomers to smooth their transition into supercomputing and parallel computing. The role of the NSF centers in making vector supercomputers accessible is noteworthy.

The third, and perhaps most critical, issue is *interest*. We are at a crossroads where a few significant applications on the Connection Machine, nCUBE 2, and Intel iPSC/860 achieve a factor of 10 or more better run-time performance than on vector supercomputers. On the other hand, there is a large body of applications, such as the typical mix of finite-element- and finite-difference-based applications, that typically achieve performance on a current massively parallel system that is comparable to that of a vector supercomputer, or at most three to five times the vector supercomputer. This level of performance is sufficient to demonstrate a price/performance advantage for the massively parallel system but not a clear raw-performance advantage. In some cases, end users are willing to buy into the newer technology on the basis of the price/performance advantage. More often, there is a great deal of reluctance on the part of potential users.

User buy-in for massive parallelism is not a vector supercomputer versus massively parallel supercomputer issue. In 1983 we faced a similar situation in vector supercomputing: many users did not want to be concerned with vector processors and how one gets optimum performance out of them. In recent years the situation has gradually improved. The bottom line is that eventually most people who are computational scientists at heart come around, and a few get left behind. In summary, I hope that someday all computational scientists in the computational science and engineering community will consider advanced computing to be part of their career and part of their job.

References

P. O. Fredericksen and O. A. McBryan, "Parallel Superconvergent Multigrid," in *Multigrid Methods*, S. McCormick, Ed., Marcel Dekker, New York (1988).

J. L. Gustafson, R. E. Benner, M. P. Sears, and T. D. Sullivan, "A Radar Simulation Program for a 1024 Processor Hypercube," in *Proceedings, Supercomputing '89*, ACM Press, New York, pp. 96–105 (1989).

J. L. Gustafson, G. R. Montry, and R. E. Benner, "Development of Parallel Methods for a 1024 Processor Hypercube," *SIAM Journal on Scientific and Statistical Computing* **9**, 609–638 (1988).

R. D. Halbgewachs, J. L. Tomkins, and John P. VanDyke, "Implementation of Midcourse Tracking and Correlation on Massively Parallel Computers," Sandia National Laboratories report SAND89-2534 (1990).

D. E. Womble, "A Time Stepping Algorithm for Parallel Computers," *SIAM Journal on Scientific and Statistical Computing* **11**, 824–837 (1990).

D. E. Womble and B. C. Young, "Multigrid on Massively Parallel Computers," in *Proceedings of the Fifth Distributed Memory Computing Conference*, D. W. Walker and Q. F. Stout, Eds., IEEE Computer Society Press, Los Alamitos, California, pp. 559–563 (1990).

The Interplay between Algorithms and Architectures: Two Examples

Duncan Buell

Duncan A. Buell received his Ph.D. degree in mathematics from the University of Illinois at Chicago in 1976. He has held academic positions in mathematics at Carleton University in Canada and in computer science at Bowling Green State University in Ohio and at Louisiana State University (LSU). In 1985 he joined the Supercomputing Research Center in Bowie, Maryland, where he is presently a Senior Research Scientist. He has done research in number theory, information retrieval, and parallel algorithms and was part of a team at LSU that built a 256-bit-wide, reconfigurable parallel integer arithmetic processor, on which he holds a patent. He has written numerous journal articles and one book.

This presentation describes briefly, using two examples, the relevance of computer architecture to the performance achievable in running different algorithms.

This session is supposed to focus on architectural features of high-performance computing systems and how those features relate to the design and use of algorithms for solving hard problems.

Table 1 presents timings from a computation-bound program. This is a benchmark that I have created, which is the computational essence of the sort of computations I normally do. As with any benchmark, there are many caveats and qualifications, but you will have to trust me that this

Table 1. An Integer Arithmetic Benchmark

	CPU Times (in Seconds) by Optimization Level	
	None	Full
SUN 4 cc compiler	92.3	80.6
SUN 4 gcc compiler	89.9	78.1
SUN 3 cc compiler	83.5	67.6
SUN 3 cc compiler	64.8	51.1
SUN 3 gcc compiler	61.6	51.4
CONVEX cc compiler	39.3	39.3 (32-bit arith.)
CONVEX cc compiler	69.0	68.2 (64-bit arith.)
CRAY-2 cc compiler	49.0	48.7 (64-bit arith.)
CRAY-2 scc compiler		48.7 (64-bit arith.)
DEC-5000/200	19.7	18.5

table is generated honestly and that in my experience, performance on this benchmark is a reasonable predictor of performance I might personally expect from a computer. This is a program written in C; not being trained in archaeology, I tend to avoid antique languages like Fortran.

Table 2 displays counts of "cell updates per second." I will get to the details in a moment. This table has largely been prepared by Craig Reese of the Supercomputing Research Center (SRC). I apologize for the fact that the tables are in inverse scales, one measuring time and the other measuring a rate.

So what's the point? As mentioned in Session 3 by Harvey Cragon, there have been some things that have been left behind as we have developed vector computing, and one of those things left behind is integer arithmetic. This is a benchmark measuring integer arithmetic performance, and on the basis of this table, one could justifiably ask why one of the machines in Table 1 is called a supercomputer and the others are not.

As an additional comment, I point out that some of the newer RISC chips are intentionally leaving out the integer instructions—this is the reason for the poor performance of the Sun Microsystems, Inc., SUN 4 relative to the Digital Equipment Corporation DEC-5000. Those in the

Table 2. The DNA Homology Problem

Cell Updates per Second
(in Thousands)

CM-2 (64K)	1,085,202	vpratio=128, algorithm 1 (8192K strings)
CM-2 (64K)	1,081,006	vpratio=32, algorithm 1 (2048K strings)
CM-2 (64K)	712,348	vpratio=1, algorithm 1 (64K strings)
CM-2 (64K)	873,813	vpratio=32, algorithm 2 (2048K strings)
CM-2 (64K)	655,360	vpratio=1, algorithm 2 (64K strings)
Splash[a]	50,000	
CRAY-2	6,400	
SUN 4/370	3,030	
SUN 4/60	2,273	
SUN 4/280	2,127	
PNAC[a]	1,099	
SUN 3/280	758	
SUN 3/60	617	
SUN 3/50	450	
CM-2[a]	212	
CRAY-2[a]	154	
CONVEX C1[a]	112	
SUN 2/140[a]	20	

[a]Computations from published services

floating-point world will not notice the absence of those instructions because they will be able to obtain floating-point performance through coprocessor chips.

The second table is a DNA string-matching benchmark. This is a computation using a dynamic programming algorithm to compare a string of characters against a great many other strings of characters. The items marked with an asterisk come from published sources; the unmarked items come from computations at SRC. Some machines are included multiple times to indicate different implementations of the same algorithm.

As for precise machines, PNAC is Dan Lopresti's special-purpose DNA string-matching machine, and Splash is a board for insertion into a SUN computer that uses Xilinx chips arranged in a linear array. The timings of the Connection Machine CM-2 (from Thinking Machines Corporation) are largely C/Paris timings from the SRC Connection Machine.

And what is the message? The DNA problem and the dynamic programming edit distance algorithm are inherently highly parallel and dominated as a computation by the act of "pushing a lot of bits around." It should, therefore, not come as a surprise that the Connection Machine, with inherent parallelism and a bit orientation, outperforms all other machines. In Table 1, we see that the absence of integer hardware drags the Cray Research CRAY-2 down to the performance level of a workstation—not even the raw speed of the machine can overcome its inherent limitations. In Table 2, we see that the fit between the computation and the architecture allows for speedups substantially beyond what one might expect to get on the basis of mundane issues of price and basic machine speed.

Now, do we care—or should we—about the problems typified in these two tables? After all, neither of these fits the mode of "traditional" supercomputing. From K. Speierman's summary of the 1983 Frontiers of Supercomputing meeting, we have the assertion that potential supercomputer applications may be far greater than current usage indicates. Speierman had begun to make my case before I even took the floor. Yes, there are hard problems out there that require enormous computation resources and that are simply not supported by the architecture of traditional high-end computers.

Two other lessons emerge. One is that enormous effort has gone into vector computing, with great improvements made in performance. But one can then argue that in the world of nonvector computing, many of those initial great improvements are yet to be made, provided the machines exist to support their kind of computation.

The second lesson is a bitter pill for those who would argue to keep the 10^{50} lines of code already written. If we're going to come up with new architectures and use them efficiently, then necessarily, code will have to be rewritten because it will take new algorithms to use the machines well. Indeed, it will be a failure of ingenuity on the part of the algorithm designers if they are unable to come up with algorithms so much better for the new and different machines that, even with the cost of rewriting code, such rewrites are considered necessary.

Linear Algebra Library for High-Performance Computers*

Jack Dongarra

Jack Dongarra is a distinguished scientist specializing in numerical algorithms in linear algebra, parallel computing, advanced computer architectures, programming methodology, and tools for parallel computers at the University of Tennessee's Department of Computer Science and at Oak Ridge National Laboratory's Mathematical Sciences Section. Other current research involves the development, testing, and documentation of high-quality mathematical software. He was involved in the design and implementation of the EISPACK, LINPACK, and LAPACK packages and of the BLAS routines and is currently involved in the design of algorithms and techniques for high-performance computer architectures.

Dr. Dongarra's other experience includes work as a visiting scientist at IBM's T. J. Watson Research Center in 1981, a consultant to Los Alamos Scientific Laboratory in 1978, a research assistant with the University of New Mexico in 1978, a visiting scientist at Los Alamos Scientific Laboratory in 1977, and a Senior Scientist at Argonne National Laboratory until 1989.

---

* This work was supported in part by the National Science Foundation, under grant ASC-8715728, and the National Science Foundation and Technology Center Cooperative Agreement No. CCR-8809615.

Dr. Dongarra received a Ph.D. in applied mathematics from the University of New Mexico in 1980, an M.S. in computer science from the Illinois Institute of Technology in 1973, and a B.S. in mathematics from Chicago State University in 1972.

Introduction

For the past 15 years, my colleagues and I have been developing linear algebra software for high-performance computers. In this presentation, I focus on five basic issues: (1) the motivation for the work, (2) the development of standards for use in linear algebra, (3) a basic library for linear algebra, (4) aspects of algorithm design, and (5) future directions for research.

LINPACK

A good starting point is LINPACK (Dongarra et al. 1979). The LINPACK project began in the mid-1970s, with the goal of producing a package of mathematical software for solving systems of linear equations. The project developers took a careful look at how one puts together a package of mathematical software and attempted to design a package that would be effective on state-of-the-art computers at that time—the Control Data Corporation (scalar) CDC 7600 and the IBM System 370. Because vector machines were just beginning to emerge, we also provided some vector routines. Specifically, we structured the package around the basic linear algebra subprograms (BLAS) (Lawson et al. 1979) for doing vector operations.

The package incorporated other features, as well. Rather than simply collecting or translating existing algorithms, we *reworked* algorithms. Instead of the traditional row orientation, we used a column orientation that provided greater efficiency. Further, we published a user's guide with directions and examples for addressing different problems. The result was a carefully designed package of mathematical software, which we released to the public in 1979.

LINPACK Benchmark

Perhaps the best-known part of that package—indeed, some people think it *is* LINPACK—is the benchmark that grew out of the documentation. The so-called LINPACK Benchmark (Dongarra 1991) appears in the appendix to the user's guide. It was intended to give users an idea of how long it would take to solve certain problems. Originally, we measured

Algorithms for High-Performance Computing   245

the time required to solve a system of equations of order 100. We listed those times and gave some guidelines for extrapolating execution times for about 20 machines.

The times were gathered for two routines from LINPACK, one (SGEFA) to factor a matrix, the other (SGESL) to solve a system of equations. These routines are called the BLAS, where most of the floating-point computation takes place. The routine that sits in the center of that computation is a SAXPY, taking a multiple of one vector and adding it to another vector:

$$y_i \leftarrow y_i + \alpha x_i \ .$$

Table 1 is a list of the timings of the LINPACK Benchmark on various high-performance computers.

The peak performance for these machines is listed here in millions of floating-point operations per second (MFLOPS), in ascending order from 16 to 3000. The question is, when we run this LINPACK Benchmark,

Table 1. LINPACK Benchmark on High-Performance Computers

Machine	Peak MFLOPS	Actual MFLOPS		System Efficiency	
Ardent Titan-1	16	7		0.44	
CONVEX C-130	62	17		0.27	
SCS-40	44	8.0		0.18	
IBM RS/6000	50	13		0.26	
CONVEX C-210	50	17		0.34	
FPS 264	54	5.6		0.10	
Multiflow 14/300	62	17		0.27	
IBM 3090/VF-180J	138	16		0.12	
CRAY-1	160	12	(27)	0.075	
Alliant FX/80	188	10	(8 proc.)	0.05	
CRAY X-MP/1	235	70		0.28	
NEC SX-1E	325	32		0.10	
ETA-10P	334	24		0.14	(0.07)
CYBER 205	400	17		0.04	
ETA-10G	644	93	(1 proc.)	0.14	
NEC SX-1	650	36		0.06	
CRAY X-MP/4	941	149	(4 proc.)	0.16	
Fujitsu VP-400	1142	20		0.018	
NEC SX-2	1300	43		0.033	
CRAY-2	1951	101	(4 proc.)	0.051	
CRAY Y-MP/8	2664	275	(8 proc.)	0.10	
Hitachi S-820/90	3000	107		0.036	

what do we actually get on these machines? The column labeled "Actual MFLOPS" gives the answer, and that answer is quite disappointing in spite of the fact that we are using an algorithm that is highly vectorized on machines that are vector architectures. The next question one might ask is, why are the results so bad? The answer has to do with the transfer rate of information from memory into the place where the computations are done. The operation—that is, a SAXPY—needs to reference three vectors and do essentially two operations on each of the elements in the vector. And the transfer rate—the maximum rate at which we are going to transfer information to or from the memory device—is the limiting factor here.

Thus, as we increase the computational power without a corresponding increase in memory, memory access can cause serious bottlenecks. The bottom line is *MFLOPS are easy, but bandwidth is difficult.*

Transfer Rate

Table 2 lists the peak MFLOPS rate for various machines, as well as the peak transfer rate (in megawords per second).

Recall that the operation we were doing requires three references and returns two operations. Hence, to run at good rates, we need a ratio of three to two. The CRAY Y-MP does not do badly in this respect. Each

Table 2. MFLOPS and Memory Bandwidth

Machine	Peak MFLOPS	Peak Transfer (megawatts/second)	Ratio
Alliant FX/80	188	22	0.12
Ardent Titan-4	64	32	0.5
CONVEX C-210	50	25	0.5
CRAY-1	160	80	0.5
CRAY X-MP/4	940	1411	1.5
CRAY Y-MP/8	2667	4000	1.5
CRAY-2S	1951	970	0.5
CYBER 205	400	600	1.5
ETA-10G	644	966	1.5
Fujitsu VP-200	533	533	1.0
Fujitsu VP-400	1066	1066	1.0
Hitachi 820/80	3000	2000	0.67
IBM 3090/600-VF	798	400	0.5
NEC SX-2	1300	2000	1.5

processor can transfer 50 million (64-bit) words per second; and the complete system, from memory into the registers, runs at four gigawords per second. But for many of the machines in the table, there is an imbalance between those two. One of the particularly bad cases is the Alliant FX/80, which has a peak rate of 188 MFLOPS but can transfer only 22 megawords from memory. It is going to be very hard to get peak performance there.

Memory Latency

Another issue affecting performance is, of course, the latency: how long (in terms of cycles) does it actually take to transfer the information after we make a request? In Table 3, we list the memory latency for seven machines. We can see that the time ranges from 14 to 50 cycles. Obviously, a memory latency of 50 cycles is going to impact the algorithm's performance.

Development of Standards

The linear algebra community has long recognized that we needed something to help us in developing our algorithms. Several years ago, as a community effort, we put together a de facto standard for identifying basic operations required in our algorithms and software. Our hope was that the standard would be implemented on the machines by many manufacturers and that we would then be able to draw on the power of having that implementation in a rather portable way.

We began with those BLAS designed for performing vector-vector operations. We now call them the Level 1 BLAS (Lawson et al. 1979). We later defined a standard for doing some rather simple matrix-vector calculations—the so-called Level 2 BLAS (Dongarra et al. 1988). Still

Table 3. Memory Latency

Machine	Latency Cycles
CRAY-1	15
CRAY X-MP	14
CRAY Y-MP	17
CRAY-2	50
CRAY-2S	35
CYBER 205	50
Fujitsu VP-400	31

later, the basic matrix-matrix operations were identified, and the Level 3 BLAS were defined (Dongarra, Du Croz, et al. 1990). In Figure 1, we show these three sets of BLAS.

Why were we so concerned about getting a handle on those three different levels? The reason lies in the fact that machines have a memory hierarchy and that the faster memory is at the top of that hierarchy (see Figure 2).

Naturally, then, we would like to keep the information at the top part to get as much reuse or as much access of that data as possible. The higher-level BLAS let us do just that. As we can see from Table 4, the Level 2 BLAS offer the potential for two floating-point operations for every reference; and with the Level 3 BLAS, we would get essentially n operations for every two accesses, or the maximum possible.

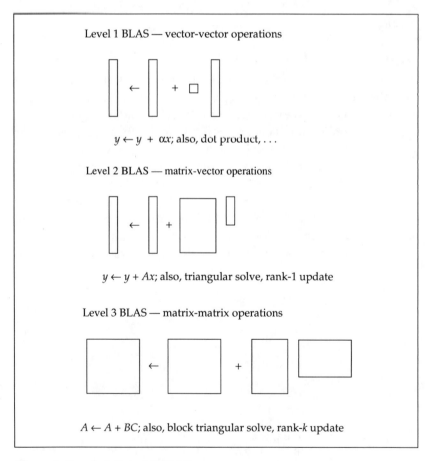

Figure 1. Levels 1, 2, and 3 BLAS.

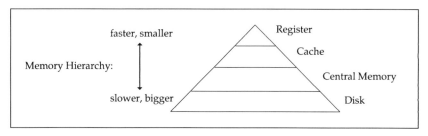

Figure 2. Memory hierarchy.

Table 4. Capabilities of Higher-Level BLAS

BLAS	Memory Reference	FLOPS	FLOPS/Memory Reference
Level 1: $y \leftarrow y + \alpha x$	$3n$	$2n$	$2/3$
Level 2: $y \leftarrow y + Ax$	n^2	$2n^2$	2
Level 3: $A \leftarrow A + BC$	$4n^2$	$2n^3$	$n/2$

These higher-level BLAS have another advantage. On some parallel machines, they give us increased granularity and the possibility for parallel operations, and they end up with lower synchronization costs.

Of course, nothing comes free. The BLAS require us to rewrite our algorithms so that we use these operations effectively. In particular, we need to develop blocked algorithms that can exploit the matrix-matrix operation.

The development of blocked algorithms is a fascinating example of history repeating itself. In the sixties, these algorithms were developed on machines having very small main memories, and so tapes of secondary storage were used as primary storage (Barron and Swinnerton-Dyer 1960, Chartres 1960, and McKellar and Coffman 1969). The programmer would reel in information from tapes, put it into memory, and get as much access as possible before sending that information out. Today people are reorganizing their algorithms with that same idea. But now instead of dealing with tapes and main memory, we are dealing with vector registers, cache, and so forth to get our access (Calahan 1979, Jordan and Fong 1977, Gallivan et al. 1990, Berry et al. 1986, Dongarra, Duff, et al. 1990, Schreiber 1988, and Bischof and Van Loan 1986). That is essentially what LAPACK is about: taking those ideas—locality of reference and data reuse—and embodying them in a new library for linear algebra.

LAPACK

Our objective in developing LAPACK is to provide a package for the solution of systems of equations and the solution of eigenvalue problems. The software is intended to be efficient across a wide range of high-performance computers. It is based on algorithms that minimize memory access to achieve locality of reference and reuse of data, and it is built on top of the Levels 1, 2, and 3 BLAS—the de facto standard that the numerical linear algebra community has given us. LAPACK is a multi-institutional project, including people from the University of Tennessee, the University of California at Berkeley, New York University's Courant Institute, Rice University, Argonne National Laboratory, and Oak Ridge National Laboratory.

We are in a testing phase at the moment and just beginning to establish world speed records, if you will, for this kind of work. To give a hint of those records, we show in Table 5 some timing results for LAPACK routines on a CRAY Y-MP.

Let us look at the LU decomposition results. This is the routine that does that work. On one processor, for a matrix of order 32, it runs at four MFLOPS; for a matrix of order 1000, it runs at 300 MFLOPS. Now if we take our LAPACK routines (which are written in Fortran), called the Level 3 BLAS (which the people from Cray have provided), and go to eight processors, we get 32 MFLOPS—a *speeddown*. Obviously, if we wish to solve the matrix, we should not use this approach!

When we go to large-order matrices, however, the execution rate is close to two GFLOPS—for code that is very portable. And for LL^T and QR factorization, we get the same effect.

Note that we are doing the same number of operations that we did when we worked with the unblocked version of the algorithms. We are not cheating in terms of the MFLOPS rate here.

One other performance set, which might be of interest for comparison, is that of the IBM RISC machine RS/6000-550 (Dongarra, Mayer, et al. 1990). In Figure 3, we plot the speed of LU decomposition for the LAPACK routine, using a Fortran implementation of the Level 3 BLAS. For the one-processor workstation, we are getting around 45 MFLOPS on larger-order matrices.

Clearly, the BLAS help, not only on the high-performance machines at the upper end but also on these RISC machines, perhaps at the lower end—for exactly the same reason: *data are being used or reused where the information is stored in its cache.*

Table 5. LAPACK Timing Results for a CRAY Y-MP (in MFLOPS)

Name	32	64	128	256	512	1024
SGETRF (*LU*)						
1 proc.	40	108	195	260	290	304
2 proc.	32	91	229	408	532	588
4 proc.	32	90	260	588	914	1097
8 proc.	32	90	205	375	1039	1974
SPOTRF (LL^T)						
1 proc.	34	95	188	259	289	301
2 proc.	29	84	221	410	539	594
4 proc.	29	84	252	598	952	1129
8 proc.	29	84	273	779	1592	2115
SGEQRF (*QR*)						
1 proc.	54	139	225	275	294	301
2 proc.	50	134	256	391	505	562
4 proc.	50	136	292	612	891	1060
8 proc.	50	133	328	807	1476	1937

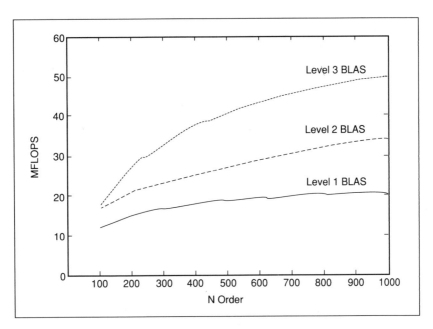

Figure 3. Variants of *LU* factorization on the IBM RISC System RS/6000-550.

Algorithm Design

Up to now, we have talked about *restructuring* algorithms; that is essentially what we did when we changed them to block form. The basic algorithm itself remains the same; we are simply affecting the locality of how we reference data and the independence of operations that we are trying to focus on—the matrix-matrix operations.

Divide-and-Conquer Approach

Let us now talk about *designing* algorithms. In this case, the basic algorithm will change. In particular, let us consider a divide-and-conquer technique for finding the eigenvalues and eigenvectors of a symmetric tridiagonal matrix (Dongarra and Sorensen 1987). The technique is also used in other fields, where it is sometimes referred to as domain decomposition. It involves tearing, or partitioning, the problem to produce small, independent pieces. Then, the eigenvalue of each piece is found independently. Finally, we "put back" the eigenvalues of these pieces, and put back the eigenvalues of these others, and so on. We were successful. In redesigning this algorithm, we ended up with one that runs in parallel very efficiently. Table 6 gives ratios of performance on a CRAY-2 for up to four processors. As we can see from the table, we are getting four times the speedup, and sometimes even better. What's more, we have an example of serendipity: the same algorithm is actually more efficient than the "best" sequential algorithm, even in the sequential setting.

Table 6. Ratio of Execution Time for TQL2 over Divide-and-Conquer Algorithm[a]

No. Proc.	100 E/(p)	100 (1)/(p)	200 E/(p)	200 (1)/(p)	300 E/(p)	300 (1)/(p)
1	1.35	1	1.45	1	1.53	1
2	2.55	1.88	2.68	1.84	2.81	1.84
3	3.39	2.51	3.71	2.55	3.79	2.48
4	4.22	3.12	4.60	3.17	50.3	3.28

[a]where E = EISPACK TQL,
   (1) = parallel divide-and-conquer algorithm on one processor, and
   (p) = parallel divide-and-conquer algorithm on p processors.

Accuracy

Working with LAPACK has given us an opportunity to go back and rethink some of the algorithms. How accurately can we solve NA problems (Demmel and Kahan 1988)? The answer depends on the accuracy of the input data and how much we are willing to spend:
- If the input data is exact, we can ask for a (nearly) correctly rounded answer—generally done only for +, *, /, , and cos.
- If the input H is uncertain in a normwise sense (true input = $H + \delta H$, where $||\delta H|| / ||H||$ is small), the conventional backward stable algorithm is suitable; it is the usual approach to linear algebra, it does not respect sparsity structure, and it does not respect scaling.
- If the input H is uncertain in a component-wise relative sense (true input = $H + \delta H$, where $\max_{i,j} |\delta H_{ij}| / |H_{ij}|$ is small), it does respect sparsity, it does respect scaling, but it does need new algorithms, perturbation theory, and error analysis.

In the end, we have new convergence criteria, better convergence criteria, and better error bounds. We also enhance performance because we now terminate the iteration in a much quicker way.

Tools

Our work in algorithm design has been supported by tool development projects throughout the country. Of particular note are the projects at Rice University and the University of Illinois. Other projects help in terms of what we might call logic or performance debugging of the algorithms—trying to understand what the algorithms are doing when they run on parallel machines. The objective here is to give the implementor a better feeling for where to focus attention and to show precisely what the algorithm is doing while it is executing on parallel machines (Dongarra, Brewer, et al. 1990).

Testing

Testing and timing have been an integral part of the LAPACK project. Software testing is required to verify new machine-specific versions. Software timing is needed to measure the efficiency of the LAPACK routines and to compare new algorithms and software. In both of these tasks, many vendors have helped us along the way by implementing basic routines on various machines and providing essential feedback (see Table 7).

The strategy we use may not be optimal for all machines. Our objective is to achieve a "best average" performance on the machines listed in Table 8. We are hoping, of course, that our strategy will also perform well

Table 7. Vendor Participation

Alliant Computer Sys.	BBN Advanced Comp.
CONVEX Computer	Cray Computer
Cray Research	Digital Equipment Corp.
Encore Computer Corp.	FPS Computing
Fujitsu	Hitachi
IBM ECSEC Italy	Intel
Kendall Square Res.	MasPar
Myrias Research Corp.	NEC
Sequent Computer Sys.	Silicon Graphics
Stardent Computer	Sun Microsystems, Inc.
Supercomputer Sys., Inc.	Thinking Machines Corp.

Table 8. Target Machines (1-100 Processors)

Alliant FX/80	IBM 3090/VF
BBN TC2000	Multiflow
CONVEX C-2	Myrias
CRAY-2	NEC SX
CRAY Y-MP	RISC machines
Encore Multimax	Sequent Symmetry
Fujitsu VP	Stardent Computer
Hitachi S-820	

on a wider range of machines, including the Intel iPSC, iWarp, MasPar, nCUBE, Thinking Machines, and Transputer-based computers.

Future Directions for Research

We have already started looking at how we can make "cosmetic changes" to the LAPACK software—adapt it in a semiautomatic fashion for distributed-memory architectures. In this effort, our current work on blocked operations will be appropriate because the operations minimize communication and provide a good surface-to-volume ratio. We also expect that this task will require defining yet another set of routines, this one based on the BLACS (basic linear algebra communication routines). Once again, we will draw on what has been done in the community for those operations.

As a preliminary piece of data, we show in Figure 4 an implementation of *LU* decomposition from LAPACK, run on a 64-processor Intel iPSC. Clearly, we are not yet achieving optimum performance, but the situation is improving daily.

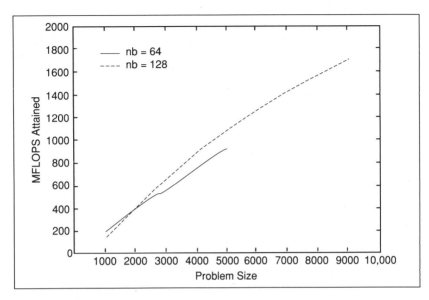

Figure 4. Pipelined *LU* factorization results for 64 and 128 nodes.

Some interest has also been expressed in developing a C implementation of the LAPACK library. And we continue to track what is happening with Fortran 90 and with the activities of the Parallel Computing Forum.

In the meantime, we are in our last round of testing of the shared-memory version of LAPACK. The package will be released to the public in 1992.

References

D. W. Barron and H. P. F. Swinnerton-Dyer, "Solution of Simultaneous Linear Equations Using a Magnetic-Tape Store," *Computer Journal* 3, 28–33 (1960).

M. Berry, K. Gallivan, W. Harrod, W. Jalby, S. Lo, U. Meier, B. Phillippe, and A. Sameh, "Parallel Algorithms on the Cedar System," Center for Supercomputing Research and Development technical report 581, University of Illinois-Urbana/Champaign (October 1986).

C. Bischof and C. Van Loan, "Computing the Singular Value Decomposition on a Ring of Array Processors," in *Large Scale Eigenvalue Problems*, J. Cullum and R. Willoughby, Eds., North-Holland, Amsterdam (1986).

D. Calahan, "A Block-Oriented Sparse Equation Solver for the CRAY-1," in *Proceedings of the 1979 International Conference Parallel Processing*, pp. 116–123 (1979).

B. Chartres, "Adoption of the Jacobi and Givens Methods for a Computer with Magnetic Tape Backup Store," technical report 8, University of Sydney, Australia (1960).

J. Demmel and W. Kahan, "Computing the Small Singular Values of Bidiagonal Matrices with Guaranteed High Relative Accuracy," Argonne National Laboratory Mathematics and Computer Science Division report ANL/MCS-TM-110 (LAPACK Working Note #3) (1988).

J. Dongarra, "Performance of Various Computers Using Standard Linear Equations Software in a Fortran Environment," technical report CS-89-85, University of Tennessee, Knoxville (1991).

J. Dongarra, O. Brewer, S. Fineberg, and J. Kohl, "A Tool to Aid in the Design, Implementation, and Understanding of Matrix Algorithms for Parallel Processors," *Parallel and Distributed Computing* **9**, 185–202 (1990).

J. Dongarra, J. Bunch, C. Moler, and G. W. Stewart, *LINPACK User's Guide*, Society for Industrial and Applied Mathematics Publications, Philadelphia, Pennsylvania (1979).

J. Dongarra, J. Du Croz, I. Duff, and S. Hammarling, "A Set of Level 3 Basic Linear Algebra Subprograms," *ACM Transactions on Mathematical Software* **16**, 1–17 (1990).

J. Dongarra, J. Du Croz, S. Hammarling, and R. Hanson, "An Extended Set of Fortran Basic Linear Algebra Subroutines," *ACM Transactions on Mathematical Software* **14**, 1–17 (1988).

J. Dongarra, I. S. Duff, D. C. Sorensen, and H. A. Van der Vorst, *Solving Linear Systems on Vector and Shared Memory Computers*, Society for Industrial and Applied Mathematics Publications, Philadelphia, Pennsylvania (1990).

J. Dongarra, P. Mayer, and G. Radicati, "The IBM RISC System 6000 and Linear Algebra Operations," Computer Science Department technical report CS-90-122, University of Tennessee, Knoxville (LAPACK Working Note #28) (December 1990).

J. Dongarra and D. Sorensen, "A Fully Parallel Algorithm for the Symmetric Eigenproblem," *SIAM Journal on Scientific and Statistical Computing* **8** (2), 139–154 (1987).

K. Gallivan, R. Plemmons, and A. Sameh, "Parallel Algorithms for Dense Linear Algebra Computations," *SIAM Review* **32** (1), 54–135 (1990).

T. Jordan and K. Fong, "Some Linear Algebraic Algorithms and Their Performance on the CRAY-1," in *High Speed Computer and Algorithm Organization*, D. Kuck, D. Lawrie, and A. Sameh, Eds., Academic Press, New York, pp. 313–316 (1977).

C. Lawson, R. Hanson, D. Kincaid, and F. Krogh, "Basic Linear Algebra Subprograms for Fortran Usage," *ACM Transactions on Mathematical Software* **5**, 308–323 (1979).

A. C. McKellar and E. G. Coffman Jr., "Organizing Matrices and Matrix Operations for Paged Memory Systems," *Communications of the ACM* **12** (3), 153–165 (1969).

R. Schreiber, "Block Algorithms for Parallel Machines," in *Volumes in Mathematics and Its Applications, Vol. 13, Numerical Algorithms for Modern Parallel Computer Architectures*, M. Schultz, Ed., Berlin, Germany, pp. 197–207 (1988).

Design of Algorithms

C. L. Liu

C. L. Liu obtained a bachelor of science degree in 1956 at Cheng Kung University in Tainan, Republic of China, Taiwan. In 1960 and 1962, respectively, he earned a master's degree and a doctorate in electrical engineering at MIT. Currently, he is Professor of Computer Science at the University of Illinois-Urbana/Champaign. Dr. Liu's principal research interests center on the design and analysis of algorithms, computer-aided design of integrated circuits, and combinatorial mathematics.

Let me begin by telling you a story I heard. There was an engineer, a physicist, and a mathematician. They all went to a conference together.

In the middle of the night, the engineer woke up in his room and discovered there was a fire. He rushed into the bathroom, got buckets and buckets of water, threw them over the fire, and the fire was extinguished. The engineer went back to sleep.

In the meantime, the physicist woke up and discovered there was a fire in his room. He jumped out of bed, opened his briefcase, took out paper, pencil, and slide rule, did some very careful computations, ran into the bathroom, measured exactly one-third of a cup of water, and poured it slowly over the fire. Of course, the fire was extinguished. And the physicist went back to sleep.

In the meantime, the mathematician woke up and also discovered there was a fire in his room. He jumped out of bed, rushed into the bathroom, turned on the faucet—and he saw water come out. He said, "Yes, the problem has a solution." And he went back to sleep.

Today we are here to talk about design of algorithms. Indeed, we sometimes use the engineer's approach, and sometimes the physicist's approach, and occasionally the mathematician's approach.

Indeed, it is a fair question that many of you can stare into the eyes of the algorithmists and say that throughout the years we have given you faster and faster and cheaper and cheaper supercomputers, more and more flexible software. What have you done in terms of algorithmic research?

Unfortunately, we are not in a position to tell you that we now have a computer capable of 10^{12} floating-point operations per second (TFLOPS) and that, therefore, there's no need to do any more research in the algorithms area. On the other hand, we cannot tell you that our understanding of algorithms has reached such a point that a "uniFLOPS" computer will solve all the problems for us. What is the problem? The problem is indeed the curse of combinatorial explosion.

I remember it was almost 30 years ago when I took my first course in combinatorics from Professor Gian-Carlo Rota at MIT, and he threw out effortlessly formulas such as n^5, $n \log n$, 2^n, 3^n, and n^n. As it turns out, these innocent-looking formulas do have some significant differences. When we measure whether an algorithm is efficient or not, we draw a line. We say an algorithm is an efficient algorithm if its computational time as a function of the size of the problem is a polynomial function. On the other hand, we say that an algorithm is inefficient if its computational time as a function of the size of the problem is an exponential function. The reason is obvious. As n increases, a polynomial function does not grow very fast. Yet on the other hand, an exponential function grows extremely rapidly.

Let me just use one example to illustrate the point. Suppose I have five different algorithms with these different complexity measures, and suppose n is equal to 60. Even if I'm given a computer that can carry out 10^9 operations per second, if your algorithm has a complexity of n, there is no problem in terms of computation time (6×10^{-8} seconds). With a complexity of n^2 or n^5, there is still no problem. There is a "small" problem with a complexity of 2^n and a "big" problem with a complexity of 3^n.

Now, of course, when the complexity is 3^n, computation time will be measured in terms of centuries. And measured in terms of cents, that adds up to a lot of dollars you cannot afford. As Professor Rota once told us, when combinatorial explosion takes place, the units don't make any difference whatsoever.

So indeed, when we design algorithms it is clear that we should strive for efficient algorithms. Consequently, if we are given a problem, it would be nice if we can always come up with efficient algorithms.

On the other hand, that is not the case. A problem is said to be *intractable* if there is known to be no efficient algorithm for solving that problem.

From the algorithmic point of view, we have paid much less attention to problems that are really intractable. Rather, a great deal of research effort has gone into the study of a class of problems that are referred to as NP-complete problems. We have tackled them for 40 or 50 years, and we cannot confirm one way or the other that there are or are not efficient algorithms for solving these problems.

So therefore, although I begin by calling this the curse of combinatorial explosion, it really is the case, as researchers, that we should look at this as the blessing of combinatorial explosion. Otherwise, we will have no fun, and we will be out of business.

So now the question is, if that is the case, what have you been doing?

Well, for the past several years, people in the area of algorithmic design tried to understand some of the fundamental issues, tried to solve real-life problems whose sizes are not big enough to give us enough trouble, or, if everything failed, tried to use approximation algorithms that would give us good, but not necessarily the best possible, results.

Let me use just a few examples to illustrate these three points.

I believe the problem of linear programming is one of the most beautiful stories one can tell about the development of algorithms.

The problem of linear programming was recognized as an important problem back during the Second World War. It was in 1947 when George Dantzig invented the simplex method. Conceptually, it is a beautiful, beautiful algorithm. In practice, people have been using it to solve linear programming problems for a long, long time. And indeed, it can handle problems of fair size; problems with thousands of variables can be handled quite nicely by the simplex method.

Unfortunately, from a theoretical point of view, it is an exponential algorithm. In other words, although we can handle problems with thousands of variables, the chance of being able to solve problems with hundreds of thousands of variables or hundreds of millions of variables is very small.

In 1979, Khachin discovered a new algorithm, and that is known as the ellipsoid algorithm. The most important feature of the ellipsoid algorithm is that it is a polynomial algorithm. The algorithm by itself is impractical because of the numerical accuracy it requires; the time it takes to run large problems will be longer than with the simplex method.

But on the other hand, because of such a discovery, other researchers got into the picture. Now they realize the ball game has moved from one arena to a different one, from the arena of looking for good exponential algorithms to the arena of looking for good polynomial algorithms.

And that, indeed, led to the birth of Karmarkar's algorithm, which he discovered in 1983. His algorithm is capable of solving hundreds of thousands of variables, and, moreover, his research has led to many other activities, such as how to design special-purpose processors, how to talk about new architectures, and how to talk about new numerical techniques. As I said, this is a good example illustrating algorithmic development. It demonstrates why it is important to have some basic understanding of various features of algorithms.

Let me talk about a second example, with which I will illustrate the following principle: when it becomes impossible to solve a problem precisely—to get the exact solution—then use a lot of FLOPS and try to get a good solution. And I want to use the example of simulated annealing to illustrate the point.

Back in 1955, Nick Metropolis, of Los Alamos National Laboratory, wrote a paper in which he proposed a mathematical formulation to model the annealing process of physical systems with a large number of particles. In other words, when you have a physical system with a large number of particles, you heat the system up to a high temperature and then slowly reduce the temperature of the system. That is referred to as annealing. Then, when the system freezes, the system will be in a state of minimum total energy. In Nick's paper he had a nice mathematical formulation describing the process.

That paper was rediscovered in 1983 by Scott Kirkpatrick, who is also a physicist. He observed that the process of annealing is very similar to the process of doing combinatorial minimization because, after all, you are given a solution space corresponding to all the possible configurations that the particles in the physical system can assume. If, somehow, you can move the configurations around so that you can reach the minimum energy state, you would have discovered the minimum point in your solution space. And that, indeed, is the global minimum of your combinatorial optimization problem.

The most important point of the annealing process is, when you reduce the temperature of your physical system, the energy of the system does not go down all the time. Rather, it goes down most of the time, but it goes up a little bit, and then it goes down, and it goes up a little bit again.

Now, in the terminology of searching a large solution space for a global minimum, that is a very reasonable strategy. In other words, most of the time you want to make a downhill move so that you can get closer and closer to the global minimum. But occasionally, to make sure that you will not get stuck in a local minimum, you need to go up a little bit so that you can jump out of this local minimum.

So therefore, as I said, to take what the physicists have developed and then use that to solve combinatorial optimization problems is conceptually extremely pleasing.

Moreover, from a mathematical point of view, Metropolis was able to prove that as T approaches infinity, the probability that the system will reach the ground state approaches 1 as a limit. For many practical problems that we want to solve, we cannot quite make that kind of assumption. But on the other hand, such a mathematical result does give us a lot of confidence in borrowing what the physicists have done and using that to solve optimization problems.

As I said, the formula I have here tells you the essence of the simulated-annealing approach to combinatorial optimization problems. If you are looking at a solution S and you look at a neighboring solution S', the question is, should you accept S' as your new solution?

This, indeed, is a step-by-step sequential search. And according to the theory of annealing, if the energy of the new state is less than the energy of the current state, the probability of going there is equal to one.

On the other hand, if the new state has an energy that is larger than the energy of the current state, then it depends on some probabilistic distribution. What I'm trying to say is, what I have here is nothing but some kind of probabilistic uphill/downhill search. And as it turns out, it will go quite well in many combinatorial-optimization problems.

And moreover, the development of such techniques will lead to a lot of interesting research questions. How about some theoretical understanding of the situation? How about possible implementation of all of these ideas? And moreover, simulated annealing basically is a very sequential search algorithm. You have to look at one solution, after that the next solution, after that the next solution. How do you parallelize these algorithms?

Let me quickly mention a rather successful experience in solving a problem from the computer-aided design of integrated circuits. That is the problem of placement of standard cells. What you are given is something very simple. You are given a large number, say about 20,000 cells. All these cells are the same height but variable widths. The problem is how to place them in rows. I do not know how many rows all together, and I do not know the relationship among these cells except that eventually there will be connections among them. And if you do it by brute force, you are talking about 20,000! possibilities, which is a lot.

Yet, on the other hand, there is a software package running on Sun Microsystems, Inc., workstations for about 10 hours that produces solutions to this problem that are quite acceptable from a practical point of view. That is my second example.

Let me talk about a third example. As I said, since we do not have enough time, enough FLOPS, to get exact solutions of many of the important problems, how about accepting what we call approximate solutions? And that is the idea of approximation algorithms.

What is an approximation algorithm? The answer is, engineers wave their hands and yell, "Just do something!"

Here is an example illustrating that some basic understanding of some of the fundamental issues would help us go a long way. Let me talk about the job-scheduling problem, since everybody here who is interested in parallel computation would be curious about that. You are given a set of jobs, and these jobs are to be scheduled—in this example, using three identical processors. And of course, when you execute these jobs, you must follow the precedence constraints.

Question: how can one come up with a schedule so that the total execution time is minimum? As it turns out, that is a problem we do not know how to do in the sense that we do not know of any polynomial algorithm for solving the problem. Therefore, if that is the case, once you have proved that the problem is NP-complete, you are given a license to use hand-waving approximation solutions.

However, if you are given a particular heuristic, a particular hand-waving approach, the question is, what is the quality of the results produced by your particular heuristics? You can say, I will run 10,000 examples and see what the result looks like. Even if you run 10,000 examples, if you do not know the best possible solution for each of these examples, you're getting nowhere.

In practice, if I give you just one particular instance and I want you to run your heuristics and tell me how good is the result that your heuristic produces, how do you answer the question? As it turns out, for this particular example, there is a rather nice answer to that question. First of all, let me mention a very obvious, very trivial heuristic we all have been using before: whenever you have a processor that is free, you will assign to it a job that can be executed.

In this case, at the very beginning, I have three processors. They are all free. Since I have only one job that can be executed, I'll execute it. After that, I have three processors that are free, but I have four jobs that can be executed. I close my eyes and make a choice; it's B, C, E. Do the rest, and so on and so forth.

In other words, the simple principle is, whenever you have jobs to be executed, execute them. Whenever your processors are free, try to do something on the processors. This is a heuristic that is not so good but not so bad, either. It can be proved that if you follow such a heuristic, then the total execution time you are going to have is never more than 1.66 times the optimal execution time. In the worst case, I will be off by 66 per cent. Now, of course, you told me in many cases 66 per cent is too much. But there's another way to make good use of this result.

Suppose you have another heuristic which is, indeed, a much better heuristic, and your total execution time is $'$ instead of . How good is $'$? Ah, I will compare it with my , since it is very easy for me to compute . If your $'$ is close to the value of , you might be off by 66 per cent.

On the other hand, if your $'$ is close to $3/5$ of , you are in pretty good shape because you must be very close to $_0$. This is an example to show you that although we do not know how to find $_0$, although we do not know how to determine a schedule that will give me an optimal schedule, on the other hand, we will be capable of estimating how good or how bad a particular heuristic is, using it on a particular example.

To conclude, let me ask the question, what is the research agenda? What are we going to do?

First point: we should pay more attention to fundamental research. And it has been said over and over again, theoretical research has been underfunded. As a college professor, I feel it is alarming that our Ph.D. students who have studied theoretical computer science very hard and have done very well face very stiff competition in getting jobs in good places. There is a danger some of our brightest, most capable computer-science students would not want to go into more theoretical, more fundamental, research but would rather spend their time doing research that is just as exciting, just as important, but fosters better job opportunities.

I think we should look into possibilities such as arranging postdoctoral appointments so that some of the people who have a strong background in theory can move over to learn something about other areas in computer science while making use of what they learned before in theoretical computer science.

Second point: with the development of fast supercomputers, there are many, many opportunities for the theoreticians to make use of what they learn about algorithms, to solve some real live application problems.

Computer-aided design is an example; image-processing, graphics, and so on are all examples illustrating that theoreticians should be encouraged to make good use of the hardware and the software that are available so that they can try to help to solve some of the important problems in computer science.

Third point: when we compare combinatorial algorithms with numerical algorithms, combinatorial algorithms are way behind in terms of their relationship with supercomputing, and we should start building a tool library—an algorithm library—so that we can let people who do not know as much, who do not care to know as much, about algorithms make good use of what we have discovered and what we have developed.

And finally, I'm not saying that as theoreticians we necessarily need TFLOPS machines, but at least give us a lot more FLOPS. Not only will we make good use of the FLOPS you can provide, but you can be sure that we'll come back and beg you for more.

Computing for Correctness

Peter Weinberger

Peter J. Weinberger is Director of the Software and Systems Research Center at AT&T Bell Laboratories. He has a Ph.D. in mathematics from the University of California at Berkeley. After teaching mathematics at the University of Michigan, he moved to Bell Labs.

He has done research in various aspects of system software, including operating systems, network file systems, compilers, performance, and databases. In addition to publishing a number of technical articles, he has coauthored the book The AWK Programming Language.

At Computer Science Research at Bell Laboratories, a part of AT&T, I have a legitimate connection with supercomputers that I'm not going to talk about, and Bell Labs has a long-standing connection with supercomputers. Instead of going over a long list of things we've done for supercomputer users, let me advertise a new service. We have this swell Fortran-to-C converter that will take your old dusty decks—or indeed, your modern dusty decks, your fancy Fortran 77—and convert from an obsolete language to perhaps an obsolescent language named ANSI C. You can even do it by electronic mail.

I'm speaking here about what I feel is a badly underrepresented group at this meeting, namely, the supercomputer nonuser. Let me explain how I picked the topic I'm going to talk about, which is, of course, not quite the topic of this session.

When you walk through the Pentagon and ask personnel what their major problem is, they say it's that their projects are late and too expensive. And if you visit industry and talk to vice presidents and ask them what their major problem is, it's the same: their stuff is late and too expensive. If you ask them why, nine out of ten times in this informal survey they will say, "software." So there's a problem with software. I don't know whether it's writing it fast or getting it correct or both. My guess is that it's both.

So I'm going to talk about a kind of computation that's becoming—to be parochial about it—more important at AT&T, but it's intended to be an example of trying to use some semiformal methods to try to get your programs out faster and make them better.

The kind of calculation that I'm going to talk about is what's described as protocol analysis. Because that's perhaps not the most obvious supercomputer-ready problem, I'll describe it in a little bit of detail.

Protocol analysis of this sort has three properties. First, it's a combinatorially unsatisfactory problem, suffering from exponential explosion. Second, it doesn't use any floating-point operations at all, and so I would be delighted at the advent of very powerful zero-FLOPS machines (i.e., machines running at zero floating-point operations per second), just supposing they are sufficiently cheap, and possibly a slow floating-point operation, as well, so I can tell how fast my program is running.

Because these problems are so big, speed is important. But what's most important is the third property, memory. Because these problems are so big, we can't do them exactly; but because they are so important, we want to do them approximately—and the more memory, the better the approximation. So what I really want is big, cheap memory. Fast is important, but cheap is better.

So what are protocols? They are important in several areas. First, there's intercomputer communications, where computers talk through networks.

They are also important, or can be made important, in understanding the insides of multiprocessors. This is a problem that was solved 20 years ago, or maybe it was only 10, but it was before the Bell System was broken up. Thus, it was during the golden age instead of the modern benighted times. We each have our own interests, right? And my interest is in having large amounts of research money at AT&T.

The problem is that there are bunches of processors working simultaneously sharing a common memory, and you want to keep them from stomping all over each other. You can model the way to do that using protocol analysis. As I said, this is a dead subject and well understood. Now let me try to tell this story without identifying the culprit.

Within the last five years, someone at Bell Labs bought a commercial multiprocessor, and this commercial multiprocessor had the unattractive feature of, every several days or so, dying. Careful study of the source, obtained with difficulty, revealed that the multiprocessor coordination was bogus. In fact, by running the protocol simulation stuff, you could see that not only was the code bogus, but the first proposed fix was bogus, too—although quite a lot less bogus in that you couldn't break it with only two processors. So that's actual protocols in multiprocessing.

There's another place where protocols are important, and that's in the phone system, which is increasingly important. Here's an intuitive idea why some sort of help with programs might be important in the phone system. You have just gone out and bought a fabulous new PBX or some other sort of electronic switchboard. It's got features: conferencing, call-waiting, all sorts of stuff. And here I am setting up a conference call, and just as I'm about to add you to the conference call, you are calling me, and we hit the last buttons at the same time. Now you're waiting for me and I'm waiting for you, and the question is, what state is the wretched switch in? Well, it doesn't much matter unless you have to reboot it to get it back.

What's interesting is that the international standards body, called the Consultative Committee in International Telegraphy and Telephony (CCITT), produces interoperability specifications for the telephone network in terms of protocol specifications. It's not exactly intuitive what the specification means, or if it's correct. In fact, there are two languages: one is a programming language, and the other is a graphical language, presumably for managers. You are supposed to use this language to do successive refinements to get a more and more detailed description of how your switch behaves. This example is in the CCITT standard language SDL (system definition language).

The trouble with international standards is that they are essentially a political process, and so you get the CCITT International Signaling System Seven Standard, and then you go around, if you happen to build switches for a living, and tell the customers that you're going to implement it. They say, "No, no, no, that's not what we use; we use CCITT Seven Standard except for all these different things."

So, what's a protocol, how are protocols modeled, and what's the point of modeling them? Take, for example, a typical computer-science pair of people. We have A, who is named Alice, and B, who is named Bob. Alice has a sequence of messages to send to Bob. The problem isn't with

Alice, Bob, or even the operating system. The problem is with the communications channel, which could have the usual bad properties—it loses messages. The whole system is correct if eventually Bob gets all the messages that Alice sends, in exactly the sequence Alice sends them.

Let me point out why this is not a trivial problem for those of you who don't believe it. The easiest thing is that Alice sends a message, and when Bob gets it, he sends an acknowledgment, and then Alice sends the next message, and so on. The problem is that the channel can lose the first message, and after a while Alice will get tired of waiting and send it again. Or the channel could lose the return message instead, in which case Alice will send the first message twice, and you will have paid your mortgage twice this month.

Now I'm going to describe what's called a sliding window, selective retransmission protocol. With each message there's a sequence number, which runs from 0 to M-1, and there can only be M/2 outstanding unacknowledged messages at once. Now you model these by what are called finite-state machines. For those of you who aren't into finite-state machines, a finite-state machine is just like a finite Markov chain without any probabilities, where you can take all the paths you're allowed to take, and on each path you might change some state variables or emit a message to one of your buddies.

The way we're going to prove this protocol correct is by modeling Alice by a finite-state machine and modeling Bob by a finite-state machine, and you can model each half of the protocol engine by a little finite-state machine. Then you just do all the transitions in the finite-state machines, and look at what happens. Now there's a sequence of three algorithms that have been used historically to study these problems.

I should point out that there's been a lot of talk here about natural parallelism, and the way you model natural parallelism in this example is by multiplying the state spaces together. You just treat the global state as the product of all the local states, and you get exponential complexity in no time at all. But that's the way it is in nature, too.

So you just build the transition graph for all these states, and you get this giant graph, and you can search around for all kinds of properties. In realistic examples you get numbers like 2^{26} reachable states. These are not the biggest cases we want to do, they're just the biggest ones we can do. There are 2^{30} edges in the graph. That's too big.

So you give up on that approach, and you look for properties that you can check on the basis of nothing other than the state you're in. And there are actually a lot of these properties.

Even in cases where we don't keep the whole graph, if each state is 64 or 128 bytes—which is not unreasonable in a big system—and if there are 2^{26} reachable states, we're still losing because there is too much memory needed. So as you're computing along, when you get to a state that you have reached before, you don't have to continue. We also have to keep track of which states we've seen, and a good way to do that is a hash table—except that even in modest examples, we can't keep track of all the stuff about all the states we've seen.

The third step is that we use only one bit per state in the hash table. If the bit is already on, you say to yourself that you've already been here. You don't actually know that unless each state has its own hash code. All you know is that you've found a state that hashes to the same bit as some other state. That means there may be part of the state space you can't explore and problems you can't find. If you think two-dimensional calculations are good enough, why should this approximation bother you?

Now what do you want to get out of this analysis? What you want are little diagrams that give messages saying that you've found a deadlock, and it's caused by this sequence of messages, and you've blown it (again) in your protocol.

Now suppose I do the example I was doing before. With a two-bit sequence number and a channel that never loses anything, there are about 90,000 reachable states, and the longest sequence of actions before it repeats in this finite thing is about 3700 long. To see the exponential explosion, suppose you allow six sequence numbers instead of four. You get 700,000 reachable states, and the longest sequence is 15,000. In four sequence numbers with the message loss, the state space is huge, but the number of reachable states is (only) 400,000. Now if the channel could duplicate messages too, then we're up to 6 million reachable states, and we're beginning to have trouble with our computer.

Now if you do the one-bit thing and hash into a 2^{29} table, you get very good coverage. Not perfect, but very good, and it runs a lot faster than with the full state storage, and it doesn't take a gigabyte to store it all.

If you break the protocol by allowing three outstanding messages, then the analyzer finds that immediately. If you change the channel so it doesn't make mistakes, then the protocol is correct, and there are 287,000 reachable states. That's three times bigger than the two-message version we started with.

There is some potential that formal methods will someday make programs a little bit better. Almost all of these methods are badly exponential in behavior. So you'll never do the problems you really want to do, but we can do better and better with bigger and bigger memories and machines. Some of this can probably be parallelized, although it's not cycles but memory that is the bottleneck in this calculation.

So that's it from the world of commercial data processing.

8

The Future Computing Environment

Panelists in this session speculated on the high-performance computing environment of the future. Discussions were about speed, memory, architectures, workstations, connectivity, distributed computing, the seamless hardware environment of networked heterogeneous computing systems, new models of computation, personalized portable interface software, and adaptive interface software, as well as audio-visual interfaces.

Session Chair

Bob Kahn, Corporation for National Research Initiatives

Interactive Steering of Supercomputer Calculations

Henry Fuchs

Henry Fuchs is the Federico Gil Professor of Computer Science and Adjunct Professor of Radiation Oncology at the University of North Carolina at Chapel Hill. His current research interests are high-performance graphics hardware, three-dimensional medical imaging, and head-mounted display and virtual environments. Dr. Fuchs is one of the principal investigators on the VISTAnet program, which is one of five gigabit network testbed projects supported by NSF and the Defense Advanced Research Projects Agency. He has been an associate editor of the Association for Computing Machinery (ACM) Transactions on Graphics (1983–88) and has chaired many conferences, including ACM's SIGGRAPH '81 (technical program chair), the 1985 Chapel Hill Conference on Advanced Research in VLSI, the 1986 Chapel Hill Workshop on Interactive 3D Graphics, and the 1990 NATO Advanced Research Workshop on 3D Imaging in Medicine (with cochairs Karl Heinz Höhne and Stephen M. Pizer). He serves on various advisory committees, including NSF's Division of Microelectronic Information Processing Systems and the ShoGraphics Technical Advisory Board.

I will discuss the aspect of the future computing environment that has to do with interactive visualization. What I mean by interactive visualization is that you can control what is happening on the supercomputer and see the results, all in an interactive loop. For instance, your Macintosh could be connected to a CRAY Y-MP, and you could have interactive visualization.

I am going to tell you about one particular application that we are pursuing in the VISTAnet project and give you some idea of where we hope to make some progress. Perhaps we could generalize so that some of the lessons we learned might be applicable to other projects.

A lot of interactive visualization has to do with getting more graphics power and seeing more than just what is on the 19-inch CRT, so I am going to emphasize that aspect of it. The VISTAnet project is pursuing, as its first application, radiation therapy treatment planning. The only way to do that right is to do some applications that you cannot do now but that you might be able to do if you had a fast enough connection.

Let us say that the treatment involves a cancer patient with a tumor. Medical practitioners decide that the way to treat the tumor is by hitting it with sufficient radiation to kill it, but they hope that there will be sufficiently low radiation to the rest of the patient's body so that it will not kill the patient. This, then, becomes an interesting computer-aided design problem that does not always have a solution. Because of the complicated anatomical structures in the human body and the erratic manner in which many tumors grow, it is almost impossible to know if there is a particular set of places where you could put radiation beams so that you can kill the tumor and not kill the patient.

This is not just an optimization problem in which you get the best answer; even the best answer may not be good enough. We are talking about the kind of problem where the window of opportunity may be 10 per cent or so. That is, if you go 10 per cent over, you may kill the patient or have very serious consequences; if you go 10 per cent under, you may not cure the patient.

Now, of course, the standard thing to do is to hit the tumor with multiple beams and then hope that at the tumor region you get lethal doses and at other places you do not get lethal doses. This is how radiation treatment is done in two dimensions (Figure 1) everywhere in the world. But, of course, the body is three dimensional, and you could aim the beams from a three-dimensional standpoint. That would give you a whole lot of places where you could aim the beam and get better treatment plans.

The problem is that if you have all these degrees of freedom, you do not know exactly where to start. Thus, the standard thing that people do is to go to protocols in which they know that for a certain kind of tumor in a certain place, they will treat it with a certain kind of beam placement.

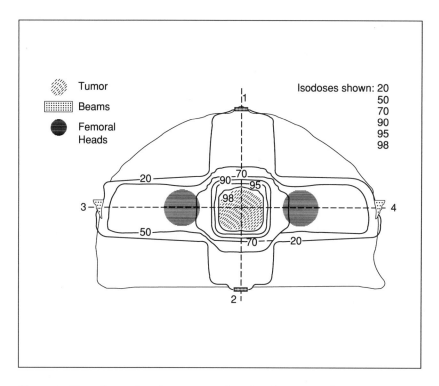

Figure 1. Two-dimensional treatment planning (courtesy of Rosenman & Chaney).

Then they look at these plots on different levels and make minor corrections when they see that there is some healthy tissue that should not get that much radiation. Because it takes a half hour to two hours on the highest-performance workstation to get a dose distribution, the typical way that this is done is that the physicist and the therapist talk about things, and then they do one particular plan and iterate a few times through over a couple of days until they are satisfied with the outcome. What we hope is, if you could do this iteration on a second-by-second basis for an hour or two hours, you could get dramatically better plans than you can with current systems.

Now I would like to discuss what kinds of visualizations people are dealing with in medical graphics. Through these graphics you could see the place where the tumor is. In digital surgery you can cut into the body, and you do have to cut into it to see what is going on inside. We hope this kind of cutting is also going to be done interactively. There are a number of different things that you have to see, all at the same time, and that you have to work with, all at the same time. When you move the beam, you have to see the new dose, and you have to compare that against the anatomy and against the tumor volume because certain kinds of tissue are more sensitive to radiation than others. A lot of patients are repeat patients, so you know that if you have treated the patient a year and a half before, certain regions are significantly more sensitive to repeat doses than they were before.

Figure 2 shows the relationship that VISTAnet plays in medical visualization. It has the CRAY Y-MP at the North Carolina Supercomputing Center, located at the Microelectronics Center for North Carolina (MCNC); Pixel-Planes 5 at the University of North Carolina (UNC), Chapel Hill; and the medical workstation, which will be at the UNC Medical School initially but which we hope to extend to Duke University and elsewhere. We work with the fastest workstations that we can get. When the patient is diagnosed, he/she gets scanned in the CT scanner and may also get other tests like magnetic resonance imaging. Then the patient can go home and return to the facility in a week. Treatment may go on for a month, perhaps twice a week. We hope at the end of six weeks that, when we do another scan, the tumor volume is reduced.

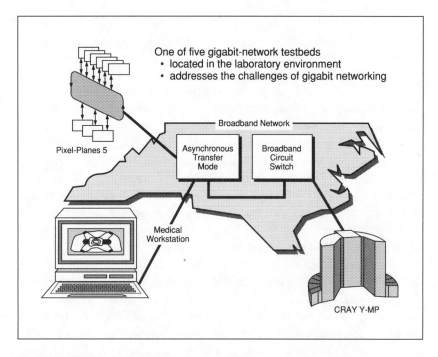

Figure 2. VISTAnet and medical networking.

The bottleneck right now in this type of treatment is the graphics because even the most powerful graphics machines cannot do those kinds of calculations and imaging at interactive rates. The problem is at the frame buffer. The way that the fastest machines currently operate is that they take the frame buffer, divide it into a large number of small frame buffers that are interleaved (the large number might be anywhere from 16 to 80), and then assign a different processing element to each one of those in an interleaved fashion so that, as you can see in Figure 3, processor A gets every fourth pixel on every fourth scan line. When a primitive comes down the pipeline, then most or all of the processors get to work at it. Figure 3 shows the kind of layout that you get when some of the video memory is assigned to each one of the processing elements and then combined together to form the video display.

There is a limit to this. The limit comes, not surprisingly, when you start getting more and more processors and smaller and smaller amounts of video RAM, and when the memory bandwidth, like in all systems, finally gets to you (Figure 4).

Figure 5 shows one of our systems that in many ways is simpler than a general-purpose one because lots of the graphics operations are totally

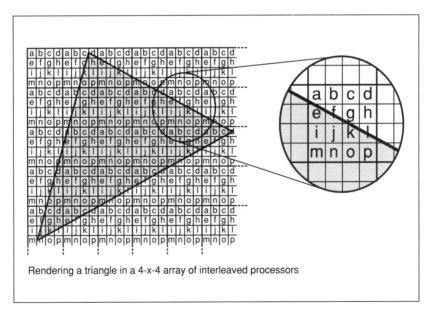

Figure 3. Layout of processing elements that eventually combines to form a video display.

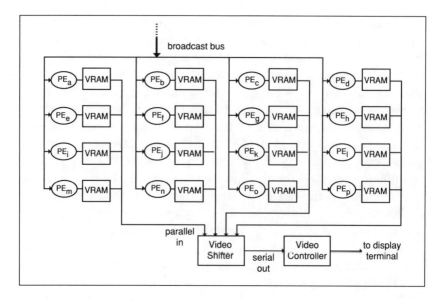

Figure 4. Interleaved image memory system.

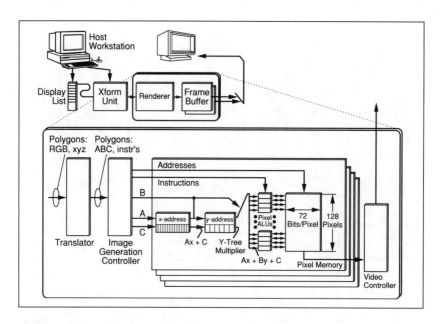

Figure 5. Layout of pixel systems.

local. That is, you do the same thing at every pixel, and you do not care what is done at the neighboring pixel.

At UNC we have been working on varieties of Pixel-Planes systems, and we are on the fifth generation. We build memory chips in which every pixel gets its own little processor. It turns out that if all you do is put a processor at every pixel, you cannot have a big enough processor to make it meaningful to get anything done. We factor out as much arithmetic as possible into a hardware linear or quadratic expression tree; in this manner we get linear and quadratic expressions essentially for free. It very fortuitously happens that almost all the rendering algorithms can be expressed as polynomials in screen space (Figure 6). Our systems basically consist of memory chips for frame buffers, and they do almost all the rendering with a pixel processor for every pixel and a global-linear and quadratic-expression evaluation. If you make these chips so that the addressing on the memory chip can change, then you could take each one, a cluster of these memory chips, and make them essentially like virtual memory so that you can assign them to different parts of the frame buffer at different times.

The ring network runs at 160 megahertz, with many boards that are general-purpose 860-based systems. Virtual pixel processors can be

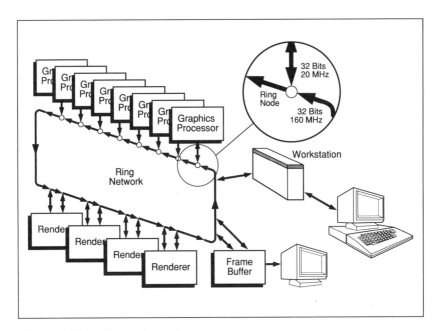

Figure 6. Pixel-Planes 5 overview.

assigned to any place on the screen. In fact, if you want to do parametric-space calculations, they work just as well in parametric space as in x,y space. Whenever you are finished, you can do block moves from the renderers to the frame buffer. In fact, some of the algorithms that people are developing use these renderers for front-end machines and use the graphics processors for the back.

It turns out that the visualization problem is a lot more than just having a fast graphics engine. If you want to see all these things at the same time well enough to be able to interact with them on a detailed basis, then you need to have a lot better perception capability than you can get with current workstation displays. A 19-inch vacuum tube is not adequate because of the complexity of cues that humans get to perceive three dimensions in the world. The solution would be to bring all human perception capabilities to bear on the problem, such as obscuration, stereopsis, kinetic depth effect, head-motion parallax, spatial memory, and so on. Our current graphics machines give us very few of these cues. The machines basically only give us obscuration. That is, we can see when something is in front and where there are other things that are in back, although we do not see the things that are in back.

It will take essentially all of our human perception capabilities to produce a sufficiently powerful visualizer to be able to work with complex, three-dimensional systems. I believe the only candidate in sight is Ivan Sutherland's pioneering work on head-mounted displays, which are currently in vogue. They are called virtual-reality systems. Basically, these are systems in which the display is on your head, your head is trapped, and you perceive some object in front of you. As you walk around, you see what is in front of you, and you can walk literally around it.

In the head-mounted display, you wear little TVs with a small tracking system to track head and hand movements. If you are looking around a room, and in the middle of the room you see molecular models, you can reach out with your three-dimensional cursor, grab the models, and move them around. The molecular-modeling work is a long-time project of Fred Brooks and is supported by the National Institutes of Health.

If eventually you want to be able to have this three-dimensional constellation in front of you because you want to see not simply obscuration, stereopsis, head-motion parallax, and so on, there is a lot more work that needs to be done, not just in image generation but in good tracking of the head and hand. You need to have something in which you can have a wide field of view and a high-resolution display.

Several kinds of tracking might become possible with the three-dimensional technology, such as mechanical, ultrasonic, inertial, magnetic, and optical. For instance, in ultrasound examinations, images could be superimposed inside the patient, and as the transducer is moved about the patient, the data are remembered so that you sweep out a three-dimensional volume of data and actually see that data image (Figure 7). Then you could do a procedure in which you could see what you were doing rather than going in blindly.

Another application we can imagine is one in which your work on an engine and the head-mounted display overlay would have three-dimensional pointers rather than two-dimensional pointers, which would give you information about the path along which an item is to be removed

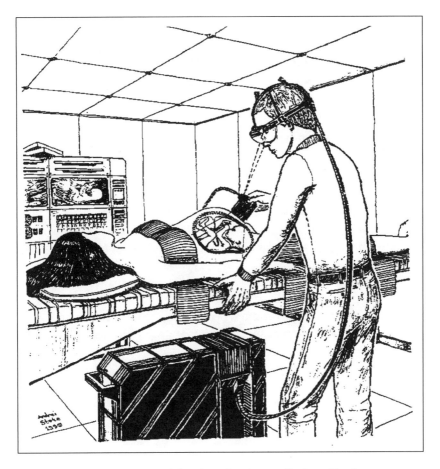

Figure 7. Three-dimensional data imaging for medical applications.

Figure 8. Three-dimensional data imaging for engineering and mechanical applications.

(Figure 8). One could imagine further applications in reconnaissance, in which information is merged from a number of different sources, or in architectural previewing, that is, viewing in three dimensions, and perhaps making changes for a building before it is actually constructed (Figure 9).

In summary, the message I have for you is that you want to think about computing environments, not just from the standpoint of program development but also of program needs. Computing environments will be heterogeneous and must include, for many applications, a very strong visualization component. Interactive visualization needs a whole lot more power than it has right now to benefit from enhanced three-dimensional perception technology.

Figure 9. Three-dimensional imaging for construction applications.

A Vision of the Future at Sun Microsystems

Bill Joy

Bill Joy is well known as a founder of Sun Microsystems, Inc., a designer of the network file system, a codesigner of Scalable Processor ARChitecture (SPARC), and a key contributor in Sun's creation of the open-systems movement. Before coming to Sun, Bill created the Berkeley version of the UNIX operating system, which became the standard for academic and scientific research in the late 1970s and early 1980s. At Berkeley, he founded the Berkeley Standard Distribution, which first distributed applications software for the PDP-11 and, later, complete systems for the VAX. He is still known as the creator of the "VI" text editor, which he wrote more than 10 years ago.

In recent years, Bill has traveled widely to speak on the future of computer technology—hardware, software, and social impacts. In the early 1980s, Bill framed what has become known as "Joy's law," which states that the performance of personal microprocessor-based systems can be calculated as MIPS = 2^{yr-84}. This prediction, made in 1984, is still widely held to be the goal for future system designs.

About 15 years ago I was at the University of Michigan working on large sparse matrix codes. Our idea was to try to decompose and "VAX-solve" a 20,000-by-20,000 sparse matrix on an IBM 370, where the computer center's charging policy charged us for virtual memory. So we, in fact, did real I/O to avoid using virtual memory. We used these same codes on early supercomputers, I think that set for me, 15 years ago, an expectation of what a powerful computer was.

In 1975 I went to the University of California-Berkeley, where everyone was getting excited about Apple computers and the notion of one person using one computer. That was an incredibly great vision. I was fortunate to participate in putting UNIX on the Digital Equipment Corporation VAX, which was meant to be a very popular machine, a very powerful machine, and also to define the unit of performance for a lot of computing simply because it didn't get any faster. Although I was exposed to the kinds of things you could do with more powerful computers, I never believed that all I needed was a VAX to do all of my computing.

Around 1982, the hottest things in Silicon Valley were the games companies. Atari had a huge R&D budget to do things that all came to nothing. In any case, if they had been successful, then kids at home would have had far better computers than scientists would, and clearly, that would have been completely unacceptable.

As a result, several other designers and I wanted to try to get on the microprocessor curve, so we started talking about the performance of a desktop machine, expressed in millions of instructions per second (MIPS), that ought to equal the quantity 2 raised to the power of the current year minus 1984. Now in fact, the whole industry has signed up for that goal. It is not exactly important whether we are on the curve. Everyone believes we should be on the curve, and it is very hard to stay on the curve. So this causes a massive investment in science, not in computer games, which is the whole goal here.

The frustrating thing in 1983 was to talk to people who thought that was enough, although it clearly was not anywhere near enough. In fact, hundreds of thousands of megabytes did not seem to me to be very much because I could not load a data set for a large scientific problem in less than 100 or 1000 megabytes. Without that much memory, I had to do I/O. I had already experienced striping sparse matrices and paging them in and out by hand, and that was not very much fun.

I think we are on target now. Enough investments have been made in the world to really get us to what I would call a 300-megapixel machine in 1991 and in 1995, a 3000-megaflops machine, i.e., a machine capable of 3000 million floating-point operations per second (FLOPS). Economics will affect the price, and different things may skew the schedule plus or minus one year, but it will not really make that much difference.

You will notice that I switched from saying megapixel to megaflops, and that is because with RISC architectures and superscalar implementations, you have the same number of MFLOPS as MIPS, if not more, in the next generation of all the RISC microprocessors. The big change in the next decade will be that we will not be tied to the desktop machine.

In the computer market now, I see an enormous installed base of software on single-CPU, single-threaded code on Macintoshes, UNIX, and DOS converging so that we can port the applications back and forth. This new class of machines will be shipped in volume with eight to 16 CPUs because that is how many I can get on a small card. In a few years, on a sheet-of-paper-size computer, I can get an eight- to 16-CPU machine with several hundred bytes or a gigabyte of memory, which is a substantial computer, quite a bit faster than the early supercomputers I benchmarked.

That creates a real problem in that I don't think we have much interesting software to run on those computers. In fact, we have a very, very small number of people on the planet who have ever had access to those kinds of computers and who really know how to write software, and they've been in a very limited set of application domains. So the question is, how do we get new software? This is the big challenge.

In 1983, I should have bought as much Microsoft Corporation stock as I could when it went public because Microsoft understood the power of what you might call the software flywheel, which is basically, once you get to 100,000 units of a compatible machine a year, the thing starts going into positive feedback and goes crazy. The reason is, as soon as you have 100,000 units a year, software companies become possible because most interesting software companies are going to be small software companies clustered around some great idea. In addition, we have a continuing flow of new ideas, but you have got to have at least 10 people to cater to the market—five people in technical fields and five in business. They cost about $100,000 apiece per year, each, which means you need $1 million just to pay them, which means you need about $2 million of revenue.

People want to pay about a couple hundred dollars for software, net, which means you need to ship 10,000 copies, which means since you really can only expect about 10 per cent penetration, you have got to ship 100,000 units a year. You can vary the numbers, but it comes out to about that order of magnitude. So the only thing you can do, if you've got a kind of computer that's shipping less than 100,000 units a year, is to run university-, research-, or government-subsidized software. That implies, in the end, sort of centralized planning as opposed to distributed innovation, and it loses.

This is why the PC has been so successful. And this is, in some sense, the big constraint. It is the real thing that prevents a new architecture, a new kind of computing platform, from taking off, if you believe that innovation will occur. I think, especially in high technology, you would be a fool not to believe that new ideas, especially for software, will come around. No matter how many bright people you have, most of them don't work for you. In addition, they're on different continents, for instance, in eastern Europe. They're well educated. They haven't had any computers there. They have lots of time to develop algorithms like the elliptical sorts of algorithms. Because there are lots of bright people out there, they are going to develop new software. They can make small companies. If they can hit a platform—that is, 100,000 units a year—they can write a business model, and they can get someone to give them money.

There are only four computing platforms today in the industry that have 100,000 units a year: DOS with Windows, Macs and UNIX on the 386, and UNIX on Scalable Processor ARChitecture (SPARC). That's it. What this tells you is that anybody who isn't on that list has got to find some way to keep new software being written for their platform. There is no mechanism to really go out and capture innovation as it occurs around the world. This includes all the supercomputers because they're equipped with way too much low volume, and they're off by orders of magnitude.

Some platforms can survive for a small amount of time by saying they're software-compatible with another one. For instance, I can have UNIX on a low-volume microprocessor, and I can port the apps from, say, SPARC or the 386 to it. But there's really no point in that because you do the economics, and you're better off putting an incremental dollar in the platform that's shipping in volume than taking on all the support costs of something that didn't make it into orbit. So this is why there's a race against time. For everyone to get their 100,000 units per year is like escaping the gravity field and not burning up on reentry.

Now, here's the goal for Sun Microsystems, Inc. We want to be the first company to ship 100,000 multiprocessors per year. This will clearly make an enormous difference because it will make it possible for people to write software that depends on having a multiprocessor to be effective. I can imagine hundreds or thousands of small software companies becoming possible.

Today we ship $5000 diskless, monochrome workstations and $10,000 standalone, color workstations; both of these are shipping at 100,000 a year. So I've got a really simple algorithm for shipping 100,000 color and 100,000 monochrome workstations a year: I simply make those multiprocessors. And by sticking in one extra chip to have two instead of one and putting the software in, people can start taking advantage of it. As you stick in more and more chips, it just gets better and better. But without this sort of a technique, and without shipping 100,000 multis a year, I don't see how you're going to get the kind of interesting new software that you need. So we may have to keep using the same 15-year-old software because we just don't have time to write any new software. Well, I don't share that belief in the past. I believe that bright new people with new languages will write new software.

The difficulty is, of course, you've got all these small companies. How are they going to get the software to the users? A 10-person company is not a Lotus or a Microsoft; they can't evangelize it as much. We have a problem in the computer industry in that the retail industry is dying. Basically, we don't have any inventory. The way you buy software these days, you call an 800 number, and you get it by the next morning. In fact, you can call until almost midnight, New York time, use your American Express card, and it will be at your door before you get up in the morning. The reason is that the people typically put the inventory at the crosspoint for, say, Federal Express, which is in Memphis, so that it only flies on one plane. They have one centralized inventory, and they cut their costs way down.

But I think there's even a cheaper way. In other words, when you want to have software, what if you already have it? This is the technique we're taking. We're giving all of our users compact-disk (CD) ROMs. If you're a small company and you write an application for a Sun, we'll put it on one of our monthly CD-ROMs for free for the first application that you do if you sign up for our software program, and we'll mail it to every installation of Sun.

So if you get a Sun magazine that has an ad for your software, you can pull a CD-ROM you already have off the shelf, boot up the demo copy of the software you like, dial an 800 number, and turn the software on with a password. Suppose there are 10 machines per site and a million potential users. That means I need 100,000 CDs, which cost about $3 apiece to manufacture. That's about $300,000. So if I put 100 applications on a CD, each company can ship its application to a million users for $3000. I could almost charge for the space in *Creative Computer Application Magazine*. The thing can fund itself because a lot of people will pay $10 for a disk that contains 100 applications that they can try, especially if it's segmented, like the magazine industry is segmented.

This is a whole new way of getting people software applications that really empowers small companies in a way that they haven't been empowered before. In fact, you can imagine if these applications were cheap enough, you could order them by dialing a 900 number where there wouldn't even have to be a human; the phone company would do the billing, and you'd just type in on your touch-tone phone the serial number of your machine, and it would read you the code back. In that case, I think you could probably support a one-person company—maybe a student in a dorm who simply pays $3000 to put a zap on the thing and arranges with some BBS-like company to do the accounting and the billing. These new ways of distributing software become possible once you spin up the flywheel, and I think they will all happen.

The workstation space I said I think will bifurcate into the machines that run the existing uniprocessor software should be shipping about a million units a year, about 100 MIPS per machine site, because that's not going to cost any more than zero MIPS. In fact, that's what you get roughly for free in that time frame. That's about $6 billion for the base hardware—maybe a $12 billion industry. I may be off by a factor of two here, but it's just a rough idea.

Then you're going to have new space made possible by this new way of letting small software companies write software, eight to 16 CPUs. That's what I can do with sort of a crossbar, some sort of simple bus that I can put in a sheet-of-paper-sized, single-board computer, in shipping at least 100,000 a year, probably at an average price of $30,000, and doing most of the graphics in software. There would not be much special-purpose hardware because that's going to depend on whether all those creative people figure out how to do all that stuff in software. And that's another, perhaps, $3 billion market.

I think what you see, though, is that these machines have to run the same software that the small multis do because that's what makes the business model possible. If you try to do this machine without having this machine to draft, you simply won't get the applications, which is why some of the early superworkstation companies have had so much trouble. It's the same reason why NeXT will ultimately fail—they don't have enough volume.

So across this section of the industry, if I had my way, it looks like we're going to ship roughly 200 TFLOPS in 1995, with lots and lots of interesting new, small software applications. The exception is that we're going to ship the 200 TFLOPS mostly as 100,000, 1000-MIP machines instead of as a few TFLOPS machines. I just have a belief that that's going to make our future change, and that's going to be where most of the difference is made—in giving 100,000 machines of that scale to 100,000 different people, which will have more impact than having 100 TFLOPS on 100 computers.

The economics are all with us. This is free-market economics and doesn't require the government to help. It will happen as soon as we can spin up the software industry.

On the Future of the Centralized Computing Environment

Karl-Heinz A. Winkler

Karl-Heinz A. Winkler worked at the Max Planck Institute before first coming to Los Alamos National Laboratory. He next went to the University of Illinois and then returned to Los Alamos, where he is a program manager in the Computing and Communications Division. Dr. Winkler's main interest is in computational science, high-speed communication, and interactive graphics, as well as in coupling numerical experiments to laboratory experiments.

Bill Joy, of Sun Microsystems, Inc., states very forcefully in the preceding paper exactly why we have to change our way of doing things. Building upon what Bill said, I would first like to discuss structural changes. By way of background, I should mention that I have been a supercomputer user all of my adult life—for at least the last 20 years. During the past year, I worked closely with Larry Smarr (see Session 10) at the University of Illinois National Center for Supercomputing Applications (NCSA), so I have learned what it is like on the other side of the fence—and what an education that was!

I think we are going through extremely exciting times in the sense that there is really a revolution going on in the way we have to do our business. In the past, if you bought yet another Cray and a hundred PCs, you could depend on not getting fired, even if you were a computer center director. That game is over. For the first time in a long time, everybody is again allowed to make disastrous investment decisions.

Having been part of a supercomputing center and working now for the Computing and Communications Division at Los Alamos National Laboratory, I find myself compelled to ask what for me is a very interesting question: if Bill is right, is there any place left for a centralized computing facility? Of course, I ask this question because my job depends on the answer.

Certainly the days are gone where you have a centralized environment and terminals around it. In a way, through X Windows, more powerful machines, and the client-host model, some of that structure will survive. If you believe in the federal High Performance Computing Initiative (HPCI), then you expect that in a few years we we will have gigabit-per-second research and education networks. And if you believe in the availability of these high-end workstations, then, of course, you envision the ultimate computer as consisting of all Bill Joy's workstations hooked together in a distributed computing environment. Obviously, that will never work globally because there are too many things going on simultaneously; you also have a lot of security concerns and a lot of scheduling problems.

Yet, in principle it is realistic to expect that the majority of the computer power in a large organization will not be in the centralized facility but in the distributed computing environment. This dramatic change that we have to react to is caused by technological advances, specifically in the microtechnology based on complementary metal oxide semiconductors—which exemplifies the smart thing to do these days. I mean, specifically, look at the forces that drive society, and bank on the technologies that address those needs rather than the needs of specialty niches. This latter point was hammered into me when we had the Supercomputer conference in Boston in 1988. At NCSA I certainly learned the value of Macintoshes and other workstations and how to use, for the first time, software I hadn't written myself.

If we look at the driving forces of society, we discover two areas we have to exploit. Take, for instance, this conference, where a relatively small number of people are convened. Consider, too, the investment base, even in CONVEX and Cray machines, combined; that base equals less than half the IBM 3090s that have been sold, and that is still less than 5000 worldwide.

There is a limit to what one can do. Referring specifically to the presentations at this conference on vector computing, faster cycle time, etc., if you want to achieve in a very few years the extreme in performance—like machines capable of 10^{12} floating-point operations per second (TFLOPS)—it's the wrong way to go, I believe. We must keep in mind the driving forces of technology. For instance, I have a Ford truck, and it has a computer chip in it. (I know that because it failed.) That kind of product is where a majority of the processing technology really goes. So one smart way of going about building a super-supercomputer is to invent an architecture based on the technological advances in, e.g., chip design that are being made anyway. These advances are primarily in the microtechnology and not in the conventional technology based on emitter-coupled logic.

Going to a massively parallel distributed machine, whether based on SIMD or MIMD architecture, allows you to exploit the driving forces of technology. What you really do is buy yourself into a technology where, because of the architectural advantage that went over the scalar and the pipelining architecture, you get away with relatively slow processors, although we see there is a tremendous speedup coming because of miniaturization. Also, you can usually get away with the cheapest, crudest memory chips. This allows you to put a machine together that, from a price/performance point of view, is extremely competitive. If you have ever opened up a Connection Machine (a Thinking Machines Corporation product), you know what I mean. There's not much in there, but it's a very, very fast machine.

Another area where one can make a similar argument is in the mass-storage arena. Unfortunately, at this conference we have no session on mass storage. I think it's one of the most neglected areas. And I think, because there is insufficient emphasis on mass storage and high-speed communication, we have an unbalanced scientific program in this country, resulting from the availability of certain components in the computing environment and the lack of others. Thus, certain problems get attention while other problems are ignored.

If you want to do, say, quantum chromodynamics, you need large memories and lots of computer time. If you want to do time-dependent, multidimensional-continuum physics, then you need not only lots of compute power and large memory but also data storage, communication, visualization, and maybe even a database so that you can make sense of it all.

One of the developments I'm most interested in is the goal of HPCI to establish a gigabit-per-second education and research network. When I had the opportunity in 1989 to testify on Senate Bill S-1067, I made it a point to talk for 15 minutes about the Library of Congress and why we don't digitize it. If you check into the problem a little closer, you find that a typical time scale on which society doubles its knowledge is about a decade—every 10 or 12 years. That would also seem to be a reasonable time scale during which we could actually accomplish the conversion from analog to digital storage. Unfortunately, even to this day, lots of knowledge is first recorded electronically and printed on paper, and then the electronic record is destroyed because the business is in the paper that is sold, not in the information.

It would be fantastic if we could use HPCI as a way to digitize Los Alamos National Laboratory's library. Then we could make it available over a huge network at very high speeds.

The supercomputing culture was established primarily at the national laboratories, and it was a very successful spinoff. One of the reasons why the NSF's Office of Advanced Scientific Computing made it off the ground so fast was because they could rely on a tremendous experience base and lots of good working software. Although I have no direct knowledge of work at the National Security Agency in digitization (because I am a foreign national and lack the necessary clearance), I nevertheless cannot imagine that there are not excellent people at the Agency and elsewhere who have not already solved the problem of converting analog data into digital form. I hear you can even take pictures in a hurry and decipher them. There's great potential here for a spinoff. Whether that's politically possible, I have no idea, but I have an excuse: I'm just a scientist.

Returning to the matter of data storage, in terms of software for a computational supercomputer environment, I must say the situation is disastrous. The common-file system was developed at Los Alamos over a decade ago, and it has served us extremely well in the role for which it was designed: processing relatively small amounts of data with very, very high quality so that you can rely on the data you get back.

Now, 10 years have passed. The Cray Timesharing System will shortly be replaced, I guess everywhere, with UNIX. It would be good to have a mass-storage system based entirely on UNIX, complete with an archival system. The most exciting recent work I'm aware of in this area was carried out at NASA Ames, with the development of MSS-2 and the UTX. But we still have a long way to go if we really want to hook the high-speed networks into a system like that.

Advances in communication also include fiber-distributed data interface, which is a marginal improvement over the Ethernet. High-performance parallel interface (better known as HIPPI) was developed at Los Alamos in the mid-1980s. But there is a tremendous lag time before technology like this shows up in commercial products.

Another question, of course, is standard communication protocols. One aspect of standard communications protocols that has always interested me is very-high-speed, real-time, interactive visualization. I realized some time ago that one could visualize for two-dimensional, time-dependent things but not for three-dimensional things, and that's why it's such a challenge. You probably need a TFLOPS machine to do the real-time calculations that Henry Fuchs mentioned earlier in this session.

Some additional problems on which I have been working are time-dependent, multidimensional-continuum physics simulations; radiation hydrodynamics; and Maxwell equations for a free-electron laser. On a Connection Machine, you can have eight gigabytes of memory right now. If you divide that by four bytes per word, you have about two gigawords. If you have 10 variables in your problem, then you have 200 million grid points. That is in principle what you could do on your machine.

If you're really interested in the dynamics of a phenomenon and do a simulation, you typically do 1000 snapshots of it, then you have your terabyte. Even at a data rate of 200 megabytes per second, it still takes 10 to 12 hours to ship the data around. A Josephson junction is only 28^2, or 128-by-128 connections. This indicates what you could do with the available machinery, assuming you could handle the data. Also, in a few years, when the earth-observing satellites will be flying, there will be a few terabytes per day being beamed down. That translates into only <100 megabits per second, but it's coming every second.

One of the things I really can appreciate concerns software. I spent a year at NCSA working with their industrial partners—about 40 per cent of their funding comes from private companies—and I found the situation very depressing. (Here at Los Alamos National Laboratory, it's even worse in a way; we hardly take advantage of any commercial software, so we miss out on many innovations coming from that quarter.) To give you an example, at NCSA there are about 120 commercial software packages on the Crays; even for their industrial customers, the fee they must pay per year is a little less than for the software you would like to get if you were operating on your Sun workstation. It's a couple of hundred thousand dollars per year.

Typically, that software is generated by a couple of people working out of one of the garage operations. Hardly anything is market tested. There's absolutely no interest in developing stuff for parallel systems. Typically you play a game of tag trying to get in touch with the software vendor, the supercomputer manufacturer, and the industrial partner you're trying to serve.

To reinforce that point, I reiterate that the standards, the software, and everything else along those lines will be determined by what comes out of personal computers, Macintoshes, and workstations because that's the most innovative sector of the environment. The implications for the rest of the environment are considerable. For example, if you don't have the same floating-point standard as you have on a workstation, I don't care what you do, you're doomed.

I would like to close with an analogy to the digital high-definition video standard. The society at large is, in fact, helping a few of us scientists solve our data-storage problem because if you have to store digital images for high-definition video, then all our terabytes will not be a big problem. Coupling real experiments to numerical experiments would provide tremendously valuable two-way feedback in the experimental world and would provide a sanity check, so to speak, both ways while you do the experiment.

Molecular Nanotechnology

Ralph Merkle

Ralph C. Merkle received his Ph.D. from Stanford University in 1979 and is best known as a coinventor of public-key cryptography. Currently, he pursues research in computational nanotechnology at the Xerox Research Center in Palo Alto, California.

We are going to discuss configurations of matter and, in particular, arrangements of atoms. Figure 1 is a Venn diagram, and the big circle with a P in it represents all possible arrangements of atoms. The smaller circle with an M in it represents the arrangements of atoms that we know how to manufacture. The circle with a U in it represents the arrangements of atoms that we can understand.

Venn diagrams let you easily look at various unions and intersections of sets, which is exactly what we're going to do. One subset is the arrangements of atoms that are physically possible, but which we can neither manufacture nor understand. There's not a lot to say about this subset, so we won't.

The next subset of interest includes those arrangements of atoms that we can manufacture but can't understand. This is actually a very popular subset and includes more than many people think, but it's not what we're going to talk about.

The subset that we can *both* manufacture *and* understand is a good, solid, worthwhile subset. This is where a good part of current research is devoted. By thinking about things that we can both understand and

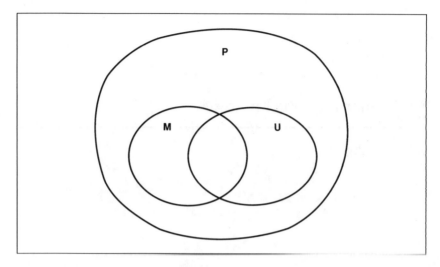

Figure 1. A Venn diagram, where P is the set of all possible arrangements of atoms, M is the set of all arrangements of atoms we can manufacture, and U is the set of all arrangements of atoms we can understand.

manufacture, we can make them better. Despite its great popularity, though, we won't be talking about *this* subset *either*.

Today, we'll talk about the subset that we *can* understand but *can't* yet manufacture. The implication is that the range of things we can manufacture will extend and gradually encroach upon the range of things that we can understand. So at some point in the future, we should be able to make most of these structures, even if we can't make them today.

There is a problem in talking about things that we can't yet manufacture: our statements are not subject to experimental verification, which is bad. This doesn't mean we *can't* think about them, and if we ever expect to build any of them we *must* think about them. But we do have to be careful. It would be a great shame if we never built any of them, because some of them are very interesting indeed. And it will be very hard to make them, especially the more complex ones, if we don't think about them first.

One thing we can do to make it easier to think about things that we can't build (and make it less likely that we'll reach the wrong conclusions) is to think about the subset of mechanical devices: machinery. This subset includes things made out of gears and knobs and levers and things. We can make a lot of mechanical machines today, and we can see how they work and how their parts interact. And we can shrink them down to smaller and smaller sizes, and they still work. At some point, they become so small that we can't make them, so they move from the subset of things that we can make to the subset of things that we can't make. But because the principles of operation are simple, we believe they would work if only we could make them that small. Of course, eventually they'll be so small that the number of atoms in each part starts to get small, and we have to worry about our simple principles of operation breaking down. But because the principles are simple, it's a lot easier to tell whether they still apply or not. And because we know the device works at a larger scale, we only need to worry about exactly how small the device can get and still work. If we make a mistake, it's a mistake in scale rather than a fundamental mistake. We just make the device a little bit bigger, and it should work. (This isn't true of some proposals for molecular devices that depend fundamentally on the fact that small things behave very differently from big things. If we propose a device that depends fundamentally on quantum effects and our analysis is wrong, then we might have a hard time making it slightly bigger to fix the problem!)

The fact remains, though, that we can't make things as small as we'd like to make them. In even the most precise modern manufacturing, we treat matter in bulk. From the viewpoint of an atom, casting involves vast liquid oceans of billions of metal atoms, grinding scrapes off great mountains of atoms, and even the finest lithography involves large numbers of atoms. The basic theme is that atoms are being dealt with in great lumbering statistical herds, not as individuals.

Richard Feynman (1961) said: "The principles of physics, as far as I can see, do not speak against the possibility of maneuvering things atom by atom." Eigler and Schweizer (1990) recently gave us experimental proof of Feynman's words when they spelled "IBM" by dragging individual xenon atoms around on a nickel surface. We have entered a new age, an age in which we can make things with atomic precision. We no longer have to deal with atoms in great statistical herds—we can deal with them as individuals.

This brings us to the basic idea of this talk, which is nanotechnology. (Different people use the term "nanotechnology" to mean very different things. It's often used to describe anything on a submicron scale, which is clearly not what we're talking about. Here, we use the term "nanotechnology" to refer to "molecular nanotechnology" or "molecular manufacturing," which is a much narrower and more precise meaning than "submicron.") Nanotechnology, basically, is the thorough, inexpensive control of the structure of matter. That means if you want to build something (and it makes chemical and physical sense), you can very likely build it. Furthermore, the individual atoms in the structure are where you want them to be, so the structure is atomically precise. And you can do this at low cost. This possibility is attracting increasing interest at this point because it looks like we'll actually be able to do it.

For example, IBM's Chief Scientist and Vice President for Science and Technology, J. A. Armstrong, said: "I believe that nanoscience and nanotechnology will be central to the next epoch of the information age, and will be as revolutionary as science and technology at the micron scale have been since the early '70's.... Indeed, we will have the ability to make electronic and mechanical devices atom-by-atom when that is appropriate to the job at hand."

To give you a feeling for the scale of what we're talking about, a single cubic nanometer holds about 176 carbon atoms (in a diamond lattice). This makes a cubic nanometer fairly big from the point of view of nanotechnology because it can hold over a hundred atoms, and if we're designing a nano device, we have to specify where each of those 176 atoms goes.

If you look in biological systems, you find some dramatic examples of what can be done. For instance, the storage capacity of DNA is roughly 1 bit per 16 atoms or so. If we can selectively remove individual atoms from a surface (as was demonstrated at IBM), we should be able to beat even that!

An even more dramatic device taken from biology is the ribosome. The ribosome is a programmable machine tool that can make almost any protein. It reads the messenger RNA (the "punched paper tape" of the biological world) and builds the protein, one amino acid at a time. All life on the planet uses this method to make proteins, and proteins are used to build almost everything else, from bacteria to whales to giant redwood trees.

There's been a growing interest in nanotechnology (Dewdney 1988, *The Economist* 1989, Pollack 1991). *Fortune Magazine* had an article about where the next major fortunes would come from (Fromson 1988), which included nanotechnology. The *Fortune Magazine* article said that *very* large fortunes would be made in the 21st century from nanotechnology and described K. Eric Drexler as the "theoretician of nanotechnology." Drexler (1981, 1986, 1988, 1992) has had a great influence on the development of this field and provided some of the figures used here.

Japan is funding research in this area (Swinbanks 1990). Their interest is understandable. Nanotechnology is a manufacturing technology, and Japan has always had a strong interest in manufacturing technologies. It will let you make incredibly small things, and Japan has always had a strong interest in miniaturization. It will let you make things where every atom is in the right place: this is the highest possible quality, and Japan has always had a strong interest in high quality. It will let you make things at low cost, and Japan has always been interested in low-cost manufacturing. And finally, the payoff from this kind of technology will come in many years to a few decades, and Japan has a planning horizon that extends to many decades. So it's not surprising that Japan is pursuing nanotechnology.

This technology won't be developed overnight. One kind of development that we might see in the next few years would be an improved scanning tunneling microscope (STM) that would be able to deposit or remove a few atoms on a surface in an atomically precise fashion, making and breaking bonds in the process. The tip would approach a surface and then withdraw from the surface, leaving a cluster of atoms in a specified location (Figure 2). We could model this kind of process today using a computational experiment. Molecular modeling of this kind of interaction is entirely feasible and would allow a fairly rapid analysis of a broad variety of tip structures and tip-surface interactions. This would let us rapidly sort through a wide range of possibilities and pick out the most useful approaches. Now, if in fact you could do something like that, you could build structures using an STM at the molecular and atomic scale.

Figure 3 shows what might be described as a scaled-down version of an STM. It is a device that gives you positional control, and it is roughly 90 nanometers tall, so it is very tiny. It has six degrees of freedom and can position its tip accurately to within something like an angstrom. We can't build it today, but it's a fairly simple design and depends on fairly simple mechanical principles, so we think it should work.

This brings us to the concept of an "assembler." If you can miniaturize an STM and if you can build structures by controlled deposition of small clusters of atoms on surfaces, then you should be able to build small structures with a small version of the STM. Of course, you'd need a small computer to control the small robotic arm. The result is something that looks like an industrial robot that is scaled down by a factor of a million. It has millionfold smaller components and millionfold faster operations.

The assembler would be programmable, like a computer-controlled robot. It would be able to use familiar chemistry: the kind of chemistry that is used in living systems to make proteins and the kind of chemistry that chemists normally use in test tubes. Just as the ribosome can bond together amino acids into a linear polypeptide, so the assembler could bond together a set of chemical building blocks into complex three-dimensional structures by directly putting the compounds in the right places. The major differences between the ribosome and the assembler

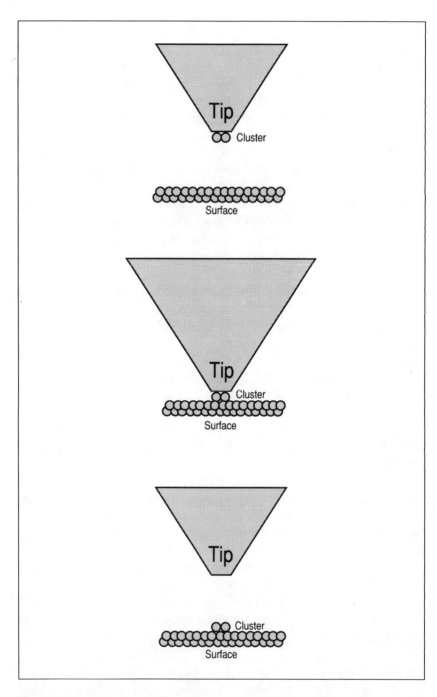

Figure 2. A scanning tunneling microscope depositing a cluster of atoms on a surface.

The Future Computing Environment 305

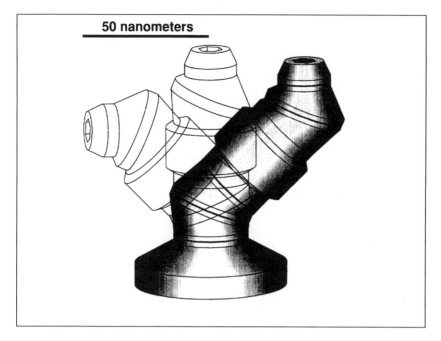

Figure 3. A scanning tunneling microscope built on a molecular scale.

are (1) the assembler has a more complex (computerized) control system (the ribosome can only follow the very simple instructions on the messenger RNA), (2) the assembler can directly move the chemical building blocks to the right place in three dimensions, and so could directly form complex three-dimensional structures (the ribosome can only form simple linear sequences and can make three-dimensional structures only by roundabout and indirect means), and (3) the assembler can form several different types of bonds (the ribosome can form just one type of bond, the bond that links adjacent amino acids).

You could also use rather exotic chemistry. Highly reactive compounds are usually of rather limited use in chemistry because they react with almost anything they touch and it's hard to keep them from touching something you don't want them to touch. If you work in a vacuum, though, and can control the positions of everything, then you can work with highly reactive compounds. They won't react with things they're not supposed to react with because they won't touch anything they're not supposed to touch. Specificity is provided by controlling the positions of reacting compounds.

There are a variety of things that assemblers could make. One of the most interesting is other assemblers. That is where you get low manufacturing cost. (At Xerox, we have a special fondness for machines that make copies of things.) The idea of assemblers making other assemblers leads to self-replicating assemblers. The concept of self-replicating machines has actually been around for some time. It was discussed by von Neumann (1966) back in the 1940s in his work on the theory of self-reproducing automata. Von Neumann's style of a self-replicating device had a Universal Computer coupled to what he called a Universal Constructor. The Universal Computer tells the Universal Constructor what to do. The Universal Constructor, following the instructions of the Universal Computer, builds a copy of both the Universal Computer and the Universal Constructor. It then copies the blueprints into the new machine, and away you go. That style of self-replicating device looks pretty interesting.

NASA (1982) did a study called "Advanced Automation for Space Missions." A large part of their study was devoted to SRSs, or Self-Replicating Systems. They concluded, among other things, that "the theoretical concept of machine duplication is well developed. There are several alternative strategies by which machine self-replication can be carried out in a practical engineering setting. An engineering demonstration project can be initiated immediately...." They commented on and discussed many of the strategies. Of course, their proposals weren't molecular in scale but were quite macroscopic. NASA's basic objective was to put a 100,000-ton, self-replicating seed module on the lunar surface. Designing it would be hard, but after it was designed, built, and installed on the lunar surface, it would manufacture more of itself. This would be much cheaper than launching the same equipment from the earth.

There are several different self-replicating systems that we can examine. Von Neumann's proposal was about 500,000 bits. The Internet Worm was also about 500,000 bits. The bacterium, *E. coli*, a self-replicating device that operates in nature, has a complexity of about 8,000,000 bits. Drexler's assembler has an estimated complexity of 100 million bits. People have a complexity of roughly 6.4 gigabits. Of course, people do things other than replicate, so it's not really fair to chalk all of this complexity up to self-replication. The proposed NASA lunar manufacturing facility was very complex: 100 to 1000 gigabits.

To summarize the basic idea: today, manufacturing limits technology. In the future we'll be able to manufacture most structures that make sense. The chief remaining limits will be physical law and design capabilities. We can't make it if it violates physical law, and we can't make it if we can't specify it.

It will take a lot of work to get there, and more than just a lot of work, it will take a lot of planning. It's likely that general-purpose molecular manufacturing systems will be complex, so complex that we won't stumble over them by accident or find that we've made one without realizing it. This is more like going to the moon: a big project with lots of complicated systems and subsystems. Before we can start such a project, though, there will have to be proposals, and analyses of proposals, and a winnowing of the proposals down to the ones that make the most sense, and a debate about which of these few best proposals is actually worth the effort to build. Computers can help a great deal here. For virtually the first time in history, we can use computational models to study structures that we can't build and use computational experiments, which are often cheap and quick, compared with physical experiments, to help us decide which path is worth following and which path isn't.

Boeing builds airplanes in a computer before they build them in the real world. They can make better airplanes, and they can make them more quickly. They can shave years off the development time. In the same way, we can model all the components of an assembler using everything from computational-chemistry software to mechanical-engineering software to system-level simulators. This will take an immense amount of computer power, but it will shave *many* years off the development schedule.

Of course, everyone wants to know how soon molecular manufacturing will be here. That's hard to say. However, there are some very interesting trends. The progress in computer technology during the past 50 years has been remarkably regular. Almost every parameter of hardware technology can be plotted as a straight line on log paper. If we extrapolate those straight lines, we find they reach interesting values somewhere around 2010 to 2020. The energy dissipation per logic operation reaches thermal noise at room temperature. The number of atoms required to store one bit of information reaches approximately one. The raw computational power of a computer starts to exceed the raw computational power of the human brain. This suggests that somewhere between 2010 and 2020, we'll be able to build computers with atomic precision. It's hard to see how we could achieve such remarkable performance otherwise, and there are no fundamental principles that prevent us from doing it. And if we can build computers with atomic precision, we'll have to have developed some sort of molecular manufacturing capability.

Feynman said: "The problems of chemistry and biology can be greatly helped if our ability to see what we are doing and to do things on an atomic level is ultimately developed, a development which, I think, cannot be avoided."

While it's hard to say exactly how long it will take to develop molecular manufacturing, it's clear that we'll get there faster if we decide that it's a worthwhile goal and deliberately set out to achieve it.

As Alan Kay said: "The best way to predict the future is to create it."

References

A. K. Dewdney, "Nanotechnology: Wherein Molecular Computers Control Tiny Circulatory Submarines," *Scientific American* **257**, 100–103 (January 1988).

K. E. Drexler, *Engines of Creation*, Anchor Press, New York (1986).

K. E. Drexler, "Molecular Engineering: An Approach to the Development of General Capabilities for Molecular Manipulation," in *Proceedings of the National Academy of Sciences of the United States of America* **78**, 5275–78 (1981).

K. E. Drexler, *Nanosystems: Molecular Machinery, Manufacturing and Computation*, John Wiley and Sons, Inc., New York (1992).

K. E. Drexler, "Rod Logic and Thermal Noise in the Mechanical Nanocomputer," in *Proceedings of the Third International Symposium on Molecular Electronic Devices*, F. Carter, Ed., Elsevier Science Publishing Co., Inc., New York (1988).

The Economist Newspaper Ltd., "The Invisible Factory," *The Economist* **313** (7632), 91 (December 9, 1989).

D. M. Eigler and E. K. Schweizer, "Positioning Single Atoms with a Scanning Tunnelling Microscope," *Nature* **344**, 524–526 (April 15, 1990).

R. Feynman, *There's Plenty of Room at the Bottom*, annual meeting of the American Physical Society, December 29, 1959. Reprinted in "Miniaturization," H. D. Gilbert, Ed., Reinhold Co., New York, pp. 282–296 (1961).

B. D. Fromson, "Where the Next Fortunes Will be Made," *Fortune Magazine*, Vol. 118, No. 13, pp. 185–196 (December 5, 1988).

NASA, "Advanced Automation for Space Missions," in *Proceedings of the 1980 NASA/ASEE Summer Study*, Robert A. Freitas, Jr. and William P. Gilbreath, Eds., National Technical Information Service (NTIS) order no. N83-15348, U.S. Department of Commerce, Springfield, Virginia (November 1982).

A. Pollack, "Atom by Atom, Scientists Build 'Invisible' Machines of the Future," *The New York Times* (science section), p. B7 (November 26, 1991).

D. Swinbanks, "MITI Heads for Inner Space," *Nature* **346**, 688–689 (August 23, 1990).

J. von Neumann, *Theory of Self Reproducing Automata*, Arthur W. Burks, Ed., University of Illinois Press, Urbana, Illinois (1966).

Supercomputing Alternatives

Gordon Bell

C. Gordon Bell, now an independent consultant, was until 1991 Chief Scientist at Stardent Computer. He was the leader of the VAX team and Vice President of R&D at Digital Equipment Corporation until 1983. In 1983, he founded Encore Computer, serving as Chief Technical Officer until 1986, when he founded and became Assistant Director of the Computing and Information Science and Engineering Directorate at NSF. Gordon is also a founder of The Computer Museum in Boston, a fellow of both the Institute of Electrical and Electronics Engineers and the American Association for the Advancement of Science, and a member of the National Academy of Engineering. He earned his B.S. and M.S. degrees at MIT.

Gordon was awarded the National Medal of Technology by the Department of Commerce in 1991 and the von Neumann Medal by the Institute of Electrical and Electronics Engineers in 1992.

Less is More

Our fixation on the supercomputer as the dominant form of technical computing is finally giving way to reality. Supers are being supplanted by a host of alternative forms of computing, including the interactive, distributed, and personal approaches that use PCs and workstations. The technical computing industry and the community it serves are poised for an exciting period of growth and change in the 1990s.

Traditional supercomputers are becoming less relevant to scientific computing, and as a result, the growth in the traditional vector supercomputer market, as defined by Cray Research, Inc., is reduced from what it was in the early 1980s. Large-system users and the government, who are concerned about the loss of U.S. technical supremacy in this last niche of computing, are the last to see the shift. The loss of supremacy in supercomputers should be of grave concern to the U.S. government, which relies on supercomputers and, thus, should worry about the loss of one more manufacturing-based technology. In the case of supercomputers, having the second-best semiconductors and high-density packaging means that U.S. supercomputers will be second.

The shift away from expensive, highly centralized, time-shared supercomputers for high-performance computing began in the 1980s. The shift is similar to the shift away from traditional mainframes and minicomputers to workstations and PCs. In response to technological advances, specialized architectures, the dictates of economies, and the growing importance of interactivity and visualization, newly formed companies challenged the conventional high-end machines by introducing a host of supersubstitutes: minisupercomputers, graphics supercomputers, superworkstations, and specialized parallel computers. Cost-effective FLOPS, that is, the floating-point operations per second essential to high-performance technical computing, come in many new forms. The compute power for demanding scientific and engineering challenges could be found across a whole spectrum of machines with a range of price/performance points. Machines as varied as a Sun Microsystems, Inc., workstation, a graphics supercomputer, a minisupercomputer, or a special-purpose computer like a Thinking Machines Corporation Connection Machine or an nCUBE Corporation Hypercube all do the same computation for five to 50 per cent of the cost of doing it on a conventional supercomputer. Evidence of this trend abounds.

An example of cost effectiveness would be the results of the PERFECT (Performance Evaluation for Cost-Effective Transformation) contest. This benchmark suite, developed at the University of Illinois Supercomputing Research and Development Center in conjunction with manufacturers and users, attempts to measure supercomputer performance and cost effectiveness.

In the 1989 contest, an eight-processor CRAY Y-MP/832 took the laurels for peak performance by achieving 22 and 120 MFLOPS (million floating-point operations per second) for the unoptimized baseline and hand-tuned, highly optimized programs, respectively. A uniprocessor Stardent Computer 3000 graphics supercomputer won the cost/performance award by a factor of 1.8 and performed at 4.2 MFLOPS with no tuning and 4.4 MFLOPS with tuning. The untuned programs on the Stardent 3000 were a factor of 27 times more cost effective than the untuned CRAY Y-MP programs. In comparison, a Sun SPARCstation 1 ran the benchmarks roughly one-third as fast as the Stardent.

The PERFECT results typify "dis-economy" of scale. When it comes to getting high-performance computation for scientific and engineering problems, the biggest machine is rarely the most cost effective. This concept runs counter to the myth created in the 1960s known as Grosch's Law, which stated that the power of a computer increased as its price squared. Many studies have shown that the power of a computer increased at most as the price raised to the 0.8 power—a dis-economy of scale.

Table 1 provides a picture of the various computing power and capacity measures for various types of computers that can substitute for supercomputers. The computer's peak power and LINPACK 1K × 1K estimate the peak power that a computer might deliver on a highly parallel application. LINPACK 100-x-100 shows the power that might be expected for a typical supercomputer application and the average speed at which a supercomputer might operate. The Livermore Fortran Kernels (LFKs) were designed to typify workload, that is, the capacity of a computer operating at Lawrence Livermore National Laboratory.

The researchers who use NSF's five supercomputing centers at no cost are insulated from cost considerations. Users get relatively little processing power per year despite the availability of the equivalent 30 CRAY

Table 1. Power of 1989 Technical Computers, in MFLOPS

Type	No. Proc. Max.	LFK per Proc.	LFK per Machine	LINPACK 100 × 100	LINPACK 1K × 1K	Peak
PC	1	–	0.1–0.5	0.1–0.5	0.1–1.0	1
Workstation	1	–	0.2–1.5	0.5–3.0	6	8
Micro/Mini	1	–	0.1–0.5	0.1–0.5	0.1–0.5	2
Supermini	6	1	4	1	6	24
Superworkstation	4	1.5–5	10	6–12	80	128
Minisuper	8	2–4.3	10	6–16	166	200
Main/Vectors	6	7.2	43	13	518	798
Supercomputer	8	19	150	84	2144	2667

X-MP processors, or 240,000 processor hours per year. When that processing power is spread out among 10,000 researchers, it averages out to just 24 hours per year, or about what a high-power PC can deliver in a month. Fortunately, a few dozen projects get 1000 hours per year. Moreover, users have to contend with a total solution time disproportionate to actual computation time, centralized management and allocation of resources, the need to understand vectorization and parallelization to utilize the processors effectively (including memory hierarchies), and other issues.

These large, central facilities are not necessarily flawed as a source of compute power unless they attempt to be a one-stop solution. They may be the best resource for the very largest users with large, highly tuned parallel programs that may require large memories, file capacity of tens or hundreds of gigabytes, the availability of archive files, and the sharing of large databases and large programs. They also suffice for the occasional user who needs only a few hours of computing a year and doesn't want to own or operate a computer.

But they're not particularly well suited to the needs of the majority of users working on a particular engineering or scientific problem that is embodied in a program model. They lack the interactive and visualization capabilities that computer-aided design requires, for example. As a result, even with free computer time, only a small fraction of the research community, between five and 10 per cent, uses the NSF centers. Instead, users are buying *smaller* computing resources to make more power available than the large, traditional, centralized supercomputer supplies. Ironic though it may seem, *less is more*.

Supersubstitutes Provide More Overall Capacity

Users can opt for a supersubstitute *if* it performs within a factor of 10 of a conventional supercomputer. That is, a viable substitute must supply up to 10 per cent the power of a super so as to deliver the same amount of computation in one day that the typical user could expect from a large, time-shared supercomputer—between a half-hour and an hour of Cray service per day and a peak of two hours. Additionally, it should be the best price performer in its class, sustain high throughput on a wide variety of jobs, and have appropriate memory and other resources.

Data compiled by the market research firm Dataquest Inc. has been arranged in Table 2 so as to show technical computer installations in 1989, along with several gauges of computational capacity: per-processor

Table 2. Installed Capacity for General-Purpose Technical Computing Environment (Source: Dataquest)

Type	Dataquest Installed	1989 Ships	1989 LFK Capacity	Companies Selling	Building	Dead
PC	3.4M	1M	1341	100s	?	?
Workstation	0.4M	145K	960	7	?	~50
Micro/Mini	0.9M	75K	30	~20	?	~100
Supermini	0.3M	7.5K	200	7	?	~10
Superworkstation	10K	10K	100	3	2	2
Minisuper	1.6K	600	32	5	>2	8
Parallel Proc.	365	250	4	24	>9	8
Main/Vectors	8.3K	100	29	3	?	3
Supercomputer	450	130	100	4	>3	3

performance on the Livermore Loops workload benchmark,* per-processor performance on LINPACK 100-x-100 and peak performance on the LINPACK 1000-x-1000 benchmark,** and total delivered capacity using the Livermore Loops workload measure, expressed as an equivalent to the CRAY Y-MP eight-processor computer's 150 MFLOPS.

How Supers Are Being Niched

Supercomputers are being niched across the board by supersubstitutes that provide a user essentially the same service but at much lower entry and use costs. In addition, all the other forms of computers, including

---

* LFKs consist of 24 inner loops that are representative of the programs run at Lawrence Livermore National Laboratory. The Spectrum of Code varies from being entirely scalar to almost perfectly vectorizable, whereby the supercomputer can run at its maximum speed. The harmonic mean is used to measure relative performances, which correspond to the time it takes to run to all 24 programs. The SPEC and PERFECT benchmarks also correlate with the Livermore benchmark.

** The LINPACK benchmark measures the computer's ability to solve a set of linear algebraic equations. These equations are the basis of a number of programs such as finite-element models used for physical systems. The small matrix size (100 × 100) benchmark corresponds to the rate at which a typical application program runs on a supercomputer. The large LINPACK corresponds to the best case that a program is likely to achieve.

mainframes with vector facilities, minis, superminis, minisupers, ordinary workstations, and PCs, offer substitutes. Thus, the supercomputer problem (i.e., the lack of the U.S.'s ability to support them in a meaningful market fashion) is based on economics as much as on competition.

Numerous machines types are contenders as supersubstitutes. Here are some observations on each category.

Workstations

Workstations from companies like Digital Equipment Corporation (DEC), the Hewlett-Packard Company, Silicon Graphics Inc., and Sun Microsystems, among others, provide up to 10 per cent of the capacity of a CRAY Y-MP processor. But they do it at speeds of less than 0.3 per cent of an eight-processor Y-MP LINPACK peak and at about two per cent the speed of a single-processor Y-MP on the LINPACK 100-x-100 benchmark. Thus, while they may achieve impressive scalar performance, they have no way to hit performance peaks for the compute-intensive programs for which the vector and parallel capabilities of supercomputers were developed. As a result, they are not ideal as supersubstitutes. Nevertheless, ordinary computers like workstations, PCs, minicomputers, and superminis together provide most of the technical computing power available today.

Minicomputers and Superminis

These machines provide up to 20 per cent of the capacity of a CRAY-MP processor. But again, with only 0.25 per cent the speed of the LINPACK peak of the Cray, they are also less-than-ideal supercomputer substitutes.

Mainframes

IBM may be the largest supplier of supercomputing power. It has installed significant computational power in its 3090 mainframes with vector-processing facilities. Dataquest has estimated that 250 of the 750 3090-processors shipped last year had vector-processing capability. Although a 3090/600 has 25 per cent of the CRAY Y-MP's LINPACK peak power, its ability to carry out a workload, as measured by Livermore Loops, is roughly one-third that of a CRAY Y-MP/8.

But we see only modest economic advantages and little or no practical benefit to be derived from substituting one centralized, time-shared resource for another. For numeric computing, mainframes are not the best performers in their price class. Although they supply plenty of computational power, they rarely hit the performance peaks that supercomputer-class applications demand. The mainframes from IBM—and even the new DEC 9000 series—suffer from the awkwardness of traditional architecture evolution. Their emitter-coupled-logic (ECL) circuit technology is costly. And the pace of improvement in ECL density lags far behind the rate of progress demonstrated by the complementary-metal-oxide-semiconductor (CMOS) circuitry employed in more cost-effective and easier-to-use supersubstitutes.

Massively Data-Parallel Computers

There is a small but growing base of special-purpose machines in two forms: multicomputers (e.g., hundreds and thousands of computers interconnected) and the SIMD (e.g., the Connection Machine, MasPar), some of which supply a peak of 10 times a CRAY Y-MP/8 with about the same peak-delivered power (1.5 GFLOPS) on selective, parallelized applications that can operate on very large data sets. This year a Connection Machine won the Bell Perfect Club Prize* for having the highest peak performance for an application. These machines are not suitable for a general scientific workload. For programs rich in data parallelism, these machines can deliver the performance. But given the need for complete reprogramming to enable applications to exploit their massively parallel architectures, they are not directly substitutable for current supercomputers. They are useful for the highly parallel programs for which the super is designed. With time, compilers should be able to better exploit these architectures that require explicitly locating data in particular memory modules and then passing messages among the modules when information needs to be shared.

The most exciting computer on the horizon is the one from Kendall Square Research (KSR), which is scalable to over 1000 processors as a large, shared-memory multiprocessor. The KSR machine functions equally well for both massive transaction processing and massively parallel computation.

Minisupercomputers

The first viable supersubstitutes, minisupercomputers, were introduced in 1983. They support a modestly interactive, distributed mode of use and exploit the gap left when DEC began in earnest to ignore its

---

* A prize of $1000 is given in each of three categories of speed and parallelism to recognize applications programs. The 1988 prizes went to a 1024-node nCUBE at Sandia and a CRAY X-MP/416 at the National Center for Atmospheric Research; in 1989 a CRAY Y-MP/832 ran the fastest.

technical user base. In terms of power and usage, their relationship to supercomputers is much like that of minicomputers to mainframes. Machines from Alliant Computer Systems and CONVEX Computer Corporation have a computational capacity approaching one CRAY Y-MP processor.

Until the introduction of graphics supercomputers in 1988, minisupers were the most cost-effective source of supercomputing capacity. But they are under both economic and technological pressure from newer classes of technical computers. The leading minisuper vendors are responding to this pressure in different ways. Alliant plans to improve performance and reduce computing costs by using a cost-effective commodity chip, Intel's i860 RISC microprocessor. CONVEX has yet to announce its next line of minisupercomputers; however, it is likely to follow the Cray path of a higher clock speed using ECL.

Superworkstations

This machine class, judging by the figures in Table 1, is the most vigorous of all technical computer categories, as it is attracting the majority of buyers and supplying the bulk of the capacity for high-performance technical computing. In 1989, superworkstation installations reached more users than the NSF centers did, delivering *four times* the computational capacity and power supplied by the CRAY Y-MP/8.

Dataquest's nomenclature for this machine class—superworkstations—actually comprises two kinds of machines: graphics supercomputers and superworkstations. Graphics supercomputers were introduced in 1988 and combine varying degrees of supercomputer capacity with integral three-dimensional graphics capabilities for project and departmental use (i.e., multiple users per system) at costs ranging between $50,000 and $200,000. Priced even more aggressively, at $25,000 to $50,000, superworkstations make similar features affordable for personal use.

Machines of this class from Apollo (Hewlett-Packard), Silicon Graphics, Stardent, and most recently from IBM all provide between 10 and 20 per cent of the computational capacity of a CRAY Y-MP processor, as characterized by the Livermore Loops workload. They also run the LINPACK 100-x-100 benchmark at about 12 per cent of the speed of a one-processor Y-MP. While the LINPACK peak of such machines is only two per cent of an eight-processor CRAY Y-MP, the distributed approach of the superworkstations is almost three times more cost effective. In other words, users spending the same amount can get three to five times as much computing from superworkstations and graphics supercomputers than from a conventional supercomputer.

In March 1990, IBM announced its RS/6000 superscalar workstation, which stands out with exceptional performance and price performance. Several researchers have reported running programs at the same speed as the CRAY Y-MP. The RS/6000's workload ability measured by the Livermore Loops is about one-third that of a CRAY Y-MP processor.

Superworkstations promise the most benefits for the decade ahead because they *conjoin more leading-edge developments than any other class of technical computer*, including technologies that improve performance and reduce costs, interactivity, personal visualization, smarter compiler technologies, and the downward migration of super applications. More importantly, superworkstations provide for interactive visualization in the same style that PCs and workstations used to stabilize mainframe and minicomputer growth. Radically new applications will spring up around this new tool that are not versions of tired 20-year-old code that ran on the supercomputer, mainframe, and minicode museums. These will come predominantly from science and engineering problems, but most financial institutions are applying supercomputers for econometric modeling, work optimization, portfolio analysis, etc.

Because these machines are all based on fast-evolving technologies, including single-chip RISC microprocessors and CMOS, we can expect performance gains to continue at the rate of over 50 per cent a year over the next five years. We'll also see continuing improvements in clock-rate growth to more than 100 megahertz by 1992. By riding the CMOS technology curve, future superworkstation architectures will likely be able to provide more power for most scientific applications than will be available from the more costly multiple-chip systems based on arrays of ECL and GaAs (gallium arsenide) gates. Of course, the bigger gains will come through the use of multiples of these low-cost processors for parallel processing.

Why Supercomputers Are Becoming Less General Purpose

Like their large mainframe and minicomputer cousins, the super is based on expensive packaging of ECL circuitry. As such, the evolution in performance is relatively slower (doubling every five years) than that of the single-chip microprocessor, which doubles every 18 months. One of the problems in building a cost-effective, conventional supercomputer is that every part—from the packaging to the processors, primary and secondary memory, and the high-performance network—typically costs more than it contributes to incremental performance gains. Supercomputers built from expensive, high-speed components have elaborate processor-memory connections, very fast transfer disks, and processing circuits that do relatively few operations per chip and per watt, and they require extensive installation procedures with high operating costs.

To get the large increases in peak MFLOPS performance, the supercomputer architecture laid down at Cray Research requires having to increase memory bandwidth to support the worst-case peak. This is partially caused by Cray's reluctance to use modem cache memory techniques to reduce cost and latency. This increase in bandwidth results in a proportional increase in memory latency, which, unfortunately, decreases the computer's scalar speed. Because workloads are dominated by scalar code, the result is a disproportionately small increase in throughput, even though the peak speed of a computer increases dramatically. Nippon Electric Corporation's (NEC's) four-processor SX-3, with a peak of 22 GFLOPS, is an example of providing maximum vector speed. In contrast, one-chip microprocessors with on-board cache memory, as typified by IBM's superscalar RS/6000 processor, are increasing in speed more rapidly than supers for scientific codes.

Thus, the supercomputer is becoming a special-purpose computer that is only really cost effective for highly parallel problems. It has about the same performance of highly specialized, parallel computers like the Connection Machine, the microprocessor-based nCUBE, and Intel's multicomputers, yet the super costs a factor of 10 more because of its expensive circuit and memory technology. In both the super and nontraditional computers, a program has to undergo significant transformations in order to get peak performance.

Now look at the situation of products available on the market and what they are doing to decrease the supercomputer market. Figure 1, which plots performance versus the degree of problem parallelism (Amdahl's Law), shows the relative competitiveness in terms of performance for supersubstitutes. Fundamentally, the figure shows that supers are being completely "bracketed" on *both* the bottom (low performance for scalar problems) and the top (highly parallel problems). The figure shows the following items:

1. The super is in the middle, and its performance ranges from a few tens of MFLOPS per processor to over two GFLOPS, depending on the degree of parallelization of the code and the number of processors

The Future Computing Environment 321

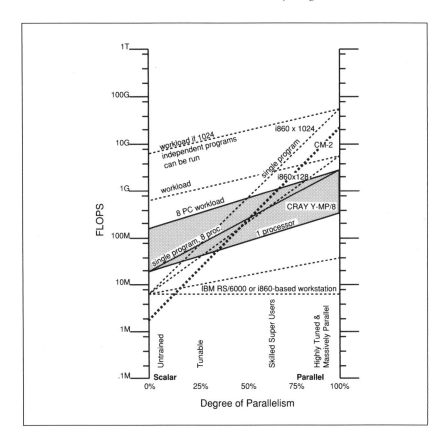

Figure 1. Performance in floating-point operations per second versus degree of parallel code for four classes of computer: the CRAY Y-MP/8; the Thinking Machines Corporation CM-2; the IBM RS/6000 and Intel i860-based workstation; and the Intel 128 and 1024 iPSC/860 multicomputers.

used. Real applications have achieved sustained performance of about 50 per cent of the peak.

2. Technical workstations supply the bulk of computing for the large, untrained user population and for code that has a low degree of vectorization (that must be tuned to run well). In 1990, the best CMOS-based technical workstations by IBM and others performed at one-third the capacity of a single-processor Y-MP on a broad range of programs and cost between $10,000 and $100,000. Thus, they are anywhere from five to 100 times more cost effective than a super costing at around $2,500,000 per processor. This situation differs from a decade ago, when supers provided over a factor of 20 greater

performance for scalar problems against all computers. The growth in clock performance for CMOS is about 50 per cent per year, whereas the growth in performance for ECL is only 10 to 15 per cent per year.
3. For programs that have a high degree of parallelization, two alternatives threaten the super in its natural habitat. The parallelization has to be done by a small user base.
   a. The Connection Machine costs about one-half the Y-MP but provides a peak of almost 10 times the Cray. One CM-2 runs a real-time application code at two times the peak of a Cray.
   b. Multicomputers can be formed from a large collection of high-volume, cost-effective CMOS microprocessors. Intel's iPSC/860 multicomputer comes in a range of sizes from typical (128 computers) to large (1K computers). IBM is offering the ability to interconnect RS/6000s. A few RS/6000s will offer any small team the processing power of a CRAY Y-MP processor for a cost of a few hundred thousand dollars.

The Supercomputer Industry

The business climate of the 1990s offers further evidence that big machines may be out of step with more cost-effective, modern computing styles. Table 2 shows the number of companies involved in the highly competitive technical computing industry. The business casualties in the high end of the computer industry last year constitute another indicator that the big-machine approach to technical computing might be flawed. No doubt, a variety of factors contributed to the demise of Control Data Corporation's Engineering Technology Associates Systems subsidiary (St. Paul, Minnesota), Chopp (San Diego, California), Cydrome Inc. (San Jose, California), Evans & Sutherland's startup supercomputer division (Sunnyvale, California), Multiflow Computer (New Haven, Connecticut), and Scientific Computer Systems (San Diego, California). But in an enormously crowded market, being out of step with the spirit of the times might have had something to do with it.

With Cray Research, Inc., having spun off Cray Computer Corporation and with several other startups designing supercomputers, it's hard to get very worried about the U.S. position vis-à-vis whether enough is being done about competitiveness. Unfortunately, less and less is being spent on the underlying circuit technologies for the highest possible speeds. It's fairly easy to predict that of the half-dozen companies attempting to build new supers, there won't be more than three viable U.S. suppliers, whereas today we only have one.

From an economic standpoint, the U.S. is fortunate that the Japanese firms are expending large development resources that supercomputers require because these same engineers could, for example, be building consumer products, robots, and workstations that would have greater impact on the U.S. computer and telecommunications markets. It would be very smart for these Japanese manufacturers to fold up their expensive efforts and leave the small but symbolically visible supercomputer market to the U.S. Japan could then continue to improve its position in the larger consumer and industrial electronics, communication, and computing sectors.

Is the Supercomputer Industry Hastening Its Own Demise?

The supercomputer industry and its patrons appear to be doing many things that hasten an early demise. Fundamentally, the market for supercomputers is only a billion dollars, and the R&D going into supers is also on the order of a billion. This simply means too many companies are attempting to build too many noncompatible machines for too small a market. Much of the R&D is redundant, and other parts are misdirected.

The basic architecture of the "true" supercomputer was clearly defined as a nonscalable, vector multiprocessor. Unfortunately, the larger it is made to get the highest peak or advertising speed, the less cost effective it becomes for real workloads. The tradeoff inherent in making a high-performance computer that is judged on the number of GFLOPS it can calculate, based on such a design, seems to be counterproductive. The supercomputer has several inconsistencies (paradoxes) in its design and use:

1. In providing the highest number of MFLOPS by using multiprocessors with multiple pipe vector units to support one to 1.5 times the number of memory accesses as the peak arithmetic speed, memory latency is increased. *However,* to have a well-balanced, general-purpose supercomputer that executes scalar code well, the memory latency needs to be low.
2. In building machines with the greatest peak MFLOPS (i.e., the advertising speed), many processors are required, raising the computer's cost and lowering per-processor performance. *However,* supercomputers are rarely used in a parallel mode with all processors; thus, supers are being built at an inherent dis-economy of scale to increase the advertising speed.

3. Having many processors entails mastering parallelism beyond that obtainable through automatic parallelization/vectorization. *However*, supercomputer suppliers aren't changing their designs to enable scaleability or to use massive parallelism.
4. In providing more than worst-case design of three pipelines to memory, or 1.5 times as many mega-accesses per second as the machine has MFLOPS, the cost effectiveness of the design is reduced at least 50 per cent. *However*, to get high computation rates, block algorithms are used that ensure memory is not accessed. The average amount of computation a super delivers over a month is only five to 10 per cent of the peak, indicating the memory switch is idle most of the time.

In addition to these paradoxes, true supers are limited in the following ways:
1. Not enough is being done to train users or to make the super substantially easier to use. Network access needs to be much faster and more transparent. The X-Terminal server interface can potentially show the super to have a Macintosh-like interface. No companies provide this at present.
2. The true supercomputer design formula seems flawed. The lack of caches, paging, and scaleability make it doomed to chase the clock. For example, paradox 4 above indicates that a super could probably deliver two to four times more power by doubling the number of processors but without increasing the memory bandwidth or the cost.
3. Cray Research describes a massively parallel attached computer. Cray is already quite busy as it attempts to enter into the minisupercomputer market. Teaming with a startup such as Thinking Machines Corporation (which has received substantial government support) or MasPar for a massively parallel facility would provide a significantly higher return on limited brain power.
4. The U.S. has enough massively parallel companies and efforts. These have to be supported in the market and through use before they perish. Because these computers are inherently specialized (note the figure), support via continued free gifts to labs and universities is not realistic in terms of establishing a real marketplace.

A Smaller, Healthier Supercomputer Industry

Let's look at what fewer companies and better R&D focus might bring:
1. The CRAY Y-MP architecture is just fine. It provides the larger address space of the CRAY-2. The CRAY-3 line, based on a new architecture, will further sap the community of skilled systems-software and applications-builder resources. Similarly, Supercomputer Systems, Inc., (Steve Chen's startup) is most likely inventing a new architecture that requires new systems software and applications. Why have three

architectures for which people have to be trained so they can support operating systems and write applications?
2. Resources could be deployed on circuits and packaging to build GaAs or more aggressive ECL-based or even chilled CMOS designs instead of more supercomputer architectures and companies.
3. Companies that have much of the world's compiler expertise, such as Burton Smith's Tera Computer Company or SSI in San Diego, could help any of the current super companies. It's unlikely that any funding will come from within the U.S. to fund these endeavors once the government is forced into some fiscal responsibility and can no longer fund them. Similarly, even if these efforts get Far-East funding, it is unlikely they will succeed.
4. Government support could be more focused. Supporting the half-dozen companies by R&D and purchase orders just has to mean higher taxes that won't be repaid. On the other hand, continuing subsidies of the parallel machines is unrealistic in the 1990s if better architectures become available. A more realistic approach is to return to the policy of making the funds available to buy parallel machines, including ordinary supers, but to not force the purchase of particular machines.

Policy Issues

Supporting Circuit and Packaging Technology

There is an impression that the Japanese manufacturers provide access to their latest and fastest high-speed circuitry to build supercomputers. For example, CONVEX gets the parts from Fujitsu for making cost-effective minisupercomputers, but these parts are not components of fast-clock, highest-speed supercomputers. The CONVEX clock is two to 10 times slower when compared with a Cray, Fujitsu, Hitachi, or NEC mainframe or super.

High-speed circuitry and interconnect packaging that involves researchers, semiconductor companies, and computer manufacturers must be supported. This effort is needed to rebuild the high-speed circuitry infrastructure. We should develop mechanisms whereby high-speed-logic R&D is supported by those who need it. Without such circuitry, traditional vector supercomputers cannot be built. Here are some things that might be done:
1. Know where the country stands vis-à-vis circuitry and packaging. Neil Lincoln described two developments at NEC in 1990—the SX-3 is running benchmark programs at a 1.9-nanosecond clock; one

processor of an immersion-cooled GaAs supercomputer is operating at a 0.9-nanosecond clock.
2. Provide strong and appropriate support for the commercial suppliers who can and will deliver in terms of quality, performance, and cost. This infrastructure must be rebuilt to be competitive with Japanese suppliers. The Department of Defense's (DoD's) de facto industrial policy appears to support a small cadre of incompetent suppliers (e.g., Honeywell, McDonnell Douglas, Rockwell, Unisys, and Westinghouse) who have repeatedly demonstrated their inability to supply industrial-quality, cost-effective, high-performance semiconductors. The VHSIC program institutionalized the policy of using bucks to support the weak suppliers.
3. Build MOSIS facilities for the research and industrial community to use to explore all the high-speed technologies, including ECL, GaAs, and Josephson junctions. This would encourage a foundry structure to form that would support both the research community and manufacturers.
4. Make all DoD-funded semiconductor facilities available *and measured* via MOSIS. Eliminate and stop supporting the poor ones.
5. Support foundries aimed at custom high-speed parts that would improve density and clock speeds. DEC's Sam Fuller (a Session 13 presenter) described a custom, 150-watt ECL microprocessor that would operate at one nanosecond. Unfortunately, this research effort's only effect is likely to be a demonstration proof for competitors.
6. Build a strong packaging infrastructure for the research and startup communities to use, including gaining access to any industrial packages from Cray, DEC, IBM, and Microelectronics and Computer Technology Corporation.
7. Convene the supercomputer makers and companies who could provide high-speed circuitry and packaging. Ask them what's needed to provide high-performance circuits.

Supers and Security

For now, the supercomputer continues to be a protected species because of its use in defense. Also, like the Harley Davidson, it has become a token symbol of trade and competitiveness, as the Japanese manufacturers have begun to make computers with peak speeds equal to or greater than those from Cray Research or Cray Computer. No doubt, nearly all the functions supers perform for defense could be carried out more cheaply by using the alternative forms of computing described above.

Supers for Competitiveness

Large U.S. corporations are painstakingly slow, reluctant shoppers when it comes to big, traditional computers like supercomputers, mainframes, and even minisupercomputers. It took three years, for example, for a leading U.S. chemical company to decide to spring for a multimillion-dollar CRAY X-MP. And the entire U.S. automotive industry, which abounds in problems like crashworthiness studies that are ideal candidates for high-performance computers, has less supercomputer power than just one of its Japanese competitors. The super is right for the Japanese organization because a facility can be installed rapidly and in a top-down fashion.

U.S. corporations are less slow to adopt distributed computing by default. A small, creative, and productive part of the organization can and does purchase small machines to enhance their productivity. Thus, the one to 10 per cent of the U.S.-based organization that is responsible for 90 to 95 per cent of a corporation's output can and does benefit. For example, today, almost all electronic CAD is done using workstations, and the product gestation time is reduced for those companies who use these modern tools. A similar revolution in design awaits other engineering disciplines such as mechanical engineering and chemistry—but they must start.

The great gain for productivity is by visualization that comes through interactive supercomputing substitutes, including the personal supercomputers that will appear in the next few years. A supercomputer is likely to increase the corporate bureaucracy and at the same time inhibit users from buying the right computer—the very users who must produce the results!

By far, the greatest limitation in the use of supercomputing is training. The computer-science community, which, by default, takes on much of the training for computer programming, is not involved in supercomputing. Only now are departments becoming interested in the highly parallel computers that will form the basis of this next (fifth) generation of computing.

Conclusions

Alternative forms for supercomputing promise the brightest decade ever, with machines that have the ability to simulate and interact with many important physical phenomena.

Large, slowly evolving central systems will continue to be supplanted by low-cost, personal, interactive, and highly distributed computing because of cost, adequate performance, significantly better performance/cost, availability, user friendliness, and all the other factors that caused users of mainframes and minis to abandon the more centralized structures for personal computing. By the year 2000, we expect nearly all personal computers to have the capability of today's supercomputer. This will enable all users to simulate the immense and varied systems that are the basis of technical computing.

The evolution of the traditional supercomputer must change to a more massively parallel and scalable structure if it is to keep up with the peak performance of evolving new machines. By 1995, specialized, massively parallel computers capable of a TFLOPS (10^{12} floating-point operations per second) will be available to simulate a much wider range of physical phenomena.

Epilogue, June 1992

Clusters of 10 to 100 workstations are emerging as a high-performance parallel processing computer—the result of economic realities. For example, Lawrence Livermore National Laboratory estimates spending three times more on workstations that are 15 per cent utilized than it does on supercomputers. Supers cost a dollar per 500 FLOPS and workstations about a dollar per 5000 FLOPS. Thus, 25 times the power is available in their unused workstations as in supers. A distributed network of workstations won the Gordon Bell Prize for parallelism in 1992.

The ability to use workstation clusters is enabled by a number of environments such as Linda, the Parallel Virtual Machine, and Parasoft Corporation's Express. HPF (Fortran) is emerging as a powerful standard to allow higher-level use of multicomputers (e.g., Intel's Paragon, Thinking Machine's CM-5), and this could also be used for workstation clusters as standardization of interfaces and clusters takes place.

The only inhibitor to natural evolution is that government, in the form of the High Performance Computing and Communications (HPCC) Initiative, and especially the Defense Advanced Research Projects Agency, is attempting to "manage" the introduction of massive parallelism by attempting to select winning multicomputers from its development-funded companies. The HPCC Initiative is focusing on the peak TFLOPS at any price, and this may require an ultracomputer (i.e., a machine costing $50 to $250 million). Purchasing such a machine would be a mistake—waiting a single three-year generation will reduce prices by a least a factor of four.

In the past, the government, specifically the Department of Energy, played the role of a demanding but patient customer, but it never funded product development—followed by managing procurement to the research community. This misbehavior means that competitors are denied the significant market of leading-edge users. Furthermore, by eliminating competition, weak companies and poor computers emerge. There is simply no need to fund computer development. This money would best be applied to attempting to use the plethora of extant machines—and with a little luck, weed out the poor machines that absorb and waste resources.

Whether traditional supercomputers or massively parallel computers provide more computing, measured in FLOPS per month by 1995, is the object of a bet between the author and Danny Hillis of Thinking Machines. Unless government continues to tinker with the evolution of computers by massive funding for massive parallelism, I believe supers will continue as the main source of FLOPS in 1995.

9

Industrial Supercomputing

Panelists in this session discussed the use of supercomputers in several industrial settings. The session focused on cultural issues and problems, support issues, experiences, efficiency versus ease of use, technology transfer, impediments to broader use, encouragement of industrial use, and industrial grand challenges.

Session Chair

Kenneth W. Neves, Boeing Computer Services

Overview of Industrial Supercomputing

Kenneth W. Neves

Kenneth W. Neves is a Technical Fellow of the Boeing Company (in the discipline of scientific computing) and Manager of Research and Development Programs for the Technology Division of Boeing Computer Services. He holds a bachelor's degree from San Jose State University, San Jose, California, and master's and doctorate degrees in mathematics from Arizona State University, Tempe, Arizona. He developed and now manages the High-Speed Computing Program dedicated to exploration of scientific computing issues in distributed/parallel computing, visualization, and multidisciplinary analysis and design.

Abstract

This paper summarizes both the author's views as panelist and chair and the views of other panelists expressed during presentations and discussions in connection with the Industrial Supercomputing Session convened at the second Frontiers of Supercomputing conference. The other panel members were Patric Savage, Senior Research Fellow, Computer Science Department, Shell Development Company; Howard E. Simmons, Vice President and Senior Advisor, du Pont Company; Myron Ginsberg, Consultant Systems Engineer, EDS Advanced Computing Center, General Motors Corporation; and Robert Hermann, Vice President for Science and Technology, United Technologies Corporation. Included in these remarks is an overview of the basic issues related to high-performance computing needs of private-sector industrial users. Discussions that ensued following the presentations of individual panel members focused on supercomputing questions from an industrial perspective in areas that include cultural issues and problems, support issues, efficiency versus ease of use, technology transfer, impediments to broader use, encouraging industrial use, and industrial grand challenges.

Introduction

The supercomputer industry is a fragile industry. In 1983, when this conference first met, we were concerned with the challenge of international competition in this market sector. In recent times, the challenge to the economic health and well-being of this industry in the U.S. has not come from foreign competition but from technology improvements at the low end and confusion in the primary market, industry. The economic viability of the supercomputing industry will depend on the acceptance by private industrial users. Traditional industrial users of supercomputing have come to understand that using computing tools at the high end of the performance spectrum provides a competitive edge in product design quality. Yet, the question is no longer one of computational power alone. The resource of "supercomputing at the highest end" is a very visible expense on most corporate ledgers.

In 1983 a case could be made that in sheer price/performance, supercomputers were leaders and if used properly, could reduce corporate computing costs. Today, this argument is no longer true. Supercomputers are at the leading edge of price/performance, but there are alternatives equally competitive in the workstation arena and in the midrange of price and performance. The issue then, is not simply accounting but one of capability. With advanced computing capability, both in memory size and computational power, the opportunity exists to improve product designs (e.g., fuel-efficient airplanes), optimize performance (e.g., enhanced oil recovery), and shorten time from conceptual design to manufacture (e.g., find a likely minimal-energy state for a new compound or medicine). Even in industries where these principles are understood, there still are impediments to the acquisition and use of high-performance computing tools. In what follows we attempt to identify these issues and look at aspects of technology transfer and collaboration among governmental, academic, and industrial sectors that could improve the economic health of the industry and the competitiveness of companies that depend on technology in their product design and manufacturing processes.

Why Use Supercomputing at All?

Before we can analyze the inhibitors to the use of supercomputing, we must have a common understanding of the need for supercomputing. First, the term supercomputer has become overused to the point of being meaningless, as was indicated in remarks by several at this conference. By a supercomputer we mean the fastest, most capable machine available by the only measure that is meaningful—sustained performance on an industrial application of competitive importance to the industry in question. The issue is not which machine is best, at this point, but that some machines or group of machines are more capable than most others, and this class we shall refer to as "supercomputers." Today this class is viewed as large vector computers with a modest amount of parallelism, but the future promises to be more complicated, since one general type of architecture probably won't dominate the market.

In the aerospace industry, there are traditional workhorse applications, such as aerodynamics, structural analysis, electromagnetics, circuit design, and a few others. Most of these programs analyze a design. One creates a geometric description of a wing, for example, and then analyzes the flow over the wing. We know that today supercomputers cannot handle this problem in its full complexity of geometry and physics. We use simplifications in the model and solve approximations as best we can. Thus, the traditional drivers for more computational power still exist. Smaller problems can be run on workstations, but "new insights" can only be achieved with increased computing power.

A new generation of computational challenges face us as well (Neves and Kowalik 1989). We need not simply analysis programs but also design programs. Let's consider three examples of challenging computing processes. First, consider a program in which one could input a desired shock wave and an initial geometric configuration of a wing and have the optimal wing geometry calculated to most closely simulate the desired shock (or pressure profile). With this capability we could greatly reduce the wing design cycle time and improve product quality. In fact, we could reduce serious flutter problems early in the design and reduce risk of failure and fatigue in the finished product. This type of computation would have today's supercomputing applications as "inner loops" of a design system requiring much more computing power than available today. A second example comes from manufacturing. It is not unusual for a finalized design to be forwarded to manufacturing just to find out that the design cannot be manufactured "as designed" for some unanticipated reason. Manufacturability, reliability, and maintainability constraints need to be "designed into" the product, not discovered downstream. This design/build concept opens a whole new aspect of computation that we can't touch with today's computing equipment or approaches. Finally, consider the combination of many disciplines that today are separate elements in design. Aerodynamics, structural analyses, thermal effects, and control systems all could and should be combined in design evaluation and not considered separately. To solve these problems, computing power of greater capability is required; in fact, the more computing power, the "better" the product! It is not a question of being able to use a workstation to solve these problems. The question is, can a corporation afford to allow products to be designed on workstations (with yesterday's techniques) while competitors are solving for optimal designs with supercomputers?

Given the rich demand for computational power to advance science and engineering research, design, and analysis as described above, it would seem that there would be no end to the rate at which supercomputers could be sold. Indeed, technically there is no end to the appetite for more power, but in reality each new quantum jump in computational power at a given location (user community) will satisfy needs for some amount of time before a new machine can be justified. The strength in the supercomputer market in the 1980s came from two sources: existing customers and "new" industries. Petrochemical industries, closely followed by the aerospace industry, were the early recruits. These industries seem to establish a direct connection between profit and/or productivity and computing power. Most companies in these industries not only bought machines but upgraded to next-generation machines within about five years. This alone established an upswing in the supercomputing market when matched by the already strong government laboratory market from whence supercomputers sprang. Industry by industry, market penetration was made by companies like Cray Research, Inc. In 1983 the Japanese entered the market, and several of their companies did well outside the U.S. New market industries worldwide included weather prediction, automobiles, chemicals, pharmaceuticals, academic research institutions (state- and NSF-supported), and biological and environmental sciences. The rapid addition of "new" markets by industries created a phenomenal growth rate.

In 1989 the pace of sales slackened at the high end. The reasons are complex and varied, partly because of the options for users with "less than supercomputer problems" to find cost-effective alternatives; but the biggest impact, in my opinion, is the inability to create new industry markets. Most of the main technically oriented industries are already involved in supercomputing, and the pace of sales has slowed to that of upgrades to support the traditional analysis computations alluded to above. This is critical to the success of these companies but has definitely slowed the rate of sales enjoyed in the 1980s. This might seem like a bleak picture if it weren't for one thing: *as important as these traditional applications are, they are but the tip of the iceberg of scientific computing opportunities in industry*. In fact, at Boeing well over a billion dollars are invested in computing hardware. Supercomputers have made a very small "dent" in this computing budget. One might say that even though supercomputers exist at almost 100 per cent penetration *by company* in aerospace, *within* companies, this penetration is less than five per cent.

Certainly supercomputers are not fit for all computing applications in large manufacturing companies. However, the acceptance of any computing tool, or research tool such as a wind tunnel, is a function of its contribution to the "bottom line." The bottom line is profit margin and market share. To gain market share you must have the "best product at the least cost." Supercomputing is often associated with design and hence, product quality. The new applications of concurrent engineering (multidisciplinary analysis) and optimal design (described above) will achieve cost reduction by ensuring that manufacturability, reliability, and maintainability are included in the design. This story needs to be technically developed and understood by both scientists and management. The real *untapped* market, however, lies in bringing high-end computation to bear on manufacturing problems ignored so far by both technologists and management in private industry.

For example, recently at Boeing we established a Computational Modeling Initiative to discover new ways in which the bottom line can be helped by computing technology. In a recent pilot study, we examined the rivet-forming process. Riveting is a critical part of airplane manufacturing. A good rivet is needed if fatigue and corrosion are to be minimized. Little is known about this process other than experimental data. By simulating the riveting process and animating it for slow-motion replay, we have utilized computing to simulate and display what cannot be seen experimentally. Improved rivet design to reduce strain during the riveting has resulted in immediate payoff during manufacturing and greatly reduced maintenance cost over the life of the plane. Note that this contributes very directly to the bottom line and is an easily understood contribution. We feel that these types of applications (which in this case required a supercomputer to handle the complex structural analysis simulation) could fill many supercomputers productively once the applications are found and implemented. This latent market for computation within the manufacturing sectors of existing supercomputer industries is potentially bigger than supercomputing use today. The list of opportunities is enormous: robotics simulation and design, factory scheduling, statistical tolerance analysis, electronic mockup (of parts, assemblies, products, and tooling), discrete simulation of assembly, spares inventory (just-in-time analysis of large, complex manufacturing systems), and a host of others.

We have identified three critical drivers for a successful supercomputing market that all are critical for U.S. industrial competitiveness: 1) traditional and more refined analysis; 2) design optimization, multidisplinary analysis, and concurrent engineering (design/build); and 3) new applications of computation to manufacturing process productivity.

The opportunities in item 3 above are so varied, even at a large company like Boeing, it is hard to be explicit. In fact, the situation requires those involved in the processes to define such opportunities. In many cases, the use of computation is traditionally foreign to the manufacturing process, which is often a "build and test" methodology, and this makes the discovery of computational opportunities difficult. What is clear, however, is that supercomputing opportunities exist (i.e., a significant contribution can be made to increased profit, market share, or quality of products through supercomputing). It is worthwhile to point out broadly where supercomputing has missed its opportunities in most industries, but certainly in the aerospace sector:

- manufacturing—e.g., rivet-forming simulation, composite material properties;
- CAD/CAM—e.g., electronic mockup, virtual reality, interference modeling, animated inspection of assembled parts;
- common product data storage—e.g., geometric-model to grid-model translation; and
- grand-challenge problems—e.g., concurrent engineering, data transfer: IGES, PDES, CALS.

In each area above, supercomputing has a role. That role is often not central to the area but critical in improving the process. For example, supercomputers today are not very good database machines, yet much of the engineering data stored in, say, the definition of an airplane is required for downstream analysis in which supercomputing can play a role. Because supercomputers are not easily interfaced to corporate data farms, much of that analysis is often done on slower equipment, to the detriment of cost and productivity.

With this as a basis, how can there be any softness in the supercomputer market? Clearly, supercomputers are fundamental to competitiveness, or are they?

Impediments to Industrial Use of Supercomputers

Supercomputers have been used to great competitive advantage throughout many industries (Erisman and Neves 1987). The road to changing a company from one that merely uses computers on routine tasks to one that employs the latest, most powerful machines as research and industrial tools to improve profit is a difficult one indeed. The barriers include technical, financial, and cultural issues that are often complex; and even more consternating, once addressed, they can often reappear over time. The solution to these issues requires both management and technologists in a cooperative effort. We begin with what are probably the most difficult issues—cultural and financial barriers.

The cultural barriers that prevent supercomputing from taking its rightful place in the computing venue abound. Topping the list is management understanding of supercomputing's impact on the bottom line. Management education in this area is sorely needed, as most managers who have wrestled with these issues attest. Dr. Hermann, one of the panelists in this session, suggested that a successful "sell" to management must include a financial-benefits story that very few people can develop. To tell this story one must be a technologist who understands the specific technical contributions computing can have on both a company's product/processes and its corporate competitive and profit goals. Of the few technologists who have this type of overview, how many would take on what could be a two-year "sell" to management? History can attest that almost every successful supercomputer placement in industry, government, or academia has rested on the shoulders of a handful of zealots or champions with that rare vision. This is often true of expensive research-tool investments, but for computing it is more difficult because of the relative infancy of the industry. Most upper-level managers have not personally experienced the effective use of research computing. When they came up through the "ranks," computing, if it existed at all, was little more than a glorified engineering calculator (slide rule). Managers in the aerospace industry fully understand the purpose of a $100 million investment in a wind tunnel, but until only in the last few years did any of them have to grapple with a $20 million investment in a "numerical" wind tunnel. Continuing with this last aerospace example, how did the culture change? An indispensable ally in the aerospace industry's education process has been the path-finding role of NASA, in both technology and collaboration with industry. We will explore government-industry collaboration further in the next section.

Cultural issues are not all management in nature. As an example, consider the increasing need for collaborative (design-build) work and multidisciplinary analysis. In these areas, supercomputing can be the most important tool in creating an environment that allows tremendous impact on the bottom line, as described above. However, quite often the disciplines that need to cooperate are represented by different (often large) organizations. Nontechnical impediments associated with change of any kind arise, such as domain protection, fear of loss of control, and career insecurities owing to unfamiliarity with computing technology. Often these concerns are never stated but exist at a subliminal level. In addition, organizations handle computing differently, often on disparate systems with incompatible geometric description models, and the technical barriers from years of cultural separation are very real indeed.

Financial barriers can be the most frustrating of all. Supercomputers, almost as part of their definition, are expensive. They cost from $10 to $30 million and thus are usually purchased at the corporate level. The expense of this kind of acquisition is often distributed by some financial mechanism that assigns the cost to those who use it. Therein lies the problem. To most users, their desk, pencils, paper, phone, desk-top computer, etc., are simply there. For example, there is no apparent charge to them, their project, or their management when they pick up the phone. Department-level minicomputers, while a visible expense, are controlled at a local level, and the expenses are well understood and accepted before purchase. Shared corporate resources, however, look completely different. They often cost real project dollars. To purchase X dollars of computer time from the company central resource costs a project X dollars of labor. This tradeoff applies pressure to use the least amount of central computing resources possible. This is like asking an astronomer to look through his telescope only when absolutely necessary for the shortest time possible while hoping he discovers a new and distant galaxy.

This same problem has another impact that is more subtle. Supercomputers like the Cray Research machines often involve multiple CPUs. Most charging formulas involve CPU time as a parameter. Consequently, if one uses a supercomputer with the mind set of keeping costs down, one would likely use only one CPU at a time. After all, a good technologist knows that if he uses eight CPUs, Amdahl's law will probably only let him get the "bang" of six or seven and then only if he is clever. What is the result? A corporation buys an eight-CPU supercomputer to finally tackle corporate grand-challenge problems, and the users immediately bring only the power of one CPU to bear on their problems for financial reasons. Well, one-eighth of a supercomputer is not a supercomputer, and one might opt for a lesser technological solution. In fact, this argument is often heard in industry today by single-CPU users grappling with financial barriers. This is particularly frustrating since the cost-reduction analysis is often well understood, and the loss in product design quality by solving problems on less competitive equipment is often not even identified!

The technological barriers are no less challenging. In fact, one should point out that the financial billing question relative to parallel processing will probably require a technological assist from vendors in their hardware and operating systems. To manage the computing resource properly, accounting "hooks" in a parallel environment need be more sophisticated. Providing the proper incentives to use parallel equipment when the overhead of parallelization is a real factor is not a simple matter. These are issues the vendors can no longer leave to the user but must become a partner in solving.

Supercomputers in industry have not really "engaged" the corporate enterprise computing scene. Computers have had a long history in most companies and are an integral part of daily processes in billing, CAD/CAM, data storage, scheduling, etc. Supercomputers have been brought into companies by a select group and for a specific need, usually in design analysis. These systems, like these organizations, are often placed "over there"—in a corner, an ivory tower, another building, another campus, or any place where they don't get in the way. Consequently, most of the life stream of the corporation, its product data, is out of reach, often electronically and culturally from the high-performance computing complex. The opportunities for supercomputing alluded to in the previous section suggest that supercomputers must be integrated into the corporate computing system. All contact with the central computing network begins at the workstation. From that point a supercomputer must be as available as any other computing resource. To accomplish this, a number of technical barriers must be overcome, such as
- transparent use,
- software-rich environment,
- visualization of results, and
- access to data.

If one delves into these broad and overlapping categories, a number of issues arise. Network topology, distributed computing strategy, and standards for data storage and transport immediately spring to mind. Anyone who has looked at any of these issues knows the solutions require management and political savvy, as well as technical solutions. At a deeper level of concern are the issues of supercomputer behavior. On the one hand, when a large analysis application is to be run, the supercomputer must bring as much of its resources to bear on the computation as possible (otherwise it is not a supercomputer). On the other hand, if it is to be an equal partner on a network, it must be responsive to the interactive user. These are conflicting goals. Perhaps supercomputers on a network need a network front end, for example, to be both responsive and powerful. Who decides this issue? The solution to this conflict is not solely the responsibility of the vendor. Yet, left unresolved, this issue alone could "kill" supercomputer usage in any industrial environment.

As supercomputer architectures become increasingly more complex, the ability to transfer existing software to them becomes a pacing issue. If existing programs do not run at all or do not run fast on new computers, these machines simply will not be purchased. This problem, of course, is a classic problem of high-end computing. Vectorization and now parallelization are processes that we know we must contend with. The issue of algorithms and the like is well understood. There is a cultural issue for technologists, however. The need to be 100 per cent efficient on a parallel machine lessens as the degree of parallelism grows. For example, if we have two $20 million computers, and one runs a problem at 90 per cent efficiency at a sustained rate of four GFLOPS (billion floating-point operations per second), and the other runs a problem at 20 per cent efficiency at 40 GFLOPS, which would you choose? I would choose the one that got the job done the cheapest! (That can not be determined from the data given! For example, at 40 GFLOPS, the second computer might be using an algorithm that requires 100 times more floating-point operations to reach the same answer. Let us assume that this is not the case and that both computers are actually using the same algorithm.) The second computer might be favored. It probably is a computer that uses many parallel CPUs. How do we charge for the computer time? How do we account for the apparently wasted cycles? I ask these two questions to emphasize that, at all times, the corporate resource must be "accounted" for with well-understood accounting practices that are consistent with corporate and government regulations!

We have paid short shrift to technological issues, owing to time and space. It is hoped that one point has become clear—that the cultural, financial, and technical issues are quite intertwined. Their resolution and the acceptance of high-end computing tools in industry will require collaboration and technology transfer among all sectors—government, industry, and academia.

Technology Transfer and Collaboration

Pending before Congress are several bills concerning tremendous potential advances in the infrastructure that supports high-performance computing. We at this meeting have a great deal of interest in cooperative efforts to further the cause of high-performance computing—to insure the technological competitiveness of our companies, our research institutions, and, indeed, our nation. To achieve these goals we must learn to work together to share fruitfully technological advances. The definition of infrastructure is perhaps a good starting point for discussing technology transfer challenges. The electronic thesaurus offers the following substitutes for infrastructure:
- chassis, framework, skeleton;
- complex, maze, network, organization, system;
- base, seat; and
- cadre, center, core, nucleus.

The legislation pending has all these characteristics. In terms of a national network that connects high-performance computing systems and large data repositories of research importance, the challenge goes well beyond simply providing connections and hardware. We want a national network that is not a maze but an organized, systematized framework to advance technology. Research support is only part of the goal, for research must be transferred to the bottom line in a sense similar to that discussed in previous sections. No single part of the infrastructure can be singled out, nor left out, for the result to be truly effective. We have spoken often in this forum of cooperative efforts among government, academia, and industry. I would like to be more explicit. If we take the three sectors one level of differentiation further, we have Figure 1.

Just as supercomputers must embrace the enterprise-wide computing establishment within large companies, the national initiatives in high-performance computing must embrace the end-user sector of industry, as well. The payoff is a more productive economy. We need a national network, just like we needed a national highway system, an analogy often used by Senator Albert Gore. Carrying this further, if we had restricted the highway system to any particular sector, we would not have seen the birth of the trucking industry, the hotel and tourism industries, and so on. Much is to be gained by cooperative efforts, and many benefits cannot be predicted in advance. Let us examine two

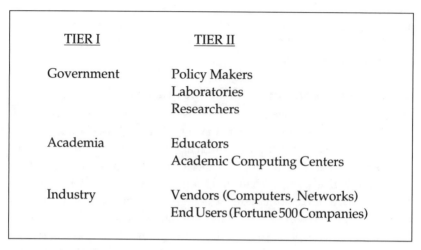

Figure 1. Technological sectors.

examples of technology transfer that came about through an investment in infrastructure, one by government and another by industry.

First is an example provided by Dr. Riaz Abdulla, from Eli Lilly Research Laboratories, in a private communication. He writes:

> For your information, supercomputing, and particularly network supercomputing at Eli Lilly became successful owing to a mutually supportive research and management position on the matter. Both the grass-roots movement here, as well as enlightened management committed to providing the best possible tools to the research staff made the enhancement of our research computer network among the best. . . . We are well on the way to establishing a network of distributed processors directly linked to the supercomputing system via high-speed links modeled after the National Center for Supercomputing Applications [at the University of Illinois, one of the NSF-funded supercomputer centers] and the vision of Professor Larry Smarr. Without the model of the NCSA, its staff of scientists, consultants, engineers and software and visualization experts, Lilly's present success in supercomputing would have been impossible.

Clearly, the government investment in supercomputing for the academic world paid off for Eli Lilly. While this was not an original goal of the NSF initiatives, it clearly has become part of the national infrastructure that NSF has become a part of in supercomputing.

In the second example, technology is transferred from the private sector to the academic and government sectors. Boeing Computer Services has been involved in supercomputing for almost two decades, from before the term was coined. We purchased Serial No. 2 of Control Data Corporation's CDC 6600, for example—a supercomputer in its day. As such, we owned and operated a national supercomputer time sales service when the NSF Advanced Scientific Computing Program was launched. We responded to a request for proposals to provide initial supercomputer time in Phase I of this program. Under contract with NSF we were able to give immediate access to supercomputing cycles. We formed a team to train over 150 research users in access to our system. This was done on location at 87 universities across the country. We provided three in-depth Supercomputing Institutes, the model of which was emulated by the centers themselves after they were established. In subsequent years we helped form, and are a member of, the Northwest Academic Computing Consortium (NWACC), along with 11 northwest universities. In collaboration we have secured NSF funding to create NWNet, the northwest regional NSF network. Boeing designed and initially operated this network but has since turned the operation over to NWACC and the University of Washington in Seattle. In other business activities, Boeing designed, installed, operates, and trains users of supercomputer centers in academia (the University of Alabama system) and government laboratories (NASA and the Department of Energy). Indeed, technology transfer is often a two-way street. The private sector is taking some very aggressive steps to advanced technology in our research laboratories, as well. (For example, see the paper following in this session, by Pat Savage, Shell Development Company, discussing Shell's leadership to the community in parallel computing tools and storage systems.)

Conclusion

We are delighted to see that much of the legislation before Congress recognizes the importance of technology transfer and collaboration among the Tier I entities of Figure 1. We are confident that all elements of Tier II will be included, but we exhort all concerned that this collaboration be well orchestrated and not left to serendipity. Transferring technology among organizations or Tier I sectors is the most difficult challenge we have, and our approach must be aggressive. The challenges of the supercomputing industry are no less difficult. They too can only be overcome by cooperation. These challenges are both technical and cultural and present an awesome management responsibility.

References

A. Erisman and K. W. Neves, "Advanced Computing for Manufacturing," *Scientific American* **257** (4), 162–169 (1987).

K. W. Neves and J. S. Kowalik, "Supercomputing: Key Issues and Challenges," in *NATO Advanced Research Workshop on Supercomputing, NATO ASI Series F, Vol. 62*, J. S. Kowalik, Ed., Springer-Verlag, New York (1989).

Shell Oil Supercomputing

Patric Savage

Patric Savage is a Senior Research Fellow in the Computer Science Department of Shell Development Company. He obtained a B.A. degree in mathematics from Rice University, Houston, in 1952 and began his career in computing in 1955, when he left graduate school to become Manager of Computer Programming at Hughes Tool Company. There, he led Hughes's pioneering efforts in the use of computers for inventory management, production control, and shop scheduling, using IBM-650, 305, and 1410 computers. Following a brief stint in the aerospace industry, he joined IBM in Los Angeles, where he designed a large grocery billing system and took part in a comprehensive study of hospital information systems.

Mr. Savage began his career with Shell in 1965 in seismic processing. This is an area that has grown into one of the world's largest nonmilitary computing endeavors, and Mr. Savage has remained active in this field since then as a computer scientist, consultant, and advisor. Since 1980 he has been very active in parallel and distributed computing systems R&D. For the past year he has been regularly attending the HIPPI and Fibre Channel Standards Working Group meetings. Recently he helped establish an Institute of Electrical and Electronics Engineers (IEEE) Standards Project that will eventually lead to a Storage System Standards protocol.

Mr. Savage is a member of the Computer Society, the IEEE, and the IEEE Mass Storage Technical Committee, which he chaired from 1986 through 1988 and for which he now chairs the Standards Subcommittee. He also chairs the Storage and Archiving Standards Subcommittee for the Society of Exploration Geophysicists and holds a life membership in Sigma Xi, the Society for Scientific Research.

I will give you a quick overview and the history of supercomputing at Shell Oil Company and then discuss our recent past in parallel computing. I will also discuss our I/O and mass-storage facility and go into what we are now doing and planning to do in parallel computing in our problem-solving environment that is under development.

Shell's involvement in high-performance computing dates from about 1963. When I arrived at Shell in 1965, seismic processing represented 95 per cent of all the scientific computing that was done in the entire company. Since then there has a been steady increase in general scientific computing at Shell. We now do a great many more reservoir simulations, and we are using codes like NASTRAN for offshore platform designs. We are also heavily into chemical engineering modeling and such.

Seismic processing has always required array processors to speed it up. So from the very beginning, we have had powerful array processors at all times. Before 1986 we used exclusively UNIVAC systems with an array processing system whose design I orchestrated. That was a machine capable of 120 million floating-point operations per second (MFLOPS) and was not a specialized device. It was a very flexible, completely programmable special processor on the UNIVAC system. We "maxed out" at 11 of those in operation. At one time we had a swing count of 13, and, for the three weeks that it lasted, we had more MFLOPS on our floor than Los Alamos National Laboratory.

In about 1986, our reservoir-simulation people were spending so much money renting time on Cray Research, Inc., machines that it was decided we could half-fund a Cray of our own. Other groups at Shell were willing to fund the other half. So that is how we got into using Cray machines. We were able and fortunate to acquire complete seismic programming codes externally and thus, we were able to jump immediately onto the Crays. Otherwise, we would have had an almost impossible conversion problem.

We began an exploratory research program in parallel computing about 1982. We formed an interdisciplinary team of seven people: three geophysicists, who were skilled at geophysical programming, and four Ph.D. computer scientists. Our goal was to enable us to make a truly giant leap ahead—to be able to develop applications that were hitherto totally unthinkable. We have not completely abandoned that goal, although we have pulled in our horns a good bit. We acquired an nCUBE 1, a 512-node research vehicle built by nCUBE Corporation, and worked with it. That was one of the very first nCUBEs sold to industry. In the process, we learned a great deal about how to make things work on a distributed-memory parallel computer.

In early 1989, we installed a single application on our nCUBE 1 at our computer center on a 256-node machine. It actually "blows away" a CRAY X-MP CPU on that same application. But the fact that it was convincingly cost effective to management is the thing that really has spurred further growth in our parallel computing effort.

To deviate somewhat, I will now discuss our I/O and mass-storage system. (The mass-storage system that many of you may be familiar with was designed and developed at Shell in conjunction with MASSTOR Corporation.) We have what we call a virtual-tape system. The tapes are in automated libraries. We do about 8000 mounts a day. We import 2000 reels and export another 2000 every day into that system. The concept is, if a program touches a tape, it has to swallow it all. So we stage entire tapes and destage entire tapes at a time. No program actually owns a tape drive; it only is able to own a virtual tape drive. We were able to have something like 27 tape drives in our system, and we were able to be dynamically executing something like 350 virtual tape units.

The records were delivered on demand from the computers over a Network Systems Hyperchannel. This system has been phased out, now that we have released all of the UNIVACs, and today our Crays access shared tape drives that are on six automated cartridge libraries. We will have 64 tape drives on those, and our Cray systems will own 32 of those tape drives. They will stage tapes on local disks. Their policy will be the same: if you touch a tape, you have to swallow the whole tape. You either have to stage it on your own local disk immediately, as fast as you can read it off of the tape, or else you have to consume it that fast.

This system was obviously limited by the number of ports that we can have. Three Crays severely strain the number of ports that you can have, which would be something like eight. Our near-term goal is to develop a tape data server that will be accessed via a switched high-performance parallel interface (HIPPI) and do our staging onto a striped-disk server that would also be accessed over a switched HIPPI. One of the problems that we see with striped-disk servers is that there is a tremendous disparity between the bandwidth of a striped-disk system and the current 3480 tape. We now suddenly come up with striped disks that will run at rates like 80 to 160 megabytes per second. You cannot handle a striped disk and do any kind of important staging or destaging using slow tape. I am working with George Michael, of Lawrence Livermore National Laboratory, on this problem. We have discussed use of striped tape that will be operating at rates like 100 megabytes per second. We believe that a prototype can be demonstrated in less than two years at a low cost.

Going back to parallel computing, I will share some observations on our nCUBE 1 experience. First, we absolutely couldn't go on very far without a whole lot more node memory, and the nCUBE 2 solved that problem for us. We absolutely have to have high-bandwidth external I/O. The reason that we were able to run only that one application was because that was a number-crunching application that was satisfied by about 100 kilobytes per second, input and output. So it was a number-cruncher. We were spoon-feeding it with data.

We have discovered that the programmers are very good at designing parallel programs. They do not need a piece of software that searches over the whole program and automatically parallelizes it. We think that the programmer should develop the strategy. However, we have found that programmer errors in parallel programs are devastating because they create some of the most obscure bugs that have ever been seen in the world of computing.

Because we felt that a parallel programming environment is essential, we enlisted the aid of Pacific-Sierra Research (PSR). They had a "nifty" product that many of you are familiar with, called FORGE. It was still in late development when we contacted them. We interested them in developing a product that they chose to call MIMDizer. It is a programmer's workbench for both kinds of parallel computers: those with distributed memories and those with shared memories. We have adopted this. The first two target machines are the Intel system and the nCUBE 2.

The thing that MIMDizer required in its development was that the target machine must be described by a set of parameters so that new target machines can be added easily. Then the analysis of your program will give a view of how the existing copy of your program will run on a given target machine and will urge you to make certain changes in it to make it run more effectively on a different target machine. I have suggested to PSR that they should develop a SIMDizer that would be applicable to other architectures, such as the Thinking Machines Corporation CM-2.

I have been seriously urging PSR to develop what I would call a PARTITIONizer. I would see a PARTITIONizer as something that would help a programmer tear a program apart and break it up so that it can be run in a distributed heterogeneous computing environment. It would be a powerful tool and a powerful adjunct to the whole package.

Our immediate plans for the nCUBE 2 are in the first quarter of 1991, when we will install a 128-node nCUBE 2, in production. For that, we will have five or six applications that will free up a lot of Cray time to run other applications that today are highly limited by lack of Cray resources.

I now want to talk about the problem-solving environment because I think there is a message here that you all should really listen to. This system was designed around 1980. Three of us in our computer science research department worked on these concepts. It actually was funded in 1986, and we will finish the system in 1992. Basically, it consists of a library of high-level primitive operations. Actually, many of these would be problem domain primitives.

The user graphically builds what we call a "flownet," or an acyclic graph. It can branch out anywhere that it wants. The user interface will not allow an illegal flownet. Every input and every output is typed and is required to attach correctly.

Every operation in the flownet is inherently parallel. Typical jobs have hundreds of operations. We know of jobs that will have thousands of operations. Some of the jobs will be bigger than you can actually run in a single machine, so we will have a facility for cutting up a superjob into real jobs that can actually be run. There will be lots of parallelism available.

We have an Ada implementation—every operation is an Ada task—and we have Fortran compute kernels. At present, until we get good vectorized compilers for Ada, we will remain with Fortran and C compute kernels. That gives us an effectiveness on the 20 per cent of the operations that really are squeaking wheels. We have run this thing on a CONVEX Computer Corporation Ada system. CONVEX, right now, is the only company we found to have a true multiprocessing Ada system. That is, you can actually run multiple processors on the CONVEX Ada system, and you will get true multiprocessing. We got linear speedup when we ran on the CONVEX system, so we know that this thing is going to work. We ran it on a four-processor CONVEX system, and it ran almost four times as fast—something like 3.96 times as fast—as it did with a single processor.

This system is designed to run on workstations and Crays and everything else in between. There has been a very recent announcement of an Ada compiler for the nCUBE 2, which is cheering to us because we did not know how we were going to port this thing to the nCUBE 2. Of course, I still do not know how we will port to any other parallel environment unless they develop some kind of an Ada capability.

Government's High Performance Computing Initiative Interface with Industry

Howard E. Simmons

Howard E. Simmons is Vice President and Senior Science Advisor in E. I. du Pont de Nemours and Company, where for the past 12 years he headed corporate research. He has a bachelor's degree from MIT in chemistry and a Ph.D. from the same institution in physical organic chemistry. He was elected to the National Academy of Sciences in 1975 and has been a Visiting Professor at Harvard University and the University of Chicago. His research interests have ranged widely from synthetic and mechanistic chemistry to quantum chemistry. Most recently, he has coauthored a book on mathematical topological methods in chemistry with R. E. Merrifield.

It is a pleasure for me to participate in this conference and share with you my perspectives on supercomputing in the industrial research, development, and engineering environments.

I will talk to you a little bit from the perspective of an industry that has not only come late to supercomputing but also to computing in general from the science standpoint. Computing from the engineering standpoint, I think, came into the chemical industry very early, and E. I. du Pont de Nemours and Company (du Pont) was one of the leaders.

Use of supercomputers at du Pont is somewhat different from the uses we see occurring in the national laboratories and academia. The differences are created to a large extent by the cultures in which we operate and the institutional needs we serve. In that context, there are three topics I will cover briefly. The first is "culture." The second is supercomputer applications software. The third is the need for interfaces to computer applications running on PCs, workstations, minicomputers, and supercomputers of differing types—for example, massively parallels.

As I mentioned in regard to culture, the industrial research, development, and engineering culture differs from that of the national laboratories and academia. I think this is because our objective is the discovery, development, manufacture, and sale of products that meet customer needs and at the same time make a profit for the company. This business orientation causes us to narrow the scope of our work and focus our efforts on solving problems of real business value in those areas in which we have chosen to operate. Here I am speaking about the bulk of industrial research, although work at AT&T's Bell Laboratories, du Pont's Central Research, and many other corporate laboratories, for instance, do follow more closely the academic pattern.

A second cultural difference is that most of the R&D effort, and consequently our staffing, has been directed toward traditional experimental sciences. We have rarely, in the past, been faced with problems that could be analyzed and solved only through computational methods. Hence, our computational science "tradition" is neither as long-standing nor as diverse as found in the other sectors or in other industries.

A significant limitation and hindrance to broad industrial supercomputer use is the vision of what is possible by the scientists and engineers who are solving problems. So long as they are satisfied with what they are doing and the way they are doing it, there is not much driving force to solve more fundamental, bigger, or more complex problems. Thus, in our industry, a lot of education is needed, not only for the supercomputer area but also in all advanced computing. We believe we are making progress in encouraging a broader world view within our technical and managerial ranks. We have been having a lot of in-house symposia, particularly in supercomputing, with the help of Cray Research, Inc. We invited not just potential users of the company but also middle managers, who are key people to convince on the possible needs that their people will have in more advanced computing.

Our company has a policy of paying for resources used. This user-based billing practice causes some difficulty for users in justifying and budgeting for supercomputer use, particularly in the middle of a fiscal cycle. A typical example is that scientists and engineers at a remote plant site—in Texas, for example—may see uses for our Cray back in Wilmington, Delaware. They have a lot of trouble convincing their middle management that this is any more than a new toy or a new gimmick. So we have done everything that we can, including forming SWAT teams that go out and try to talk to managers throughout the corporation in the research area and give them some sort of a reasonable perspective of what the corporation's total advanced computing capabilities are.

The cultural differences between the national laboratories, universities, and industry are certainly many, but they should not preclude mutually beneficial interactions. The diversity of backgrounds and differences in research objectives can and should be complementary if we understand each other's needs.

The second major topic I will discuss is software, specifically applications software. We presume for the present time that operating systems, communications software, and the like will be largely provided by vendors, at least certainly in our industry. There are several ways to look at the applications software issue. The simplest is to describe our strategies for acquiring needed analysis capabilities involving large "codes." In priority, the questions we need to ask are as follows:

- Has someone else already developed the software to solve the problem or class of problems of interest to us? If the answer is yes, then we need to take the appropriate steps to acquire the software. In general, acquisition produces results faster and at lower cost than developing our own programs.
- Is there a consortium or partnership that exists or might be put together to develop the needed software tools? If so, we should seriously consider buying in. This type of partnering is not without some pitfalls, but it is one that appeals to us.
- Do we have the basic expertise and tools to develop our own special-purpose programs in advanced computing? The answer here is almost always yes, but rarely is it a better business proposition than the first two options. This alternative is taken only when there is no other viable option.

To illustrate what's been happening, our engineers have used computers for problem solving since the late 1950s. Since we were early starters, we developed our own programs and our own computer expertise. Today, commercial programs are replacing many of our "home-grown" codes. We can no longer economically justify the resources required to develop and maintain in-house versions of generic software products. Our engineers must concentrate on applying the available computational tools faster and at lower life-cycle costs than our competition.

Our applications in the basic sciences came later and continue to undergo strong growth. Many of our scientists write their own code for their original work, but here, too, we face a growing need for purchased software, particularly in the molecular-dynamics area.

Applications software is an ever-present need for us. It needs to be reasonably priced, reliable, robust, and have good documentation. In addition, high-quality training and support should be readily available. As we look forward to parallel computers, the severity of the need for good applications software will only increase, since the old and proven software developed for serial machines is becoming increasingly inadequate for us.

Finally, integration of supercomputers into the infrastructure or fabric of our R&D and engineering processes is not easy. I believe it to be one of the primary causes for the slow rate of penetration in their use.

For the sake of argument, assume that we have an organization that believes in and supports computational science, that we have capable scientists and engineers who can use the tools effectively, that the computers are available at a reasonable cost, and that the needed software tools are available. All of these are necessary conditions, but they are not sufficient.

Supercomputers historically have not fit easily into the established computer-based problem-solving environments, which include personal computers, workstations, and minicomputers. In this context, the supercomputer is best viewed as a compute engine that should be almost transparent to the user. To make the supercomputer transparent requires interfaces to these other computing platforms and the applications running on them. Development of interfaces is an imperative if we are going to make substantial inroads into the existing base of scientific and engineering applications. The current trend toward UNIX-based operating systems greatly facilitates this development. However, industry tends to have substantial investments in computer systems running proprietary operating systems (e.g., IBM/MVS, VAX/VMS, etc.).

Three brief examples of our supercomputer applications might help to illustrate the sort of things we are doing and a little bit about our needs. In the first example, we design steel structures for our manufacturing plants using computer-aided design tools on our Digital Equipment Corporation VAX computers. Central to this design process is analysis of the structure using the NASTRAN finite-element analysis program. This piece of the design process is, of course, very time consuming and compute intensive. To break that bottleneck, we have interfaced the design programs to our CRAY X-MP, where we do the structural analyses. It is faster by a factor of 20 to 40 in our hands, a lot lower in cost, and it permits us to do a better design job. We can do a better job with greater compute power so that we can do seismic load analyses even when the structures are not in high-risk areas. This, simply for economic reasons, we did not always do in the past. This capability and vision lead to new approaches to some old problems.

The second example is—as we explore new compounds for application and new products—part of the discovery process that requires the determination of bulk physical properties. In selected cases we are computing these in order to expedite the design and development of commercial manufacturing facilities. We find high value in areas ranging from drug design to structural property relations in polymers. A good example is the computation of basic thermodynamic properties of such small halocarbons as Freon chlorofluorocarbon replacements. This effort is critical to our future and the viability of some of our businesses. It is very interesting to note that these are ab initio quantum mechanical calculations that are being used directly in design of both products and plants. So in this case we have had no problem in convincing the upper management in one of the most traditional businesses that we have of the great value of supercomputers because this is necessary to get some of these jobs done. We gain a substantial competitive advantage by being able to develop such data via computational methodologies and not just experimentally. Experimental determination of these properties can take much longer and cost more.

A third example, atmospheric chemistry modeling to understand and to assess the impact of particular compounds in the ozone-depletion problem—and now global warming—is another area where we have had a significant supercomputer effort over many years. This effort is also critical to the future and viability of some of our businesses. As a consequence, this is an area where we chose to develop our own special-purpose programs, which are recognized as being state of the art.

In looking forward, what can we do together? What would be of help to us in industry?

One answer is to explore alternatives for integrating supercomputers into a heterogeneous network of computer applications and workstations so they can be easily accessed and utilized to solve problems where high-performance computing is either required or highly desirable.

Second, we could develop hardware and software to solve the grand challenges of science. Although it may not be directly applicable to our problems, the development of new and novel machines and algorithms will benefit us, particularly in the vectorization and parallelization of algorithms.

Third, we might develop applications software of commercial quality that exploits the capabilities of highly parallel supercomputers.

Fourth, we could develop visualization hardware and software tools that could be used effectively and simply by our scientists and engineers to enhance their projects. We would be anxious to cooperate with others jointly in any of these sorts of areas.

The bottom line is that we, in our innocence, believe we are getting real business value from the use of supercomputers in research, development, and engineering work. However, to exploit this technology fully, we need people with a vision of what is possible, we need more high-quality software applications—especially for highly parallel machines—and we need the capability to easily integrate supercomputers into diverse problem-solving environments. Some of those things, like the latter point, are really our job. Yet we really need the help of others and would be very eager, I think, to work with a national laboratory in solving some problems for the chemical industry.

An Overview of Supercomputing at General Motors Corporation

Myron Ginsberg

Myron Ginsberg currently serves as Consultant Systems Engineer at the Electronic Data Systems Advanced Computing Center, General Motors Research and Environmental Staff, Warren, Michigan. Until May 1992, he was Staff Research Scientist at General Motors Research Laboratories.

During a 13-year tenure at General Motors, Dr. Ginsberg was significantly involved in GM's initial and continuing supercomputer efforts, which led to the first installation of a Cray supercomputer in the worldwide auto industry. He is also Adjunct Associate Professor in the Electrical Engineering and Computer Science Department, College of Engineering, at the University of Michigan. He has edited four volumes on vector/parallel computing applications in the auto industry. He has three times been the recipient of the Society of Automotive Engineers' (SAE's) Award for Excellence in Oral Presentation and has earned the SAE Distinguished Speaker Plaque, as well. Dr. Ginsberg serves on the Editorial Board of Computing Systems in Engineering *and on the Cray Research, Inc., Fortran Advisory Board. He has also been a Distinguished National Lecturer for the American Society of Mechanical Engineers, the Society for Industrial and Applied Mathematics, and the Association for Computing Machinery.*

Abstract

The use of supercomputers at General Motors Corporation (GM) began in the GM Research Laboratories (GMR) and has continued there, spreading to GM Divisions and Staffs, as well. Topics covered in this paper include a review of the computing environment at GM, a brief history of GM supercomputing, worldwide automotive use of supercomputers, primary GM applications, long-term benefits, and the challenges for the future.

Introduction

In this paper, we will review the computing environment at GM, give a brief history of corporate supercomputing, indicate worldwide automotive utilization of supercomputers, list primary applications, describe the long-term benefits, and discuss the needs and challenges for the future.

People and the Machine Environment

Supercomputing activities at GM have been focused primarily on projects in GMR and/or cooperative activities between GMR and one or more GM Divisions or Staffs.

There are approximately 900 GMR employees, with about 50 per cent of these being R&D professionals. In this latter group, 79 per cent have a Ph.D., 18 per cent an M.S., and 3 per cent a B.S. as their highest degree. In addition, there are Electronic Data Systems (EDS) personnel serving in support roles throughout GM.

General Motors was the first automotive company to obtain its own in-house Cray Research supercomputer, which was a CRAY 1S/2300 delivered to GMR in late 1983. Today, GM has a CRAY Y-MP4/364 at GMR, a CRAY Y-MP4/232 at an EDS center in Auburn Hills, Michigan, and a CRAY X-MP/18 at Adam Opel in Germany. Throughout GM, there is a proliferation of smaller machines, including a CONVEX Computer Corporation C-210 minisuper at B-O-C Flint, Michigan, IBM mainframes, Digital Equipment Corporation (DEC) minis, a Stardent 2000 graphics super at C-P-C Engineering, numerous Silicon Graphics high-end workstations, and a large collection of workstations from IBM, Sun Microsystems, Inc., Apollo (Hewlett-Packard), and DEC. There is extensive networking among most of the machines to promote access across GM sites.

History of Supercomputing at GM

Table 1 summarizes GM's involvement with supercomputers. In 1968, GMR entered into a joint effort with Control Data Corporation (CDC) to explore the potential use of the STAR-100 to support graphics consoles. A prototype of that machine, the STAR 1-B, was installed at GMR. This project was terminated in 1972.

GM next started looking at supercomputers in late 1979. At that time the GM computing environment was dominated by several IBM mainframes (IBM 3033). Scientists and engineers developed an intuitive feel with respect to sizing their programs. They were aware that if they exceeded certain combinations of memory size and CPU time, then their job would not be completed the same day. They tried to stay within those bounds, but that became extremely difficult to do as the physical problems being considered grew increasingly complex and as they sought to develop two- and three-dimensional models.

In 1981, benchmarks were gathered both from GMR and GM Staffs and Divisions for testing on the CRAY-1 and on the CDC CYBER 205. These benchmarks included representative current and anticipated future work that would require very-large-scale computations. The results indicated that the CRAY-1 would best satisfy our needs. To get initial experience of our employees on that machine, we began to use a CRAY-1 at Boeing Computer Services and tried to ramp up our usage until such time as we could economically utilize our own in-house CRAY. Finally, in late 1983, a CRAY-1S/2300 was delivered to GMR and was in general use in early 1984. The utilization of that machine steadily grew until it was replaced by a CRAY X-MP/24 in 1986, and then that machine was replaced by a two-processor CRAY Y-MP in late 1989, with an additional CPU upgrade in early 1991. Other Cray supercomputers at GM were introduced at Adam Opel in 1985 and at EDS in 1991.

Automotive Industry Interest in Supercomputers

At about the same time GM acquired its own Cray supercomputer in late 1983, Chrysler obtained a CDC CYBER 205 supercomputer. Then in early 1985, Ford obtained a CRAY X-MP/11. As of late 1991, there were approximately 25 Cray supercomputers worldwide in automotive companies in addition to several nonautomotive Crays used by auto companies.

Table 1. Summary of the History of Supercomputing at GM

1968–72	GMR-Control Data cooperative work on STAR-100 project with STAR 1-B prototype at GMR
1979–80	Investigate research and production needs for corporate supercomputer
1981–82	Benchmarking CDC CYBER 205 and CRAY-1S
1982–83	Use Boeing Computer Services CRAY-1S
1984–86	CRAY-1S/2300 at GMR
1985	CRAY-1S/1000 at Adam Opel in Germany
1986–89	CRAY X-MP/24 replaces CRAY-1S/2300 at GMR
1988	CRAY X-MP/14 replaces CRAY-1S/1000 at Adam Opel in Germany
1989–90	CRAY Y-MP4/232 replaces CRAY X-MP/24 at GMR
1990	CRAY X-MP/14 upgraded to X-MP/18 at Adam Opel in Germany
1991	CRAY Y-MP4/332, upgrade of one additional CPU at GMR CRAY Y-MP4E/232, EDS machine at their Auburn Hills, Michigan, center
1992	CRAY Y-MP4/364, upgrade of 32 million words at GMR

Figure 1 portrays an interesting trend in the growth of supercomputers within the worldwide automotive community. It depicts the number of Cray CPUs (not machines), including both X-MP and Y-MP processors, in the U.S., Europe, and the Far East in 1985, 1988, and 1991. In 1985, no automotive Cray CPUs existed in the Far East, and only two were in use in the U.S. (GM and Ford). In sharp contrast, at the end of 1991, there were 26 Cray CPUs (13 machines) in the Far East, compared with a total of 14 (four machines) in all U.S. auto companies! The specific breakdown by machines is given in Table 2; the ranking used is approximately by total CPU computational power and memory. We note that the Far East, specifically Japanese, auto companies occupy five of the top 10 positions. Their dominance would be even more obvious in Figure 1 if Japanese supercomputer CPUs were included; several of the Far East auto companies own or have access to one or more such machines in addition to their Crays.

It is interesting to note that once the first supercomputer was delivered to the automotive industry in late 1983, just about every major car company in the world began to acquire one or more such machines for in-house use within the following eight years, as evidenced by Figure 1 and Table 2.

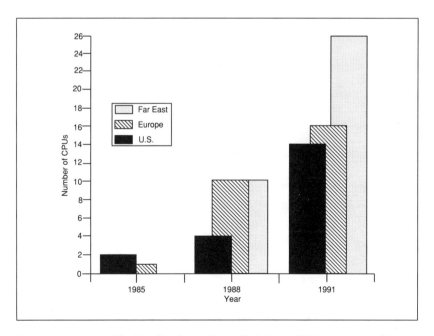

Figure 1. Geographic distribution of installed Cray CPUs in the world auto industry, 1985–91.

Table 2. Cray Supercomputers in the Worldwide Auto Industry as of 1991[a]

Ford	Y-MP8/7128 (Y-MP C90, 4Q 92)
Nissan	Y-MP8/664, X-MP/12
GM/EDS	Y-MP4/364, Y-MP4E/232, X-MP/18
Honda	Y-MP8/364
Volkswagen AG	Y-MP8/364
Mercedes	Y-MP4/232, X-MP/28
Mazda	Y-MP2E/232, X-MP/216
Toyota	Y-MP8/232, X-MP/116
BMW	Y-MP8/232
Mitsubishi	Y-MP4/216
Fiat	Y-MP4/216
Peugeot	Y-MP2/232
Isuzu	Y-MP2E/232
Chrysler	Y-MP2E/232 (Y-MP8i/464, 2Q 92)
Daihatsu	Y-MP2/216
Kia Motors	Y-MP4/116
Hino Motors	Y-MP2E/132
Yamaha Motors	Y-MP2E/116
Renault	X-MP/216

[a] In addition to the above, Saab Scandia uses a CRAY X-MP/48 at the Swedish National Supercomputer Center, and Porsche uses the CRAY-2 at the University of Stuttgart.

One of the reasons for the initial delay to introduce supercomputers in the auto industry was a significant economic downturn in the early 1980s, combined with the high cost of supercomputers at that time ($5 million to $10 million range). There was also a reluctance to acquire a machine that might not be filled to capacity for quite a while after acquisition. Nevertheless, U.S., European, and the Far East auto companies began experimenting with supercomputers at service bureaus during the early 1980s.

The early acquisition of supercomputers by U.S. government labs, such as Los Alamos and Livermore, helped to spearhead the future use of supercomputers by auto companies, as well as by other businesses in private industry. The experience gained with adapting programs to supercomputers was reported in the open literature, as well as at professional meetings where people from the automotive industry could interact with personnel from the national laboratories. Furthermore, many of the programs developed at those labs became available in the public domain. Also, some joint cooperative projects began to develop between the national labs and U.S. auto companies.

Applications

Table 3 summarizes many of the supercomputer applications currently running at GM.

Most of the supercomputer applications represent finite element or finite difference two- or three-dimensional mathematical models of physical phenomena. Both early and current applications at GM have been dominated by work in the aerodynamics area (computational fluid dynamics), combustion modeling, and structural analysis (including crashworthiness analysis); see, for example, Hammond (1985), Meintjes

---

Table 3. A Sampling of GM Supercomputer Applications

Flows, Sprays, and Combustion on Two-Stroke Engines
Front-End Airflow System Design
Engine Combustion Model Development
Crashworthiness Simulation
Simulation of Passenger-Compartment Heating and Cooling
Sheet-Metal-Forming Analysis
Fundamental Research on Sprays
Internal Flow Passage Research
Underhood Cooling
Aerosol Dynamics
Biological Modeling
Structural Analysis on Integrated Systems
Turbomachinery Flow Analysis
VLSI Design and Simulation
Vehicle Interior Acoustics Modeling
Ignition Chemistry
Exterior Vehicle Aerodynamics (CFD Problems)

(1986), Grubbs (1985), Haworth and El Tahry (1990), Haworth et al. (1990), El Tahry and Haworth (1991), Ginsberg (1988, 1989), Ginsberg and Johnson (1989), Ginsberg and Katnik (1990), Johnson and Skynar (1989), Khalil and Vander Lugt (1989), and Shkolnikov et al. (1989). This work involves both software developed in-house (primarily by GMR personnel) and use of commercial packages (used primarily by personnel in GM Divisions and Staffs). Within the past several years, additional applications have utilized the GMR supercomputer; see, for example, sheet-metal-forming applications as discussed by Chen (May 1991, July 1991), Chen and Waugh (1990), and Stoughton and Arlinghaus (1990). A most recent application by the newly formed Saturn Corporation is using the GMR Cray and simulation software to design strategically placed "crush zones" to help dissipate the energy of a crash before it reaches vehicle occupants (General Motors Corporation 1991).

In addition to the use of the Cray supercomputer, GMR scientists and engineers have been experimenting with other high-performance computers, such as hypercube and transputer-based architectures. Such machines provide a low-cost, distributed parallel computing facility. Recent work in this area on such machines includes that described by Baum and McMillan (1988, 1989), Malone (1988, 1989, 1990), Malone and Johnson (1991a, 1991b), and Morgan and Watson (1986, 1987). A more complete list of GM applications of high-performance computers is given by Ginsberg (1991).

Long-Term Benefits

There are several factors that justify the use of supercomputers for automotive applications. For example, the speed of such machines makes it possible to perform parameter studies early in the design cycle, when there is only a computer representation of the vehicle, and even a physical prototype may not yet exist; at that stage, a scientist or engineer can ask "what if" questions to try to observe what happens to the design as specific parameters or combination of parameters are changed. Such observations lead to discarding certain design approaches and adopting others, depending upon the results of the computer simulations. This can reduce the amount of physical prototyping that has to be done and can lead to significant improvements in quality of the final product. Other long-term benefits include improved product safety via crashworthiness modeling and greater fuel economy via aerodynamics simulations.

Accurate computer simulations have the potential to save money by reducing both the number of physical experiments that need to be performed and the time to prepare for the physical testing. For example, in the crashworthiness area, each physical crash involves a custom, handmade car that can only be used once and may take several months to build. Furthermore, the typical auto industry cost of performing one such physical crash on a prototype vehicle can be upwards of $750,000 to $1,000,000 per test! It thus becomes apparent that realistic computer simulations have the potential to produce substantial cost savings.

The successful application of supercomputers in all phases of car design and manufacturing can hopefully lead to significant reductions in the lead time necessary to bring a new product to market. The use of supercomputers in the auto industry is still in its infancy. Creative scientists and engineers are just beginning to explore the possibilities for future automotive applications.

Needs and Challenges

Los Alamos National Laboratory received its first Cray in 1976, but the American automotive community did not begin acquiring in-house supercomputers until over seven years later. The American automobile industry needs more immediate access to new supercomputer technologies in order to rapidly utilize such machines for its specific applications. This will require growth in cooperative efforts with both government laboratories and universities to explore new architectures, to create highly efficient computational algorithms for such architectures, and to develop the necessary software support tools.

Another challenge for the future is in the networking area. Supercomputers must be able to communicate with a diverse collection of computer resources, including other supercomputers, and this requires very high bandwidth communication networks, particularly if visualization systems are to be developed that allow real-time interaction with supercomputer simulations.

The demand for faster and more realistic simulations is already pushing the capabilities of even the most sophisticated uniprocessor architectures. Thus, we must increase our investigation of parallel architectures and algorithms. We must assess the tradeoffs in using supercomputers, minisupers, and graphics supers. We must determine where massively parallel machines are appropriate. We must be able to develop hybrid approaches where portions of large problems are assigned to a variety of architectures, depending upon which machine is the most efficient for dealing with a specific section of computation. This again requires cooperative efforts among private industry, government labs, and universities; commercial tools must be developed to assist scientists and engineers in producing highly efficient parallel programs with a minimal amount of user effort.

Realistic simulations demand visualization rather than stacks of computer paper. Making videos should become routine for scientists and engineers; it should not be necessary for such persons to become graphics experts to produce high-quality, realistic videos. In the automotive industry, videos are being produced, particularly in the crashworthiness (both side-impact and frontal-barrier simulations) and aerodynamics areas.

The challenges above are not unique to the auto industry alone. Rapid U.S. solutions to these needs could help the American automotive industry to increase its competitiveness in the world marketplace.

References

A. M. Baum and D. J. McMillan, "Message Passing in Parallel Real-Time Continuous System Simulations," General Motors Research Laboratories publication GMR-6146, Warren, Michigan (January 27, 1988).

A. M. Baum and D. J. McMillan, "Automated Parallelization of Serial Simulations for Hypercube Parallel Processors," in *Proceedings, Eastern Multiconference on Distributed Simulation*, Society for Computer Simulation, San Diego, California, pp. 131–136 (1989).

K. K. Chen, "Analysis of Binder Wrap Forming with Punch-Blank Contact," General Motors Research Laboratories publication GMR-7330, Warren, Michigan (May 1991).

K. K. Chen, "A Calculation Method for Binder Wrap with Punch Blank Contact," General Motors Research Laboratories publication GMR-7410, Warren, Michigan (July 1991).

K. K. Chen and T. G. Waugh, "Application of a Binder Wrap Calculation Model to Layout of Autobody Sheet Steel Stamping Dies," Society of Automotive Engineers paper 900278, Warrendale, Pennsylvania (1990).

S. H. El Tahry and D. C. Haworth, "A Critical Review of Turbulence Models for Applications in the Automotive Industry," American Institute of Aeronautics and Astronautics paper 91-0516, Washington, DC (January 1991).

General Motors Corporation, "Saturn Sales Brochure," S02 00025 1090 (1991).

M. Ginsberg, "Analyzing the Performance of Physical Impact Simulation Software on Vector and Parallel Processors," in *Third International Conference on Supercomputing: Supercomputing 88, Vol. 1, Supercomputer Applications*, L. P. Kartashev and S. I. Kartashev, Eds., International Supercomputer Institute, Inc., St. Petersburg, Florida, pp. 394–402 (1988).

M. Ginsberg, "Computational Environmental Influences on the Performance of Crashworthiness Programs," in *Crashworthiness and Occupant Protection in Transportation Systems*, T. B. Khalil and A. I. King, Eds., American Society of Mechanical Engineers, New York, pp. 11–21 (1989).

M. Ginsberg, "The Importance of Supercomputers in Car Design/Engineering," in *Proceedings, Supercomputing USA/Pacific 91*, Meridian Pacific Group, Inc., Mill Valley, California, pp. 14–17 (1991).

M. Ginsberg and J. P. Johnson, "Benchmarking the Performance of Physical Impact Simulation Software on Vector and Parallel Computers," in *Supercomputing '88, Vol. II, Science and Applications*, J. L. Martin and S. F. Lundstrom, Eds., Institute of Electrical and Electronics Engineers Computer Society Press, Washington, D.C., pp. 180–190 (1989).

M. Ginsberg and R. B. Katnik, "Improving Vectorization of a Crashworthiness Code," Society of Automotive Engineers paper 891985, Warrendale, Pennsylvania; also in *SAE Transactions*, Sec. 3, Vol. 97, Society of Automotive Engineers, Warrendale, Pennsylvania (September 1990).

D. Grubbs, "Computational Analysis in Automotive Design," *Cray Channels* 7 (3), 12–15 (1985).

D. C. Hammond Jr., "Use of a Supercomputer in Aerodynamics Computations at General Motors Research Laboratories," in *Supercomputers in the Automotive Industry*, M. Ginsberg, Ed., special publication SP-624, Society of Automotive Engineers, Warrendale, Pennsylvania, pp. 45–51 (July 1985).

D. C. Haworth and S. H. El Tahry, "A PDF Approach for Multidimensional Turbulent Flow Calculations with Application to In-Cylinder Flows in Reciprocating Engines," General Motors Research Laboratories publication GMR-6844, Warren, Michigan (1990).

D. C. Haworth, S. H. El Tahry, M. S. Huebler, and S. Chang, "Multidimensional Port-and-Cylinder Flow Calculations for Two- and Four-Valve-per-Cylinder Engines: Influence of Intake Configuration on Flow Structure," Society of Automotive Engineers paper 900257, Warrendale, Pennsylvania (February 1990).

J. P. Johnson and M. J. Skynar, "Automotive Crash Analysis Using the Explicit Integration Finite Element Method," in *Crashworthiness and Occupant Protection in Transportation Systems*, T. B. Khalil and A. I. King, Eds., American Society of Mechanical Engineers, New York, pp. 27–32 (1989).

T. B. Khalil and D. A. Vander Lugt, "Identification of Vehicle Front Structure Crashworthiness by Experiments and Finite Element Analysis," in *Crashworthiness and Occupant Protection in Transportation Systems*, T. B. Khalil and A. I. King, Eds., American Society of Mechanical Engineers, New York, pp. 41–51 (1989).

J. G. Malone, "Automated Mesh Decomposition and Concurrent Finite Element Analysis for Hypercube Multiprocessor Computers," *Computer Methods in Applied Mechanics and Engineering* **70** (1), 27–58 (1988).

J.G. Malone, "High Performance Using a Hypercube Architecture for Parallel Nonlinear Dynamic Finite Element Analysis," in *Proceedings, Fourth International Conference on Supercomputing: Supercomputing 89, Vol. 2, Supercomputer Applications*, L. P. Kartashev and S. I. Kartashev, Eds., International Supercomputer Institute, Inc., St. Petersburg, Florida, pp. 434–438 (1989).

J. G. Malone, "Parallel Nonlinear Dynamic Finite Element Analysis of Three-Dimensional Shell Structures," *Computers and Structures* **35** (5), 523–539 (1990).

J. G. Malone and N. L. Johnson, "A Parallel Finite Element Contact/ Impact Algorithm for Nonlinear Explicit Transient Analysis: Part I, The Search Algorithm and Contact Mechanics," General Motors Research Laboratories publication GMR-7478, Warren, Michigan (1991a).

J. G. Malone and N. L. Johnson, "A Parallel Finite Element Contact/ Impact Algorithm for Nonlinear Explicit Transient Analysis: Part II, Parallel Implementation," General Motors Research Laboratories publication GMR-7479, Warren, Michigan (1991b).

K. Meintjes, "Engine Combustion Modeling: Prospects and Challenges," *Cray Channels* **8** (4), 12–15 (1987); extended version in *Supercomputer Applications in Automotive Research and Engineering Development*, C. Marino, Ed., Computational Mechanics Publications, Southhampton, United Kingdom, pp. 291–366 (1986).

A. P. Morgan and L. T. Watson, "Solving Nonlinear Equations on a Hypercube," in *ASCE Structures Congress '86: Super and Parallel Computers and Their Impact on Civil Engineering*, M. P. Kamat, Ed., American Society of Civil Engineers, New Orleans, Louisiana, pp. 1–15 (1986).

A. P. Morgan and L. T. Watson, "Solving Polynomial Systems of Equations on a Hypercube," in *Hypercube Multiprocessors*, M. T. Heath, Ed., Society for Industrial and Applied Mathematics, Philadelphia, Pennsylvania, pp. 501–511 (1987).

M. B. Shkolnikov, D. M. Bhalsod, and B. Tzeng, "Barrier Impact Test Simulation Using DYNA3D," in *Crashworthiness and Occupant Protection in Transportation Systems*, T. B. Khalil and A. I. King, Eds., American Society of Mechanical Engineers, New York, pp. 33–39 (1989).

T. Stoughton and F. J. Arlinghaus, "Sheet Metal Forming Simulation Using Finite Elements," *Cray Channels* **12** (1), 6–11 (1990).

Barriers to Use of Supercomputers in the Industrial Environment

Robert Hermann

Robert J. Hermann was elected Vice President, Science and Technology, at United Technologies Corporation (UTC) in March 1987. In this position, Dr. Hermann is responsible for assuring the development of the company's technical resources and the full exploitation of science and technology by the corporation. He also has responsibility for the United Technologies Research Center and the United Technologies Microelectronics Center. Dr. Hermann joined UTC in 1982 as Vice President, Systems Technology, in the electronics sector. He was named Vice President, Advanced Systems, in the Defense Systems Group in 1984.

Dr. Hermann served 20 years with the National Security Agency, with assignments in research and development, operations, and NATO. In 1977 he was appointed principal Deputy Assistant Secretary of Defense for Communications, Command, Control, and Intelligence. He was named Assistant Secretary of the Air Force for Research, Development, and Logistics in 1979 and Special Assistant for Intelligence to the Undersecretary of Defense for Research and Engineering in 1981.

He received B.S., M.S., and Ph.D. degrees in electrical engineering from Iowa State University, Ames, Iowa. Dr. Hermann is a member of the National Academy of Engineering, the Defense Science Board, and the National Society of Professional Engineers' Industry Advisory Group. He is also

Chairman of the Naval Studies Board and of the Executive Committee of the Navy League's Industrial Executive Board.

I will discuss my point of view, not as a creator of supercomputing-relevant material or even as a user. I have a half-step in that primitive class called management, and so I will probably reflect most of that point of view.

United Technologies Corporation (UTC) makes jet engines under the name of Pratt and Whitney. We make air conditioners under the name of Carrier. We make elevators under the name of Otis. We make a very large amount of automobile parts under our own name. We make helicopters under the name of Sikorsky and radars under the name of Norden.

There is a rich diversity between making elevators and jet engines. At UTC we are believers in supercomputation—that is, the ability to manage the computational advantages that are qualitatively different today than they were five years ago; and they will probably be qualitatively different five years from now.

The people in Pratt and Whitney and in the United Technologies Research Center who deal with jet engines have to deal with high-temperature, high-Mach-number, computational fluid dynamics where the medium is a plasma. These are nontrivial technical problems, and the researchers are interested in three-dimensional Navier-Stokes equations, and so on. It is in an industry where being advanced has visible, crucial leverage, which in turn results in motivation. Thus, there are pockets in UTC where I would say we really do believe, in an analytic sense, in design, process, simulation, and visualization.

It seems to me that when I use the term "supercomputation," I have to be in some sense connoting doing things super—doing things that are unthinkable or, at least, unprecedented. You have to be able to do something that you just would not have even tried before. Thus, an important barrier in "supercomputation" is that it requires people who can think the unthinkable, or at least the unprecedented. They have to have time, they have to have motivation, and they have to have access.

Also, those same people clearly have to have hardware, software, math, physics, application, and business perspectives in their head. The critical ingredient is that you need, in one intellect, somebody who understands the software, the hardware, the mathematics to apply it, the physics to understand the principles, and the business application. This is a single-intellect problem or, at least, a small-group problem. If you do not have this unity, you probably cannot go off and do something that was either unthinkable or unprecedented. Getting such individuals and groups together is indeed a barrier.

A business point of view will uncover another big barrier in the way we organize our businesses and the way that businesses are practiced routinely. The popular way of doing business is that the total business responsibility for some activity is placed in the hands of a manager. Total business responsibility means that there are many opportunities to invest various kinds of resources: time, money, management. Supercomputation is certainly not the thing that leaps to mind the first time when someone in most businesses is asked, "What are some of the big, burning problems you have?"

In our environment, you legitimately have to get the attention of the people who have the whole business equation in their heads and in their responsibility packages. One thing that does get attention is to say that small purchases are easier to make than large purchases. UTC is a very large corporation. At $20 billion and 200,000 employees, you would think that at that level you could afford to make large purchases. However, we have broken the company down in such a way that there are no large outfits. It is a collection of small outfits such that it is more than ten times easier to make ten $100,000 purchases than one $1 million purchase. That equation causes difficulty for the general problem of pulling in the thing called supercomputation because in some sense, supercomputation cannot be bought in small packages. Otherwise, it isn't super.

It is also true that the past experiences of the people who have grown up in business are hard to apply to supercomputation. It is not like building a factory. A factory, they know, makes things.

UTC is an old-line manufacturing outfit. We are one of thousands of old-line manufacturing outfits that exist on a global basis. We are the class of folks who make the money in the world that supports all the research, development, and investment.

The people who are in charge do not naturally think in terms of supercomputation because it is moving too fast. We have to educate that set of people. It is not an issue of pointing fingers in blame, although we are representative. But I would also say to someone who is trying to promote either the application of supercomputation as a field or national competitiveness through the use of supercomputation, "This is a barrier that has to be overcome." It will probably not be overcome totally on the basis of the motivation of the structure of the corporation itself.

We need to be educated, and I have tried to figure out what is inhibiting our using supercomputers. Several possible answers come to mind.

First, we do not know how to relate the advantage to our business. And we do not have time to do it, because our nose is so pressed to the grindstone trying to make money or cash flow or some other financial equation. The dominance of the financial equation is complete as it is, and it is fundamental to the existence of the economic entity. But somehow or another, there has to be some movement toward making people know more about the application of supercomputers to their business advantage.

Another issue is the question of how to pay for supercomputers. If you purchase large items that are larger than the normal business element can buy, you have to cooperate with somebody else, which is a real big barrier because cooperating with somebody else is difficult.

Also, how you do the cost accounting is a nontrivial business. Indeed, we at UTC probably would not have a Cray if it had not been forced on us by the National Aerospace Plane Program Office. When we got it, we tried to figure out how to make it useful across the corporation, and we were hindered, obstructed, and eventually deterred in every way by the cost-accounting standards applied by the government.

Now, what are we doing about it? I would say we are trying to do something about it, although we may be way behind as a culture and as a group. We are trying to build on those niche areas where we have some capability, we are trying to use our own examples as precedents, we are surveying ourselves to try to understand what is meaningful, and we are trying to benchmark ourselves against others.

In 1989 we participated in some self-examination that we did over the course of the year. We have agreed that we are going to establish a network in which we can do scientific computation in a joint venture with laboratories, etc., to transport the necessary technology.

This is also a national issue. The national competitiveness issue must be somewhere out there at the forefront. In the national competitiveness area, to become a patriot, supercomputation is important—as infrastructure, not as a subsidy. I would think that some notion of an infrastructure, which has some geographic preference to it, is likely to be needed. I would therefore argue that networked data highways and attached supercomputation networks have some national competitiveness advantages, which are a little bit different from the totally distributed minicomputer that you can ship anywhere and that does not have a particular geographic or national preference associated with it.

From a national point of view, and as a participant in national affairs, I can have one view. But from a corporate point of view, I am somewhat neutral on the subject: if we do not do it in the U.S., the Europeans probably will, and the Japanese probably will; and we will then have to use the European-Japanese network because it is available as a multinational corporation.

10

Government Supercomputing

This panel included users from various organs of government—voracious consumers of computers and computing since the 1940s. National laboratories, the intelligence community, the federal bureaucracy, and NSF computing centers all make use of supercomputers. The panel concentrated on the future and referred to the past to give the future some perspective. Panelists agreed that all the problems identified during the first Frontiers of Supercomputing conference have not been solved. Thus, the panelists focused on continuing challenges related to cultural issues, support, efficiency versus user friendliness, technology transfer, impediments to accessibility, and government policy.

Session Chair

**George Michael,
Lawrence Livermore National Laboratory**

Planning for a Supercomputing Future*

Norm Morse

For most of the past decade, Norman R. Morse has served as leader of the Computing and Communications Division at Los Alamos National Laboratory. He has a bachelor's degree in physics from Texas A&I University, Kingsville, and a master of science in electrical engineering and computer science from the University of New Mexico, Albuquerque. Under his leadership, the Laboratory's computing facilities were expanded to include the newly constructed Data Communications Center and the Advanced Computing Laboratory. In addition, he promoted the use of massively parallel computers, and through his efforts, the facilities now house three Connection Machines—two CM-2s and one CM-5. Norm recently returned to a staff position, where he plans to pursue research in clustered workstation paradigms for high-performance computing.

Over the past few years I, together with other people from Los Alamos National Laboratory, have been examining the state of computing at Los Alamos and have been thinking about what our desired future state of computing might be. I would like to share with you some of the insights

---

* The author wishes to acknowledge the contributions of the Computing and Communications Initiative Project Team at Los Alamos National Laboratory and the numerous Laboratory staff members who worked with the team.

that we have had, as well as some thoughts on the forces that may shape our future.

We used a strategic planning model to guide our thinking. Figure 1 shows that model. From a particular current state of computing, there are many possible future states into which institutional computing can evolve, some of them more desirable than others. There are drivers that determine which one of these possible future states will result. We have been interested in examining this process and trying to understand the various drivers and how to use them to ensure that we arrive at what we perceive to be our desired state of computing.

In the late 1960s and early 1970s, the current state of computing was batch. We had evolved from single-user systems into a batch-processing environment. The future state was going to evolve from that state of technology and from the influences of that time.

The evolution from batch computing to timesharing computing came from a variety of drivers (Figure 2). One major driver was technology: terminals were invented about that time, and some rudimentary network capabilities were developed to support the needs for remote computing. The mainframes essentially didn't change—they were von Neumann central supercomputers. Software was developed to support a timesharing model of computing. And an important, nontechnical factor was that the money for computing came in through the applications that

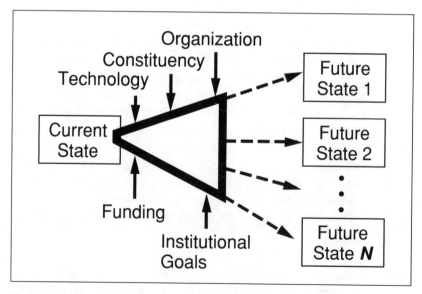

Figure 1. General model of the evolution of institutional computing.

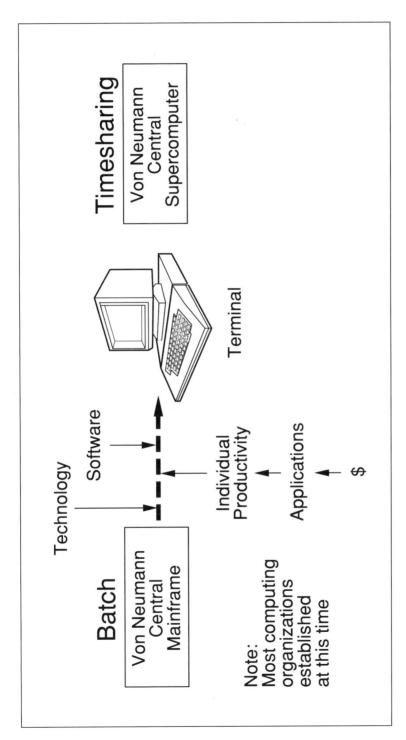

Figure 2. The evolution from batch to timesharing computing in the 1970s.

people were interested in solving. The people who were developing and running those applications were interested in individual productivity.

It may have been in this period of time that we stopped thinking that machine productivity was the most important issue that we had to deal with, and we began to think that the productivity of the individuals who were using the machine should be maximized. So we evolved into a timesharing environment in the 1970s. We recognized the value of a centrally managed network and central services to support timesharing. Mass storage and high-quality print services became an important part of the network.

In the 1980s we went from a timesharing environment to a distributed environment (Figure 3). And again, the important influences that drove us from timesharing to distributed included advances in technology. But there were also other factors: a large user community required more control of their computing resource, and they valued the great increase in interactivity that came from having a dedicated computer on their desks.

The 1980s became the era of workstations—a lot of computational power that sat on your desk. Networks became more reliable and universal. We began to understand that networks were more than just wires that tied computers together. Users needed increased functionality, as well as more bandwidth, to handle both the applications and the user interfaces. Many of the centralized services began to migrate and were managed on user networks. We started thinking about doing visualization. Von Neumann central supercomputers, along with departmental-class mainframes, were still the workhorses of this environment. Massively parallel supercomputers were being developed.

The next environment hasn't sorted itself out yet. The future picture, from a hardware and software technology viewpoint, is becoming much more complicated. We're calling the next environment the high-performance computing environment (Figure 4).

Again, there are N possible future states into which we could evolve. The drivers or enablers that are driving the vector from where we are now to where we want to be in the future are getting more complicated, and they're not, in all cases, intuitively obvious.

The high-performance computing model that I see evolving, at least at Los Alamos, is one composed of three major parts: parallel workstations, networks, and supercomputers. I think that general-purpose computing is going to be done on workstations. The supercomputers are going to end up being special-purpose devices for the numerically intensive

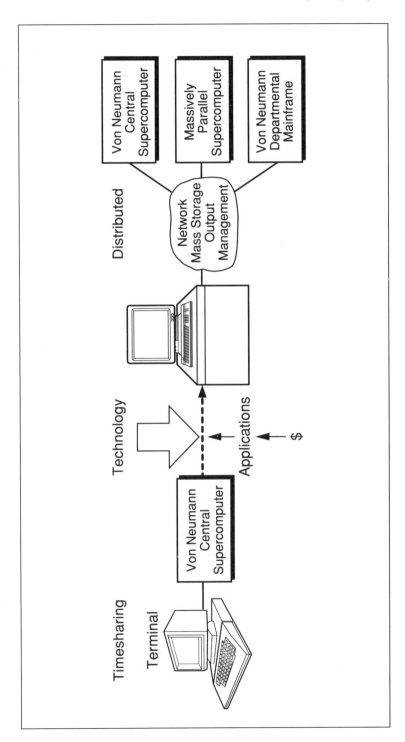

Figure 3. The evolution from a timesharing- to a distributed-computing environment in the 1980s.

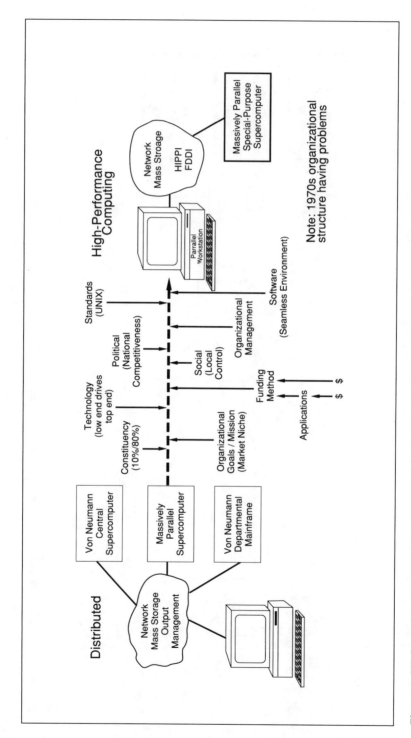

Figure 4. High-performance computing emerging from a distributed-computing environment in the 1990s.

portion of client tasks. In fact, I personally think they've always been special purpose, regardless of how much work we've put in to try to make them general-purpose devices. I think in the long term, supercomputers have to be massively parallel to achieve the speeds required to solve the grand-challenge problems. That's not to say that the workhorses of today will go away; the von Neumann vector pipeline machines and the departmental-class machines have been around for a long time and will remain essential for a large class of problems.

This high-performance computing environment will evolve and will consist of these three major elements. The real questions are, what will be the balance among them, and how well does that balance satisfy the needs of a particular organization?

Constituency is one of the important drivers. The workstations sitting on people's desks and the computers that people typically learn to use in universities are a very important part of the high-performance computing environment. Of the 8000 clients who use our computing center, virtually every one of them uses a personal computer. Somewhere around 250 people use 90 per cent of our supercomputing cycles in any given month. So when we looked for people to argue for various parts of this high-performance computing environment, we could find 8000 people who would argue for the workstation part. The Laboratory internet is becoming increasingly important to a broad population in the Laboratory because of the need to communicate with colleagues locally and internationally. So we could find 8000 people who would argue for networks. But on the other hand, there are only a few hundred who will argue vehemently for the supercomputing environment. This kind of support imbalance can shift the future state to one in which there is a strong workstation and network environment but a very weak supercomputing capability.

I would guess that the statistics at most sites are similar to this. There are a small number of people who dominate the use of the supercomputing resources, doing problems that are important to the mission of the institution. And if the institution is in the business of addressing grand-challenge problems, it takes a lot of supercomputing cycles to address those problems.

Given the environment described, I feel that the low-end technology will drive this evolution and is going to drive the top end. That is, the massively parallel supercomputers of the future will be made up of building blocks (hardware and software) developed for the workstation market. There are many reasons driving this trend, but one of the most important is the fact that the workstation market is huge compared with the supercomputer market. A tremendous effort is under way to develop hardware and software for the workstation market. If supercomputers are made from workstation building blocks, the remaining question is whether the supercomputing capability will be closely coupled/closely integrated or loosely coupled/closely integrated. The marketplace will shake out the answer in the next few years.

Standards are going to be even more important in this new environment. For example, UNIX is going to run across the whole environment. It should be easy for people to do as much of their work on workstations as possible, and, when they run out of the power to do their work there, they will be able to use other, more powerful or less heavily used resources in the network to finish their jobs. This means that the UNIX systems must be compatible across a large variety of computing platforms. Computing vendors need to cooperate to build software systems that make this easy.

Another important driver, the funding method, may be different from what we've seen in the past. Traditionally, the money has come in through the applications, driving the future state of computing. The applications people drive the capabilities that they need. With the High Performance Computing Initiative, there is the potential, at least, for money to come directly into building a computing capability. And I think we need to be very careful that we understand what this capability is going to be used for. If we end up building a monument to computing that goes unused, I think we will not have been very successful in the High Performance Computing Initiative.

One last issue that I'll mention is that there are a lot of social issues pushing us toward our next state of computing. Local control seems to be the most important of those. People like to eliminate all dependencies on other folks to get their jobs done, so local control is important. We need to make the individual projects in an organization cognizant of the mission of the organization as a whole and to maintain capabilities that the organization needs to secure its market niche.

High-Performance Computing at the National Security Agency

George Cotter

George R. Cotter currently serves as Chief Scientist for the National Security Agency (NSA). From June 1988 to April 1990, he was the Chairman of the Director's Senior Council, a study group that examined broad NSA and community issues. From June 1983 to June 1988, Mr. Cotter served as Deputy Director of Telecommunications and Computer Services at NSA, in which position he was responsible for implementing and managing worldwide cryptologic communications and computer systems.

Mr. Cotter has a B.A. from George Washington University, Washington, DC, and an M.S. in numerical science from Johns Hopkins University, Baltimore, Maryland. He has been awarded the Meritorious and Exceptional Civilian Service medals at NSA and in 1984 received the Presidential Rank of Meritorious Cryptologic Executive. Also in 1984, he received the Department of Defense Distinguished Civilian Service Award.

Introduction

High-performance computing (HPC) at the National Security Agency (NSA) is multilevel and widely distributed among users. NSA has six major HPC complexes that serve communities having common interests. Anywhere from 50 to several hundred individuals are served by any one complex. HPC is dominated by a full line of systems from Cray Research,

Inc., supplemented by a few other systems. During the past decade, NSA has been driving toward a high level of standardization among the computing complexes to improve support and software portability. Nevertheless, the standardization effort is still in transition. In this talk I will describe the HPC system at NSA. Certain goals of NSA, as well as the problems involved in accomplishing them, will also be discussed.

Characterization of HPC

NSA's HPC can handle enormous input data volumes. For the division between scalar and vector operations, 30 per cent scalar to 70 per cent vector is typical, although vector operations sometimes approach 90 per cent. Little natural parallelism is found in much of the code we are running because the roots of the code come from design and implementations on serial systems. The code has been ported and patched across 6600s, 7600s, CRAY-1s, X-MPs, and right up the line. We would like to redo much of that code, but that would present a real challenge.

An important characteristic of our implementation is that both batch and interactive operations are done concurrently in each complex with much of the software development. Some of these operations are permanent and long-term, whereas others are experimental. The complexes support a large research community. Although interactive efforts are basically day operations, many batch activities require operating the systems 24 hours a day, seven days a week.

HPC Architecture

At NSA, the HPC operating environment is split between UNIX and our home-grown operating system, Folklore, and its higher-level language, IMP. The latter is still in use on some systems and will only disappear when the systems disappear.

The HPC architecture in a complex consists of the elements shown in Figure 1. As stated before, both Folklore and UNIX are in use. About five or six years ago, NSA detached users from direct connection to supercomputers by giving the users a rich variety of support systems and more powerful workstations. Thus, HPC is characterized as a distributed system because of the amount of work that is carried out at the workstation level and on user-support systems, such as CONVEX Computer Corporation machines and others, and across robust networks into supercomputers.

NSA has had a long history of building special-purpose devices that can be viewed as massively parallel processors because most of them do

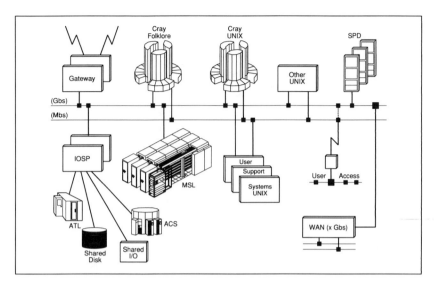

Figure 1. NSA's HPC architecture in the 1990s.

very singular things on thousands of processors. Over the past few years, NSA has invested a great deal of effort to upgrade networking and storage capacity of the HPC complexes. At present, a major effort is under way to improve the mass-storage system supporting these complexes. Problems abound in network support. Progress has been slow in bringing new network technology into this environment because of the need to work with a large number of systems, with new protocols, and with new interfaces. A great deal of work remains to be done in this field.

Software Environment

IMP, Fortran, and C are the main languages used in HPC at NSA. Although a general Ada support function is running in the agency (in compliance with Department of Defense requirements to support Ada), HPC users are not enthusiastic about bringing up Ada compilers on these systems. NSA plans to eliminate IMP because it has little vectorizing capability, and the user has to deal with vectorizing.

Faster compilers are needed, particularly a parallelizing C compiler. HPC also requires full-screen editors, as well as interactive debuggers that allow partial debugging of software. Upgrading network support is a slow process because of the number of systems involved and new protocols and interfaces. Upgrading on-line documentation, likewise, has been slow. Software support lags three to five years behind the introduction of new hardware technology, and we don't seem to be gaining ground.

Mass-Storage Requirements

A large number of automatic tape libraries, known as near-line (1012-bit) storage, have deteriorated and cannot be repaired much longer. Mass-storage systems must be updated to an acceptable level. Key items in the list of storage requirements are capacity, footprint, availability, and bit-error rate, and these cannot be overemphasized. In the implementation of new mass-storage systems, NSA has been driven by the need for standardization and by the use of commercial, supportable hardware, but the effort has not always been completely successful.

One terabyte of data can be stored in any one of the ways shown graphically in Figure 2. If stacked, the nine-track tape reels would reach a height 500 feet, almost as high as the Washington Monument. Clearly, the size and cost of storage on nine-track tapes is intolerable if large amounts of data are to be fed into users' hands or into their applications. Therefore, this type of storage is not a solution.

NSA is working toward an affordable mass-storage system, known as R1/R2, because the size is manageable and the media compact (see Figure 3). This goal should be achieved in the middle 1990s. Central to the system will be data management and a data-management system, database system, and storage manager for this kind of capability, all being considered as a server to a set of clients (Cray, CONVEX, Unisys). The mass-storage system also includes Storage Tek silos having capabilities approaching a terabyte in full 16-silo configuration. In addition, E-Systems is developing (funded by NSA) a very large system consisting of a D2 tape, eight-millimeter helical-scan technology, and 1.2×10^{15} bits in a box that has a relatively small footprint. Unfortunately, seconds to minutes are required for data transfer calls through this system to clients being served, but nevertheless the system represents a fairly robust near-line storage capacity.

Why is this kind of storage necessary? Because one HPC complex receives 40 megabits of data per second, 24 hours a day, seven days a week—so one of these systems would be full in two days. Why is the government paying for the development of the box? Why is industry not developing it so that NSA might purchase it? Because the storage technology industry is far from robust, sometimes close to bankruptcy.

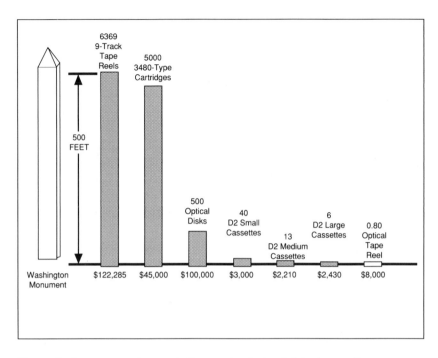

Figure 2. Storage requirements for one terabyte of data, by medium.

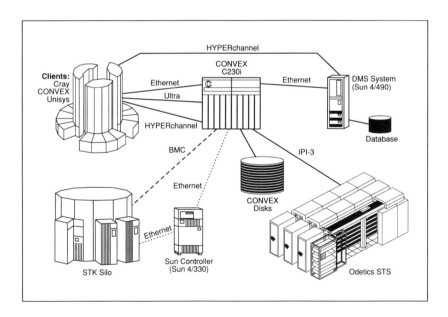

Figure 3. Mass-storage system for R1/R2 architecture.

Summary of Issues

I have addressed the following important issues:
- cost/performance relationships;
- large memory;
- mass storage;
- software environments;
- network support; and
- new architecture.

NSA is driven by computing requirements that would demand a 40 per cent improvement each year in cost/performance if annual investment were to be held steady. Since we are far from getting that improvement—even though cost/performance has improved a great deal over the years—the complexes are growing. We have problems that are intractable today because sufficiently large memories are not available on the systems. Mass storage and software environments have been thoroughly discussed. Network support, which is lagging behind, has not worked well with the storage industry or with the HPC industry. A much tighter integration of developments in the networking area is necessary to satisfy the needs of NSA.

HPC facilities issues include space, power, and cooling. We are seriously considering building an environmentally stable building that will allow the import of 40 kilowatts of power to the systems. However, such outrageous numbers should drive the computer industry toward cooler systems, new technology, and into the direction of superconductivity.

The High Performance Computing Initiative: A Way to Meet NASA's Supercomputing Requirements for Aerospace

Vic Peterson

Victor L. Peterson is Deputy Director of the National Aeronautics and Space Administration (NASA) Ames Research Center. He has a bachelor's degree in aeronautical engineering from Oregon State University, a master's degree in aeronautic and astronautic sciences from Stanford University, and a master's degree in management from the Alfred P. Sloan Fellow's Program at MIT. For over 15 years, he has directed programs to advance the use of supercomputers in various fields of science and engineering. He was one of the founders of NASA's Numerical Aerodynamic Simulation System Program.

Supercomputers are being used to solve a wide range of aerospace problems and to provide new scientific insights and physical understanding. They are, in fact, becoming indispensable in providing solutions to a variety of problems. In the engineering field, such problems include aerodynamics, aerothermodynamics, structures, propulsion systems, and controls. In the scientific field, supercomputers are tackling problems in turbulence physics, chemistry, atmospheric sciences, astrophysics, and human modeling. Examples of applications in the engineering field relate to the design of the next-generation high-speed civil transports, high-performance military aircraft, the National

Aerospace Plane, Aeroassisted Orbital Transfer vehicles, and a variety of problems related to enhancing the performance of the Space Shuttle. Example applications involving scientific inquiry include providing new insights into the physics and control of turbulence, determination of physical properties of gases, solids, and gas-solid interactions, evolution of planetary atmospheres—both with and without human intervention—evolution of the universe, and modeling of human functions such as vision.

Future computer requirements in terms of speed and memory have been estimated for most of the aerospace engineering and scientific fields in which supercomputers are widely used (Peterson 1989). For example, requirements for aircraft design studies in which the disciplines of aerodynamics, structures, propulsion, and controls are treated simultaneously for purposes of vehicle optimization can exceed 10^{15} floating-point operations per second and 10^{11} words of memory if computer runs are not to exceed about two hours (Figure 1). Of course, these requirements can be reduced if the complexity of the problem geometry and/or the level of physical modeling are reduced. These speed and memory requirements are not atypical of those needed in the other engineering and scientific fields (Peterson 1989).

Advancements in the computational sciences require more than more powerful computers (Figure 2). As the power of the supercomputer grows, so must the speed and capacity of scientific workstations and both fast-access online storage and slower-access archive storage. Network bandwidths must increase. Methods for numerically representing problem geometries and generating computational grids, as well as solution algorithms, must be improved. Finally, more scientists and engineers must be trained to meet the growing need stimulated by more capable computer systems.

The need for advancements in the computational sciences is not limited to the field of aerospace. Therefore, both the executive and legislative branches of the federal government have been promoting programs to accelerate the development and application of high-performance computing technologies to meet science and engineering requirements for continued U.S. leadership. The thrust in the executive branch is an outgrowth of studies leading to the federal High Performance Computing Initiative (HPCI) described in the September 8, 1989, report of the Office of Science and Technology Policy. The thrust in the legislative branch is summarized in draft legislation in both houses of Congress (S. 1067, S. 1976, and H. R. 3131, considered during the second session of

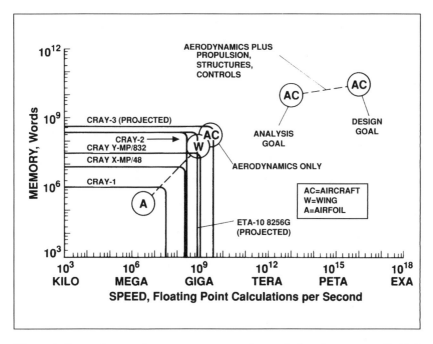

Figure 1. Computer speed versus memory requirements (two-hour runs with 1988 methods; aerodynamics from Reynolds-averaged Navier-Stokes equations).

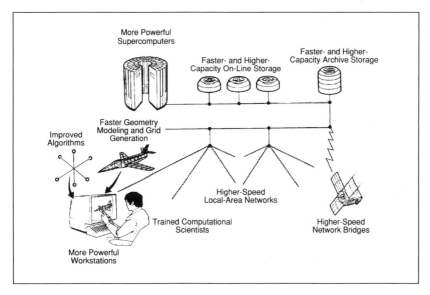

Figure 2. Advancements in computational sciences require more powerful supercomputers.

the 101st Congress). Some differences between the executive and legislative programs currently exist, but both programs have similar goals, and they both identify the Defense Advanced Research Projects Agency (DARPA), the Department of Energy (DOE), NASA, and NSF as principal implementing agencies. Participating organizations include the Environmental Protection Agency, the National Institutes of Health (NIH), the National Institute of Standards and Technology, and the National Oceanic and Atmospheric Agency.

Roles of the four principal agencies, together with lead-agency designations, have been assigned in the executive-branch version of HPCI (Table 1). Four areas of activity have been defined as (1) high-performance computing systems (with DARPA as the lead agency); (2) advanced software technology and algorithms (NASA to lead); (3) the National Research and Education Network (DARPA to lead for network research, and NSF to lead for network deployment); and (4) basic

Table 1. Responsibilities of Principal Agencies Involved in HPCI[a]

Activity	NASA	DARPA	NSF	DOE
High-Performance Computing Systems	• Application Testbeds • Architecture Evaluation	• Parallel Systems • Systems Software • Microsystems	• Basic Architecture Research	• Application Testbeds • Architecture Evaluation
Advanced Software Technology and Algorithms	• Software Coordination • Visualization • Data Management	• Software Tools • Parallel Algorithms	• Software Tools • Databases • Access	• Software and Computing Research
National Research and Education Network	• Network Interconnect	• Gigabit Research	• NREN Deployment	• Network Interconnect
Basic Research and Human Resources	• Universities • Institutes and Centers	• Universities • Industry	• Universities • Engineering Research and Science and Technology Centers	• Universities • National Labs

[a]Shading indentifies lead agency or agencies.

research and human resources (no lead agency). The participating organizations will undertake efforts to solve grand-challenge computational problems appropriate to their missions.

Objectives of NASA involvement in HPCI are threefold: (1) develop algorithm and architecture testbeds capable of fully utilizing massively parallel concepts and increasing end-to-end performance, (2) develop massively parallel architectures scalable to 10^{12} floating-point operations per second, and (3) demonstrate technologies on NASA research challenges.

NASA applications or grand-challenge problems will be undertaken in three distinct areas: (1) computational aerosciences, (2) earth and space sciences, and (3) remote exploration and experimentation. The Ames Research Center will lead in the computational-aerosciences area, and the problems will relate to integrated multidisciplinary simulations of aerospace vehicles throughout their mission profiles. The Goddard Spaceflight Center will lead in the earth- and space-sciences area, and the problems will relate to multidisciplinary modeling and monitoring of the earth and its global changes and assessments of their impact on the future environment. Finally, the Jet Propulsion Laboratory will lead in the remote-exploration and experimentation area, and the problems will relate to extended-duration human exploration missions and remote exploration and experimentation.

In summary, supercomputing has become integral with and necessary to advancements in many fields of science and engineering. Approaches to making further advancements are known, so the performance of supercomputing systems is pacing the rate of progress. Supercomputer performance requirements for making specific advancements have been estimated, and they range over seven or eight orders of magnitude in speed and two orders of magnitude in main-memory capacity beyond current capabilities. A major new thrust in high-performance computing is being planned to help meet these requirements and assure continued U.S. leadership in the computational sciences into the 21st century.

Reference
Victor L. Peterson, "Computational Challenges in Aerospace," *Future Generation Computer Systems* **5** (2–3), 243–258 (1989).

The Role of Computing in National Defense Technology

Bob Selden

Bob Selden received his B.A. degree from Pomona College, Claremont, California, and his Ph.D. in physics from the University of Wisconsin, Madison.

He worked at Los Alamos National Laboratory from 1979 to 1988, then served as science advisor to the Air Force Chief of Staff and to the Secretary of the Air Force from 1988 to 1991. Subsequently, he returned to Los Alamos in his current position as an Associate Director for Laboratory Development. In this capacity, his principal responsibilities include providing the strategic interface with the Department of Defense.

Bob Selden has received both the Air Force Association's Theodore von Karman Award for outstanding contributions to defense science and technology and the Air Force Decoration for exceptional civilian service.

The focus of my presentation is on the use of computers and computational methods in the research and development of defense technology and on their subsequent use in the technology, itself. I approach this subject from a strategic standpoint.

Technology for defense is developed to provide the capabilities required to carry out tasks in support of our defense policy and strategy. Considering the broad scope of defense policy and strategy, and the capabilities needed to support them, makes the fundamental role of computers and computational methods self-evident.

Our national security policy can be simply stated as our commitment to protect our own freedom and our way of life and to protect those same things for our friends and allies around the world. National security strategy can be put into two broad categories. The first category is deterring nuclear war. The second is deterring or dissuading conventional war and, failing that, maintaining the capability to conduct those actions that are necessary at the time and place you need to conduct them.

These simply stated strategy objectives provide the basis for the defense forces that exist today. As we look into the future, the characteristics of the kinds of systems that the military has to have are forces that are mobile and have speed, flexibility, and a lot of lethality. You may have to attack tank armies in the desert and go in and shoot weapons at the enemy one at a time with an airplane, against an enemy that has sophisticated defenses, which can result in many losses of airplanes and pilots. You also need accuracy. For instance, suppose the U.S. has to go in and take out all of the facilities related only to chemical warfare. In that case, chemical storage sites and the means of delivering chemical weapons would have to be exactly targeted. To have the fundamental capability to do any of those things, we need systems that provide information, communications, and command and control, as well as the ability to make all the elements tie together so that we know where those elements are, when they are going to be there, how to organize our forces, and how to make the best use of those forces.

Now, let us look at the enabling technologies that allow such complexities of modern warfare to take place successfully. Many of the key enabling technologies depend on the exploitation of electronics and electromagnetics. In short, a major part of the ball game is the information revolution—computing, sensors, communication systems, and so forth, in which very, very dramatic changes are already under way and more dramatic changes are yet to come.

Supercomputing as a research methodology has not truly come of age inside the Department of Defense (DoD). As a whole, the DoD laboratories, as opposed to the national laboratories, also have not been involved with computing as a methodology. It is true that part of the problem is cost, as well as procurement regulations, etc. But the real issue is that there is not a supercomputing culture—a set of people like there is in many of the organizations we have heard from during this conference, to push for computing as a major methodology for doing R&D. Being able to recognize the significance of the broad supercomputing culture will result in a tremendous payoff in investments in large-scale computation as a part of the research process within DoD.

Despite these comments, we are seeing an absolutely unprecedented use of data processing, computing, and computing applications in military hardware, operations, simulations, and training. This is a revolution in the kinds of military equipment we use and in the way we train. It is also true that the number-one logistics problem for maintenance cited in military systems today is software.

Now I would like to discuss some of the impact and applications of computing in military systems and military operations. Computing is a fairly unique kind of technology in that it is both enabling and operational in end products. It is an enabling technology because you do research with computers, and it is an operational technology because you use it with real equipment in real time to do things in the analysis and management, as well as in the systems, themselves. Computing and computational methods are pervasive from the very beginning of the research, all the way to the end equipment.

In operations, real-time computing is an extremely challenging problem. For instance, to be able to solve a problem in a fighter airplane by doing the data processing from a complex electronic warning system and a synthetic-aperture radar, the computational data processing and analysis must be accomplished in near real time (perhaps seconds) and a display or other solutions presented to the pilots so that they will be able to make a decision and act on it. This complex interaction is one of the hardest computational problems around. In fact, it is every bit as challenging as any problem that is put on a Cray Research, Inc., computer at Los Alamos National Laboratory.

Another area of application is in the area of simulation, which includes training, simulation, and analysis. This is going to be an area that is just on the verge of exploding into the big time, partly because of the funding restrictions imposed on the use of real systems and partly because the training simulators, themselves, are so powerful. We already have cockpit simulators for pilots, tanks, training, war games, and so on. The National Testbed that the Strategic Defense Initiative is sponsoring in Colorado Springs is also an example of those kinds of large-scale computer simulations.

The world of computing has changed a great deal over the past decade. A look at Figure 1, the distribution of installed supercomputing capability in the U.S. in 1989, shows the leadership of defense-related organizations in supercomputing capabilities. It also shows a growing capability within DoD.

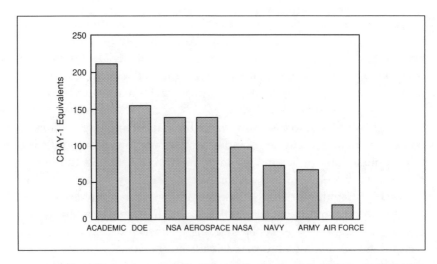

Figure 1. Installed supercomputing capability in the U.S. in 1989. CYBER 205-class or above (source: Cray Research, Inc.).

In conclusion, computing is coming of age in both the development and operation of defense technology. Future capabilities are going to rely even more on computation and computational methodology, and this will also be a time of planning, training, and analysis. Computing is a pervasive enabling technology.

NSF Supercomputing Program

Larry Smarr

> *Larry Smarr is currently a professor of physics and astronomy at the University of Illinois-Urbana/Champaign and since 1985 has also been the Director of the National Center for Supercomputing Applications.*
>
> *He received his Ph.D. in physics from the University of Texas at Austin. After a postdoctoral appointment at Princeton University, Dr. Smarr was a Junior Fellow in the Harvard University Society of Fellows. His research has resulted in the publication of over 50 scientific papers.*
>
> *Dr. Smarr was the 1990 recipient of the Franklin Institute's Delmer S. Fahrney Medal for Leadership in Science or Technology.*

I attended the 1983 Frontiers of Supercomputing conference at Los Alamos National Laboratory, when the subject of university researchers regaining access to supercomputers—after a 15-year hiatus—was first broached. There was a lot of skepticism as to whether such access would be useful to the nation. That attitude was quite understandable at the time. The university community is the new kid on the block, so far as participants at this conference are concerned, and we were not substantially represented at the meeting in 1983.

Today, the attitude is quite different. Part of my presentation will be devoted to what has changed since 1983.

As you know, in 1985–86 the National Science Foundation (NSF) set up five supercomputing centers, one of which has since closed (see Al

Brenner's paper, Session 12). The four remaining centers are funded through 1995. Three of the four supercomputer center directors are attending this conference. Apart from myself, representing the National Center for Supercomputing Applications (NCSA) at the University of Illinois, there is Sid Karin from San Diego and Michael Levine from Pittsburgh, as well as the entire NSF hierarchy—Rich Hirsh, Mel Ciment, Tom Weber, and Chuck Brownstein, right up to the NSF Director, Erich Bloch (a Session 1 presenter).

During the period 1983–86, we started with no network. The need to get access to the supercomputer centers was the major thing that drove the establishment of the NSF network. The current rate of usage of that network is increasing at 25 per cent, compounded, per month. So it's a tremendous thing.

There were three universities that had supercomputers when the program started; there are now well over 20. So the capacity in universities has expanded by orders of magnitude during this brief period. During those five years, alone, we've been able to provide some 11,000 academic users, who are working on almost 5000 different projects, access to supercomputers, out of which some 4000 scientific papers have come. We've trained an enormous number of people, organized scientific symposia, and sponsored visiting scientists on a scale unimagined in 1983.

What I think is probably, in the end, most important for the country is that affiliates have grown up—universities, industries, and vendors—with these centers. In fact, there are some 46 industrial partners of the sort discussed extensively in Session 9, that is, the consumers of computers and communications services. Every major computer/communications-services vendor is also a working partner with the centers and, therefore, getting feedback about what we need in the future.

If I had to choose one aspect, one formerly pervasive attitude, that has changed, it's the politics of inclusion. Until the NSF centers were set up, I would say most supercomputer centers were operated by exclusion, that is, inside of laboratories that were fairly well closed. There was no access to them, except for, say, the Department of Energy Magnetic Energy Facility and the NSF National Center for Atmospheric Research. In contrast, the NSF centers' goal is to provide access to anyone in the country that has a good idea and the capability of trying it out.

Also, unlike almost all the other science entities in the country, instead of being focused on a particular science and engineering mission, we are open to all aspects of human knowledge. That's not just the natural sciences. As you know, many exciting breakthroughs in computer art and music and in the social sciences have emerged from the NSF centers.

If you imagine supercomputer-center capacity represented by a pie chart (Figure 1), the NSF directorate serves up the biggest portion to the physical sciences. Perhaps three-quarters of our cycles are going to quantum science. I find it very interesting to recall being at Livermore in the 1970s, and it was all continuum field theory, fluid dynamics, and the like. So the whole notion of which kind of basic science these machines should be working on has flip-flopped in a decade, and that's a very profound change.

The centers distribute far more than cycles. They're becoming major research centers in computational science and engineering. We have our own internal researchers—world-class people in many specialties—that work with the scientific community, nationwide; some of the most important workshops in the field are being sponsored by the centers. You're also seeing us develop software tools that are specific to particular disciplines: chemistry, genome sequencing, and so forth. That will be a significant area of growth in the future.

There's no preexisting organizational structure in our way of doing science because the number of individuals who do computing in any field of science is still tiny. Their computational comrades are from biology,

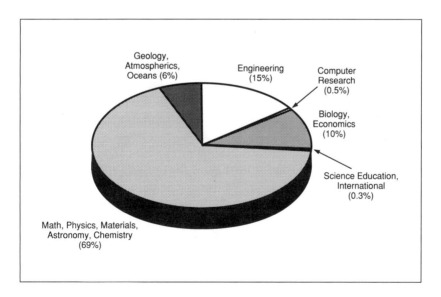

Figure 1. Usage, by discipline, at NSF supercomputing centers.

chemistry, engineering—you name it—and there are no national meetings and no common structure that holds them together culturally. So the centers are becoming a major socializing force in this country.

What we are seeing, as the centers emerge from their first five-year period of existence and enter the next five-year period, is a move from more or less off-the-shelf, commercially available supercomputers to a very wide diversity of architectures. Gary Montry, in his paper (see Session 6), represents the divisions of parallel architecture as a branching tree. My guess is that you, the user, will have access to virtually every one of those little branches in one of the four centers during the next few years.

Now, with respect to the killer-micro issue (also discussed by Ben Barker in Session 12), in the four extant centers we have about 1000 workstations and personal computers, and at each center we have two or three supercomputers. Just like all of the other centers represented here, we at NCSA have focused heavily on the liberating and enabling aspect of the desktop. In fact, I would say that at the NSF centers from the beginning, the focus has been on getting the best desktop machine in the hands of the user and getting the best network in place—which in turn drives more and more use of supercomputers. If you don't have a good desktop machine, you can't expect to do supercomputing in this day and age. So workstations and supercomputers form much more of a symbiosis than a conflict. Furthermore, virtually every major workstation manufacturer has a close relationship with one or more of the centers.

The software tools that are developed at our centers in collaboration with scientists and then released into the public domain are now being used by over 100,000 researchers in this country, on their desktops. Of those, maybe 4000 use the supercomputer centers. So we have at least a 25-to-one ratio of people that we've served on the desktop, compared with the ones that we've served on the supercomputers, and I think that's very important. The Defense Advanced Research Projects Agency has, as you may know, entered into a partnership with NSF to help get some of these alternate architectures into the centers. In the future, you're going to see a lot of growth as a result of this partnership.

The total number of CRAY X-MP-equivalent processor hours that people used at all five centers (Figure 2) has steadily increased, and there is no sign of that trend tapering off. What I think is more interesting is the number of users who actually sign on in a given month and do something on the machines (Figure 3). There is sustained growth, apart from a period in late 1988, when the capacity didn't grow very fast and the machines became saturated, discouraging some of the users. That was a very clear warning to us: once you tell the scientific community that

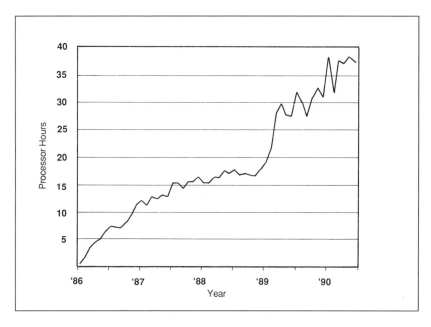

Figure 2. Total CRAY X-MP-equivalent processor hours used in five NSF supercomputing centers.

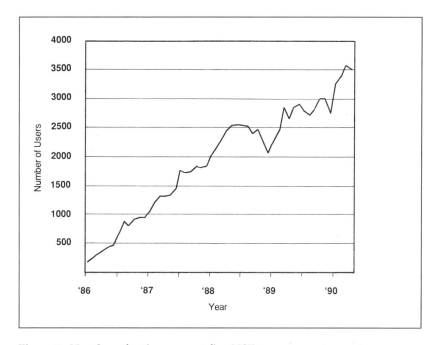

Figure 3. Number of active users at five NSF supercomputer centers.

you're going to provide a new and essential tool, you've made a pact. You have to continue upgrading on a regular, and rapid, basis, else the user will become disenchanted and do some other sort of science that doesn't require supercomputers. We think that this growth will extend well into the future.

I am especially excited about the fact that users, in many cases, are for the first time getting access to advanced computers and that the number of first-time users grew during the time that desktops became populated with ever-more-powerful computers. Instead of seeing the demand curve dip, you're seeing it rise even more sharply. Increasingly, you will see that the postprocessing, the code development, etc., will take place at the workstation, with clients then throwing their codes across the network for the large uses when needed.

Who, in fact, uses these centers? A few of our accounts range upwards of 5000 CPU hours per year, but 95 per cent of our clients consume less than 100 hours per year (Figure 4). The implication is that much of the work being done at the centers could be done on desktop machines. Yet, these small users go to the trouble to write a proposal, go through peer review, experience uncertainty over periods of weeks to months as to whether and when they'll actually get on the supercomputer, and then have to go through what in many cases is only a 9600-Baud connect by the time we get down to the end of the regional net.

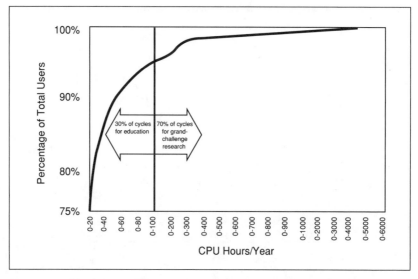

Figure 4. Percentage of total users versus annual CPU-hour consumption, January FY 1988 through April FY 1990: 95 per cent of all users consume less than 100 CPU hours per year.

It's like salmon swimming upstream: you can't hold them back. We turn down probably 50 per cent of the people who want to get on the machine, for lack of capacity.

What has happened here is that the national centers perform two very different functions. First, a great many of the users, 95 per cent of them, are being educated in computational science and engineering, and they are using their workstations simultaneously with the supercomputers. In fact, day to day, they're probably spending 90 per cent of their working hours on their desktop machines. Second, because of the software our centers have developed, the Crays, the Connection Machines, the Intel Hypercubes are just windows on their workstations. That's where they are, that's how they feel.

You live on your workstation. The most important computer to you is the one at your fingertips. And the point is, with the network, and with modern windowing software, everything else in the country is on your desktop. It cuts and pastes right into the other windows, into a word processor for your electronic notebook.

For instance, at our center, what's really amazing to me is that roughly 20 per cent of our users on a monthly basis are enrolled in courses offered by dozens of universities—courses requiring the student to have access to a supercomputer through a desktop Mac. That percentage has gone up from zero in the last few years.

These and the other small users, representing 95 per cent of our clients, consume only 30 per cent of the cycles. So 70 per cent of the cycles, the vast majority of the cycles, are left for a very few clients who are attacking the grand-challenge problems.

I think this pattern will persist for a long time, except that the middle will drop out. Those users who figure it out, who know what software they want to do, will simply work on their RISC workstations. That will constitute a very big area of growth. And that's wonderful. We've done our job. We got them started.

You can't get an NSF grant for a $50,000 workstation unless you've got a reputation. You can't get a reputation unless you can get started. What the country lacked before and what it has now is a leveraging tool. Increasing the human-resource pool in our universities by two orders of magnitude is what the NSF centers have accomplished.

But let me return to what I think is the central issue. We've heard great success stories about advances in supercomputing from every agency, laboratory, and industry—but they're islands. There is no United States of Computational Science and Engineering. There are still umpteen colonies or city-states. The network gives us the physical wherewithal to change that. However, things won't change by themselves. Change requires political will and social organization. The NSF centers are becoming a credible model for the kind of integration that's needed because, just in terms of the dollars, alone (not the equipment-in-kind and everything else—just real, fundable dollars), this is the way the pie looks in, say, fiscal year 1989 (FY 1989) (Figure 5).

There is a great deal of cost sharing among each of the key sectors—the state, the regional areas, the consumers of the computers, the producers. NSF is becoming the catalyst pulling these components together. Next, we need to do something similar, agency to agency. The High Performance Computing Initiative (HPCI) is radical because it is a prearranged, multiagency, cooperative approach. The country has never seen that happen before. Global Change is the only thing that comes close. But that program hasn't had the benefit of a decade of detailed planning the way HPCI has. It's probably the first time our country has tried anything like it.

I would like to hear suggestions on how we might mobilize the people attending this conference—the leaders in all these islands—and, using the political and financial framework afforded by HPCI over the rest of this decade, change our way of doing business. That's our challenge. If we can meet that challenge, we won't need to worry about competitiveness in any form. Americans prevail when they work together. What we are not good at is making that happen spontaneously.

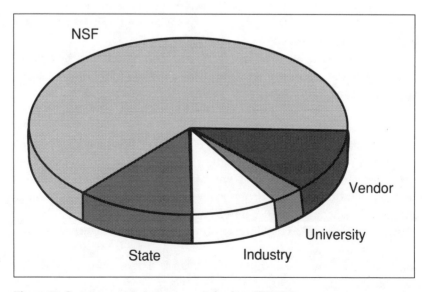

Figure 5. Supercomputer-center cost sharing, FY 1989.

11

International Activity

In this session, panelists discussed international supercomputing technical developments, including future prospects, systems, and components, as well as opportunities for international cooperation and competition. The role of government assistance at all levels was also examined.

Session Chair

Seymour Goodman, University of Arizona

A Look at Worldwide High-Performance Computing and Its Economic Implications for the U.S.*

Robert Borchers, Seymour Goodman, Michael Harrison, Alan McAdams, Emilio Millán, and Peter Wolcott

Robert R. Borchers is currently the Assistant to the Director for University Relations at Lawrence Livermore National Laboratory (LLNL). Before accepting this assignment in 1992, he was the Associate Director for Computation at LLNL. In that role, his responsibilities included overseeing all large-scale computing at LLNL. He has been active in the supercomputing community as a member of the Supercomputing Conference Steering Committee, the Program Advisory Committee for NSF Centers, and numerous other panels and committees. He is the immediate past and founding editor of Computers in Physics, *published by the American Institute of Physics. Before coming to LLNL in 1979, Bob held professional and administrative positions at the University of Wisconsin, Madison, and the University of Colorado, Boulder.*

---

* This paper is based on the presentations of several Session 11 panelists. The authors wish to thank Karyn R. Ames and Gary D. Doolen of Los Alamos National Laboratory and conference transcriber Steven T. Brenner for their help in compiling the information.

Seymour E. Goodman is Professor of Management Information Systems and Policy and head of the Mosaic research group at the University of Arizona. He studies international developments in information technology and related public policy issues. Professor Goodman has chaired several national-advisory and study groups concerned with international computing, including the Committee to Study International Developments in Computer Science and Technology and the Computer Subpanel of the Panel on the Future Design and Implementation of U.S. National Security Export Controls, both under the National Research Council of the National Academy of Sciences. He is a contributing editor of International Perspectives of the Communications of the ACM *and editor of technology and transnational issues for* International Information Systems. *Professor Goodman was an undergraduate at Columbia University and received his Ph.D. from Caltech.*

Michael A. Harrison is Professor of Computer Science at the University of California at Berkeley. He received a B.S. and M.S. in electrical engineering from Case Institute of Technology in Cleveland in 1958 and 1959, respectively, and a Ph.D. from the University of Michigan in Ann Arbor in 1963. His activities on behalf of professional societies is extensive. He is currently a Director of the American Federation of Information Processing Societies and has served for four years on the Computer Science and Technology Board of the National Academy of Sciences. Professor Harrison is a consulting editor for Addison Wesley Publishing Co. and is an editor of Discrete Mathematics, Discrete Applied Mathematics, Information Processing Letters, Theoretical Computer Science, *and the* Journal of Computer and System Science. *He has written five books and well over a hundred technical papers. Areas of research in which he specializes include switching theory, automata, formal language theory, protection in operating systems, electronic document systems, and programming environments. Currently, his work centers upon the creation of multimedia systems. Professor Harrison is the founder and Chairman of the Board of Gain Technology Inc.*

Alan K. McAdams received his B.A. from Yale University and his M.B.A. and Ph.D. from Stanford University. He has taught at Cornell University throughout his academic career. Professor McAdams was a Senior Staff Economist with the President's Council of Economic Advisors from 1971 through 1972, with areas of responsibility in the economics of science and technology policy. From 1972 to 1978, he was a member of the NRC/NAS Advisory Panel for the Institute for Computer Sciences and Technology of the National Bureau of Standards. Professor McAdams chaired the Office of Technology Assessment Advisory Panel for the study U.S. Industrial Competitiveness: Steel, Electronics, and Automobiles *(1981). He is a member of the American Economic Association and the Institute for Electrical and Electronics Engineers. His publications include "The Computer Industry," in* Structure of American Industry, *6th edition (1982);* Economic Benefits and Public Support of a National Education and Research Network *(1988); several monographs on electronic networks (1987–1988); and* HDTV and the Information Age *(1991).*

Emilio Millán graduated with a bachelor's degree in science and Russian from Dartmouth College, Hanover, New Hampshire. He is now a master's degree candidate at the Department of Computer Science at the University of Illinois-Urbana/Champaign, where he pursues his interests in machine translation and language analysis. Research programs that he currently pursues are being carried out in conjunction with Seymour Goodman and Peter Wolcott of the University of Arizona, Tucson.

Peter Wolcott received his B.A. (magna cum laude) in computer science and Russian from Dartmouth College in 1984. He is a Ph.D. candidate in the Management Information Systems Department at the University of Arizona. His specialties are the development of software and high-performance computing systems in the former Soviet Union and Eastern Europe. He is a member of the Institute of Electrical and Electronics Engineers Computer Society, the Association for Computing Machinery, and the Dobro Slovo Slavic-studies honor society.

Abstract

The Japanese are mounting a threat to the American position as preeminent producer of high technology. This threat has substantial implications not only for American high-technology industries and the high-performance computing industry in particular but for national security, as well. This paper examines the worldwide high-performance computing market in an attempt to place the U.S., Japan, and a number of other countries in a global context. The reasons for the erosion of American dominance are considered and remedies suggested.

A Brief Technical Overview of the Present-Day Landscape

The United States has historically been the dominant country in the world in terms of both supercomputer development and application. The U.S. has the lead in both vector and parallel processing, and Cray Research, Inc., continues to be the preeminent company in the high-performance system industry. Moreover, the wide spectrum of approaches employed by U.S. supercomputer developers has resulted in an extremely fertile research domain from which a number of commercially successful companies have emerged—CONVEX Computer Corporation, Thinking Machines Corporation, and nCUBE Corporation among them. The U.S. high-performance-system user base can claim a sophistication exceeding or roughly equal to that in any other country.

However, we are not alone. A number of countries have undertaken extensive research efforts in the high-performance computing arena, including the Soviet Union, Japan, and some in Western Europe. Others, such as Bulgaria, Israel, and China, have initiated research in this area, and many countries now employ supercomputers. In this section we examine some of the more substantial efforts worldwide.

The Soviet Union

The Soviets have a long history of high-performance computing. The USSR began research into computing shortly after World War II and produced functional digital computers in the early 1950s. The first efforts in parallel processing began in the early 1960s, and research in this area has continued steadily since then.

Soviet scientists have explored a wide spectrum of approaches in developing high-performance systems but with little depth in any one. Consequently, the Soviets have yet to make a discernible impact on the global corpus of supercomputing research. The Soviets to date have neither put into serial production a computer of CRAY-1 performance or greater—only within the last few years have they prototyped a machine at that level—nor have they yet entered the worldwide supercomputer market. However, Soviet high-performance computing efforts conducted within the Academy of Sciences have exhibited higher levels of innovation than have their efforts to develop mainframes, minicomputers, and microcomputers.*

The BESM-6, a machine that is capable of a million instructions per second (MIPS) and was in serial production from 1965 to 1984, has been, until recently, the workhorse of the Soviet scientific community. The concept of a recursive-architecture machine with a recursive internal language, recursive memory structure, recursive interconnects, etc., was reported by Glushkov et al. (1974). The ES-2704, which only recently entered limited production, is a machine embodying these architectural and data-flow features. Computation is represented as a computational node in a graph. The graph expands as nodes are decomposed and contracts as results are combined into final results.

The ES-2701, developed at the Institute of Cybernetics in Kiev, like the ES-2704, incorporates distributed-memory flexible interconnects but is based on a different computational paradigm—there called a macropipeline computation—in which pipelining occurs at the algorithm level. Computation, under some problems, progresses as a wave across the processor field as data and intermediate results are passed from one processor to the next.

The ES-2703 is promoted as a programmable-architecture machine. The architecture is based on a set of so-called macroprocessors connected by a crossbar switch that may be tuned by the programmer. The "macro" designation denotes microcode or hardware implementation of complex mathematical instructions.

The El'brus project is the most heavily funded in the Soviet Union. The El'brus-1 and -2 were strongly influenced by the Burroughs 700-series architecture, with its large-grain parallelism, multiple processors sharing banks of common memory, and stack-based architecture for the individual processors. A distinguishing feature of this first El'brus machine stemmed from the designers' decision to use, in lieu of an assembly language, an Algol-like, high-level procedural language with underlying hardware support. This compelled the El'brus design team to

---

* Development of these latter computers has been confined largely to imitation of Western, primarily IBM and DEC, architectures.

maintain software compatibility across the El'brus family at the level of a high-level language, which in turn enabled them to use very different architectures for some of their later models (e.g., the El'brus-3 and mini-El'brus, both very-long-instruction-word machines).

Most of the more successful machines, from the point of view of production, have been developed through close cooperation between the Academy of Sciences and industry organizations. One such machine, the PS-2000, was built by an organization in the eastern Ukraine—the Impul's Scientific Production Association. The PS-2000 could have up to 64 processors operating in a SIMD fashion, and its successor, the PS-2100, combines 10 groupings of the 64 processors, with the whole complex then being able to operate in a MIMD fashion. Although now out of production, 200 PS-2000s were produced in various configurations and now are actively used primarily in seismic and other energy-related applications. Series production of the PS-2100 began in 1990.

The development of high-performance computing in the Soviet Union is hindered by a number of problems. For one, the supply of components, both from indigenous suppliers and from the West, is inconsistent. Moreover, the state of mass storage is very weak. The 317-megabyte disks, which not long ago represented the Soviet state of the art, continue to be quite rare. Further, *perestroika*-related changes have caused sharp reductions in funding of several novel architecture projects, and a number have been terminated.

Western Europe

In Western Europe, while there has been no prominent commercial attempt to build vector processors, much attention has been paid to developing distributed processing and massively parallel, primarily Transputer-based, processors. Efforts in this realm have resulted in predominantly T-800 Transputer-based machines claiming processing rates of 1.5 million floating-point operations per second (MFLOPS) per processor, with up to 1000 processors and with RISC-based chips promising to play a sizable role in the future. To date, however, the Europeans have been low-volume producers, with few companies having shipped more than a handful of machines. Two such exceptions are the U.K.'s Meiko and Germany's Parsytec.

Meiko and Parsytec have proved to be the two most commercially successful European supercomputer manufacturers, with over 300 and 600 customers worldwide, respectively. Meiko produces two scalable, massively parallel dynamic-architecture machines—the Engineer's Computing Surface and the Embedded Real-Time Computing Surface—

with no inherent architectural limit on the number of processors. Among Meiko's clients are several branches of the U.S. military and the National Security Agency. Parsytec's two Transputer-based MIMD systems, the MultiCluster and SuperCluster, are available in configurations with maximums of 64 and 400 processors, respectively.

Lesser manufacturers of high-performance computing include Parsys, Active Memory Technology (AMT), and ESPRIT—the European Strategic Program for Research and Development in Information Technology.* The U.K.-based Parsys is the producer of the SuperNode 1000, another Transputer-based parallel processor, with 16 to 1024 processors in hierarchical, reconfigurable arrays. AMT's massively parallel DAP/CP8 510C (1024 processors) and 610C (4096 processors) boast processing speeds of 5000 MIPS (140 MFLOPS) and 20,000 MIPS (560 MFLOPS), respectively. Spearheaded by the Germans, ESPRIT's SUPRENUM project has produced the four-GFLOPS, MIMD SUPRENUM-1 and is continuing development of the more powerful SUPRENUM-2.

The Europeans have proved themselves as experts in utilizing vector processors as workhorses. Vector processors can be found in use in Germany, France, and England. Though the Europeans have been extensive users of U.S.-made machines, Japanese machines have recently started to penetrate the European market.

Japan

Japan is maturing in its use and production of high-performance systems. The Japanese have elevated vector processing to a fine art, both in the case of hardware and software, and are producing world-class systems that rival those of Cray. Moreover, the installed base of supercomputers in Japan has climbed to over 150, the number of Japanese researchers working in the realm of computational science and engineering is growing, and the quality of their work is improving.

The first vector processors to emerge from Japan, such as the Fujitsu VP-200, generated a lot of excitement. Initial benchmarks indicated that these early supercomputers, with lots of vector pipelines—characteristic of the Japanese machines—were very fast. The Fujitsu machine was followed by the Hitachi S-820 and then the Nippon Electric Company (NEC) SX-2, which was, at that time, the fastest single processor in the

---

* Still others include the British Computer Systems Architecture, Real World Graphics, and Dolphin Server Technology, whose Orion project has as its aim the production, by 1993, of a 1000-MIPS multiprocessing server based on Motorola processors. ESPRIT is a joint venture of the European Community.

world.* These machines also boasted many vector pipes, as well as automatic interactive vectorizing tools of high quality.

Recent Japanese announcements indicate that the trend toward greater vectorization will continue. The NEC SX-3, for example, employs a processor that can produce 16 floating-point results every three-nanosecond clock cycle, a performance that amounts to more than five GFLOPS per processor.

It merits mention, however, that while Japanese high-performance computers compete well in the "megaflop derby," their sustained performance on production workloads remains unknown. Huge memory bandwidth hides behind the caches of these Japanese machines, and the memories are a fairly long distance from the processors, which probably inhibits their short vector performance.

Parallel processing is not, however, being ignored in Japan. The Japanese have a number of production parallel processors now to which they are devoting much attention. In at least two areas of parallel processing, the Japanese have made significant progress. Most, if not all, Japanese semiconductor manufacturers are using massively parallel circuit simulators, and the NEC fingerprint identification machine, used in police departments worldwide, represents one of the largest-selling massively parallel processors in the world.

The Japanese recently have begun showing signs of accommodating U.S. markets. For one thing, Japanese manufacturers are exhibiting some willingness to accommodate the IEEE and Cray floating-point arithmetic formats, in addition to the IBM format their machines currently support. Secondly, some machines, notably the SX-3, now run UNIX. These and other existing signs indicate that the Japanese seek not only to accommodate the American market but to aggressively enter it.

The software products available on Japanese supercomputers and the monitoring tools available to scientific applications programmers from Japanese vendors appear to be as good as or better than those available from Cray Research. Consequently, applications software being developed in Japan may be better vectorized as a result of the better tools and vendor-supplied software. Further, Japanese supercomputer centers seem to be having little, if any, difficulty obtaining access to the best U.S.-developed applications software.

While the U.S. appears to be preeminent in all basic research areas of computational science and engineering, the Japanese are making

---

* The Hitachi S-280/80 is now considered to be the fastest single-processor supercomputer for most applications.

significant strides as the current generation of researchers matures in its use of supercomputers and a younger generation is trained in computational science and engineering. The environment in which Japanese researchers work is also improving, with supercomputer time and better software tools being made increasingly available. Networking within the Japanese supercomputing community, however, remains underdeveloped.

The American, Soviet, European, and Japanese machines and their parameters are compared in Table 1.

The Japanese Challenge and "McAdams's Laws"

We now shift focus and tone to take up a number of the economic and political issues associated with high-performance computing by employing "McAdams's Laws" to examine the nature and possible impact of the Japanese challenge in the high-technology market.

Introduction

At the end of World War II, the U.S. gross national product equaled over half of the gross product of the entire world. During the post-World War II period of American economic and military hegemony, the U.S. pursued a national policy that favored activities designed to contain "world communism" over the interests of its domestic economy. Starting from the position of overwhelming predominance, these choices seemed necessary and obvious.

Since that time, much has happened in the world to clarify our perceptions. World communism was not only successfully contained over the last 40 years, but today, in many nations, communism and socialism are being abandoned in favor of democracy and capitalism. The fall of communism in Eastern Europe and elsewhere is viewed by many as a harbinger of a "victory" over world communism and a demonstration of the superiority of American-style laissez faire capitalism to other economic systems. However, there are also many who believe that in its efforts to contain communism, the U.S. may have brought its economy—especially its high-technology sectors—to a position close to ruin.

Law 1: that which is currently taking place is not impossible.

The perception of U.S. dominance as assured and perpetual is severely flawed. The U.S. may soon cease to be the world's commercial leader in the field of supercomputers and has rapidly lost ground in other areas,

Table 1. Parameters of Various High-Performance Systems

Machine	Peak Performance	Number of Processors	Year of First Production[a]	
American				
CRAY X-MP	0.87 GFLOPS	4	1983	
CRAY X-MP	2.7 GFLOPS	8	1988	
CRAY-2	2 GFLOPS	4	1984	
CRAY-3[b]	16 GFLOPS	16	N/A	
CRAY C90	16 GFLOPS	16		
Soviet				
BESM-6	1 MIPS	1	1965	(1964)
ES-2701	530 MIPS	48	N/A	(1984)
ES-2703	1 GIPS (32-bit)	64 macroprocessors	N/A	(1985)
ES-2704	100 MIPS	24 computational 48 communications 12 switching	1990	(1980)
El'brus-1	12–15 MIPS	10	N/A	(1979)
El'brus-2	94 MFLOPS	10	1985	(1984)
El'brus-3[b]	6.4 GFLOPS	16	N/A	(N/A)
El'brus-MKP	560 MFLOPS	1	1991	(1988)
Electronika-SSBIS	450 MFLOPS	2	1991?	(1990)
PS-2000	200 MIPS (24-bit)	64	1981	(1980)
PS-2100	1.5 GIPS (32-bit)	640	1990	(1987)
European				
Parsytec MultiCluster		64 (max.)		
Parsytec SuperCluster		400 (max.)		
AMT DAP/CP8 510C	5,000 MIPS (140 MFLOPS)	1,024		
AMT DAP/CP8 610C	20,000 MIPS (560 MFLOPS)	4,096		
ESPRIT SUPRENUM-1	4 GFLOPS			
Japanese				
Fujitsu VP-200	4 GFLOPS	1	1983	
Hitachi S-810/20	0.63 GFLOPS	1	1983	
Hitachi S-820/80	3 GFLOPS	1	1988	
NEC SX-2	1.3 GFLOPS	1	1985	
NEC SX-3	22 GFLOPS	4	1990	
Fujitsu VP-400E	1.7 GFLOPS	1	1987	
Fujitsu VP-2600	5 GFLOPS	1	1990	

[a] For Soviet machines, "year of first production" is not necessarily a good benchmark, so in parentheses appears the year that prototype testing and refinement began. N/A indicates that the machine never entered serial production.

[b] Projected values.

as well, including machine tools, consumer electronics, semiconductor-manufacturing equipment, and high-performance semiconductors. Even areas of U.S. strength, such as aircraft and computer hardware and software, may soon be at risk unless strong action is taken. American competitiveness, much less dominance, in these and other high-technology areas can no longer be assumed.

Alarms have been sounded at many levels. The National Advisory Committee on Semiconductors, in its recently released second annual report, refers to the semiconductor industry as "an industry in crisis" and urges the federal government to act immediately or risk losing the semiconductor industry in its entirety and with it, the computer industry, as well. The Administration itself has just identified 22 critical technologies vital to U.S. military and economic security, a list of technologies virtually identical to those identified earlier and individually by the Departments of Defense and Commerce as vital to the future of the U.S. in world geopolitical and economic competition.

A concerted effort on the part of the Japanese, combined with complacency on the part of American industry and unfavorable trade conditions between the U.S. and Japan, have brought about this situation in which spheres of U.S. industry have lost former dominance and competitiveness in certain international markets. The world has changed. The U.S. is no longer predominant.

Japan: Vertical Integration, *Keiretsu*, and Government Coordination

In 1952, when the Japanese became independent, they set a goal, embodied in the motto, "We will match the standard of living in the West." At that time, over half of the Japanese population was engaged in subsistence agriculture, and yet, by improving their output-to-input ratio, they were able to improve their productivity. Since then, the Japanese have moved from subsistence agriculture into light manufacturing, into heavy and chemical goods, and into the higher-technology areas. Today, as a result of an innovative corporate structure and governmental industrial orchestration, the Japanese have positioned themselves to become the dominant suppliers of information technologies to the world.

Japan today has a different, more sophisticated structure to its economy than our own. The major Japanese firms producing computers are all vertically integrated, meaning that a strong presence is maintained across the spectrum of computing machinery—from micros to supercomputers—and in all allied technologies: microelectronics, networking, consumer electronics, etc. In contrast, there is only one vertically integrated company in the computer field in the U.S.—IBM.

Table 2 illustrates this situation. The three leading Japanese supercomputer firms—NEC, Fujitsu, and Hitachi—are integrated all the way from consumer electronics to microcomputers, minis, intermediates, mainframes, and supercomputers. All are large-scale producers of semiconductors. In contrast, U.S. firms producing semiconductors are either "merchant" suppliers, with the bulk of their sales and earnings coming from the sale of semiconductors on the open, merchant market, or "captive" suppliers, such as IBM and AT&T, which produce semiconductors only to satisfy internal demand. Further, when previously successful merchant suppliers have been purchased and merged into large U.S. companies to become both captive and merchant suppliers, they have uniformly gone out of the merchant business.* Usually they have shut down completely. No U.S. captive supplier has become a successful merchant supplier to the market. Japanese firms, however, do both successfully. They are captive suppliers to themselves, and they are merchant suppliers to the market. This suggests something amiss in our system or between our system and that of the Japanese.

Table 2. Market-Segment Participation by Selected U.S. and Japanese Manufacturers of Supercomputers and/or Semiconductors

	Cray	DCD/ETA	IT, Intel, Motorola	AT&T	IBM	NEC	Fujitsu	Hitachi
Supercomputers	x	x			x[a]	x	x	x
Mainframe Computers		x			x	x	x	x
Intermediate Computers					x	x	x	x
Minicomputers		x		x	x	x	x	x
Microcomputers					x	x	x	
Consumer Electronics						x	x	x
Semiconductors (Merchant)			x			x	x	x
Semiconductors (Captive)				x	x	x	x	x

[a] IBM is reentering the supercomputer market.

* AT&T has since withdrawn from the manufacture of mass-market semiconductors, and many U.S. firms were driven out of the dynamic random-access memory chip business between 1985 and 1987.

Japanese firms are not only integrated across virtually every aspect of semiconductors and computers but are also prominent members of Japanese industrial conglomerates, or *keiretsu*. The major Japanese *keiretsu* groups are each structured around a major bank.* The Bank of Japan supports these banks with differential interest rates for loans to "target" industries.

From their *keiretsu* structure, Japanese corporations get diversification, risk reduction, and a lower cost of capital, which allows them to maintain a favorable output-to-input ratio. Additionally, while competition in their home market among the *keiretsu* forces reduced costs and improved quality, these firms will cooperate when operating in foreign markets or when facing foreign firms in their home market. Should they temporarily fail to cooperate, the Japanese government steps in and reestablishes "harmony," especially in their dealings with outsiders, or *gaijin*. Thus, the Japanese have not only a team but a team with a strategy.

The U.S.: Rugged Individualism and Trade-War Losses

In contrast, American industry has rejected a close working relationship with the government and insists on that truly American concept of rugged individualism. Whereas this arrangement has at times resulted in extraordinarily rapid growth of the American high-technology industries, it has also resulted in an uncoordinated industrial environment in which poor decisions have been made.

Economic policy decisions must be made with respect to certain economic relationships, which can be easily illustrated formally. Four variables are required for this somewhat oversimplified example:
- q = quantity of output;
- p = price of products;
- w = wages per hour; and
- i = number of hours worked.

If five computers are sold ($q = 5$) for $1000 each ($p = \$1000$), total revenue will be $5000 ($qp = \$5000$). If the wage rate of the work force is $10 per hour ($w = \$10$) and the input of work hours is 500 ($i = 500$), the cost to produce the machines is also $5000 ($iw = \$5000$), and no profit is realized:

$$qp = \$5000 ;$$

$$iw = \$5000 .$$

---

* The *keiretsu* banks servicing the top six *keiretsu* are all among the world's ten largest banks.

More generally,

$$qp = iw.$$

If we now divide both sides of this relationship by ip, we get

$$\frac{qp}{ip} = \frac{iw}{ip},$$

and canceling, we get

$$\frac{q}{i} = \frac{w}{p}.$$

This ratio, in a nutshell, illustrates what is needed to break even. On the right-hand side of the equation are two factors expressed in dollars—w, the hourly wage and p, the price of the computers—whereas on the left is the ratio of q, machines produced, to i, input hours. The ratio of q/i is the rate of output per unit of input and represents what economists call the (average) production function of the process. It is a relationship determined by the technology. If the technology is the same in two countries, this ratio will be (roughly) the same for those countries. A country with lower hourly wages, w (e.g., $5), will be able to charge a lower price, p (e.g., $500) for its product and still break even. That is, with the same plant and equipment (or production function) on the left-hand side of the equation and low wage rates on the right-hand side, low prices will be possible for the products from the low-wage countries. If, however, wages are very high (say $20), then the ratio of w/p will require that p, the product price, also be high (in our example, $2000).

International competition is as simple as that. It is neither "good" nor "bad," it is simply inexorable. The simple relationship demonstrates why so many jobs are being lost by the U.S. to developing countries. Their low wages make it possible for them to produce products from relatively stable technologies more cheaply and thus charge lower prices than can we in the U.S.

There are solutions to this problem, but the U.S. has generally failed to implement them. For example, it may be possible to improve U.S. plants and equipment (and/or its management) so that employees are more productive and thereby achieve greater output (q) per unit of input (i). This would justify a higher wage (w) in relation to a given unit price (p) for the output. This has been a major tenet of Japanese strategy for decades. Another solution involves producing a higher-quality product for which consumers will be willing to pay more, thus justifying the higher wages paid to the work force, since a higher p can justify a higher w.

Because the U.S. has as its long-term objective to maintain or increase its relative standard of living, then one or both of these strategies are required. Even then, a way must be found to inhibit the rate of diffusion to low-wage economies of an innovative, highly productive production process and/or of product quality innovations. Only in these ways can higher wages—and thus a reasonable standard of living—be sustained in our economy over the long haul. The implications of these facts are pretty clear; they are very much a part of our day-to-day experience.

A reasonable economic development strategy for the U.S. must be in the context of these major forces influencing outcomes worldwide. The major forces aren't definitive of final outcomes, but they do establish the limits within which policies, plans, and strategies can be successful.

The need to respond to the low prices offered by other countries has been recognized in this country for many years. The solutions attempted have largely been ineffectual quick fixes, in essence trying to catch up without catching up rather than facing up to the imperatives of improving product quality and productivity in general.

Law 2: you don't catch up without catching up.

Our response to the present challenge has, to date, included a miscellany of wishful thinking, "concession bargaining," and manipulating monetary factors. Concession bargaining sought to cut the wages of U.S. workers producing high-technology goods—and thus their standard of living—so that the U.S. could match the prices that low-wage countries are offering. When "concession wages" didn't work, the U.S. then decided to find another financial gimmick that would permit us to lower the price of our goods in world markets.

The U.S. decided to cut the exchange rate in half. At the bottom, the dollar was worth 120 yen, while before it had been worth 240 yen and more. In effect, this introduces a factor (in this case 1/2) between the world price before the change and the world price after, while leaving the U.S. domestic equation unchanged. Given that we have over a $5 trillion economy today, cutting our exchange rate with the world in half amounts to giving away $2.5 trillion in the relative value of our economy. The $30 billion improvement in the balance of trade due to the lower world price of our goods yielded a return on our investment of only a little over one cent on each dollar of value lost. This is not an intelligent way to run an economy.

Trade: "Successful" Negotiations and "Potato Chips"

Another major difficulty for U.S. high-technology manufacturers is the asymmetry in U.S. and Japanese market accesses; U.S. markets are wide open through our ideology, whereas Japanese markets are not. The U.S. government, which historically has shunned "intervention," has been reluctant to "start a trade war" and insists that U.S. firms are better off standing alone. The government, whose policies might be stated as, "It doesn't matter if we export computer chips or potato chips,"* has been practicing what can only be called unilateral disarmament in international trade battles.

Law 3: when two countries are in a trade war and one does not realize it, that country is unlikely to win.

The Japanese markets in target industries are, have been, and are likely to remain closed. In the U.S., we refer to such a phenomenon as "protectionism." This market was protected when the Japanese were behind. This market was protected while the Japanese caught up. This market remains protected even now after the Japanese have achieved substantial superiority in many of its products.

These facts violate conventional wisdom that equates protectionism with sloth. Clearly, that has not been the case for the Japanese. Protectionism for the Japanese can't be all bad; it is possible for protectionism to work to the benefit of a nation; it has for Japan. Today, the U.S. has a weekly trade deficit with Japan of about one billion dollars—almost half of which is spent on automobiles—which feeds into the Japanese R&D cycle rather than our own.

Protectionism has long been the name of the game for Japan, and the U.S. has an extremely poor track record in trade negotiations with the Japanese. There are general trade principles understood by both the U.S. and Japan, some of which are embodied in specific, written trade agreements. An important one is that each side should buy those technological

---

* The exact quote—"Potato chips, semiconductor chips, what's the difference? They are all chips."—was made by Michael Boskin, Chairman of the Council of Economic Advisors to the President.

products that the other side can produce more economically. The Japanese, however, routinely violate these principles whenever it is convenient to do so, and the U.S. does not pressure the Japanese to meet their obligations. Seventeen negotiations between the United States and Japan in semiconductors have been almost completely ineffectual. These general comments are illustrated in Figure 1, which is also known to the industry as the "worm chart."

Figure 1 shows the share of the Japanese semiconductor market held by U.S. producers over the period 1973–1986 in relation to a series of "successful" negotiations between the United States and Japan to open the Japanese semiconductor market to competition by U.S. firms. It is startling to note that the U.S. market share has dropped by approximately one per cent—from 10 to nine per cent—in the wake of these "successful" negotiations.

During the early period, Japanese firms were no match for their American rivals. By the early 1980s, they had caught up in many areas and were already ahead in some. By the end of the period, Japanese firms had established general superiority. Yet, throughout the entire period, and irrespective of the relative quality of the U.S. versus Japanese products,

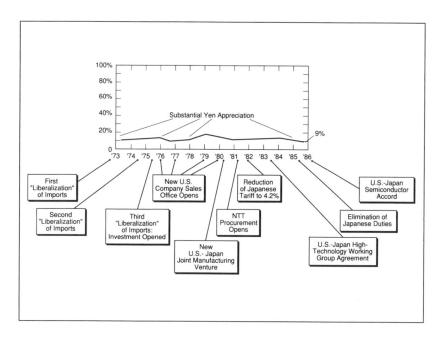

Figure 1. The Worm Chart: U.S. share of Japanese semiconductor market.

the U.S. market share has remained virtually the same. It could be characterized as a "worm," lying across the chart at the 10 per cent level.

There's something about the way the Japanese manage their economy that has led to a constant U.S. share of its semiconductor business in the range of approximately 10 per cent. Given the multitude of variables involved, it is a virtual impossibility that markets alone could have brought about such a result.

Remedies
Law 4: an important aspect of change is that things are different afterward.

It is not difficult to see that no panaceas exist. Various remedies to ameliorate the situation, however, do.

The present-day situation must be recognized as a crisis. The U.S. government and, to a lesser degree, industry have failed to recognize this as a crisis. Fixing that which is wrong now is many times more difficult than would have been the case just a few years ago.

The Japanese market will have to be pried open for American products. When President Bush appointed Carla Hills to the post of U.S. Trade Representative, he gave her a crowbar to emphasize that very point. To do so, the Japanese must be held to their obligations under current trade agreements, and legislation intended to bring about a more equitable trade situation, perhaps along the lines of the High Performance Computing Initiative, should be passed. Should these measures fail, the U.S. might be wise to consider assuming some vestiges of protectionism, at least as the means to pry open the Japanese market.*

More qualified people are needed in Washington to attend to the problem. Both NSF and the Defense Advanced Research Projects Agency are trying to recruit such individuals, with both experiencing difficulty.

The industry has changed, and American industry must change with it. Many of the lessons to be learned in this case come directly from the Japanese. For example, to survive these days, a supercomputer company must be vertically integrated and must generate money to be invested in R&D.

---

* The U.S. and Japan agreed in May 1991, once again, that Japan will take those actions necessary to assure that U.S. firms' share of their market reaches 20 per cent. This represents the third time the two countries have agreed to reach the same goal that was not reached as a result of either of the prior agreements.

More money must be invested in R&D. Another key to industrial survival today is a market share to generate money for investment in research and development. R&D spending in Japan went up 14 per cent during one recent 12-month period, alone, and four or five Japanese companies have R&D budgets that exceed the entire budget of NSF, including NEC, whose budget exceeds NSF's by one-third.

U.S. industry must improve its productivity, change its values, and technologically catch up. A top-down approach to the solution of eroding U.S. leadership in the area of high-performance computing will not work. Fooling with the exchange rate won't work, nor will any number of additional stopgap measures. Changing our values, above all else, means abandoning our short-term view in terms of industrial planning, education, consumer buying habits, and government involvement in industry. Incentives must be introduced if industry is to be expected to assume a long-term view. Unfortunately, industry in this country has a very short-term view. They won't take the technology that is available in the universities. The good technology that is developed there is being siphoned off to Japan, where there are interested people. Meanwhile, industry occupies itself worrying about satisfying the investors at the next shareholders' meeting.

Education—at all levels—must be improved in this country. The decline of public elementary and secondary education is well-documented and demands both increased governmental spending and fundamental changes in values. Similar reforms are necessary at the university level, as well. Improving education in America, however, can not be accomplished in short order. Sy Goodman:

> The educational issue, long-term as it is, is still absolutely critical. The U.S. university community, in my opinion, is overly complacent about what it thinks it is, relative to the rest of the world.

Industry should investigate the possibility of government involvement, perhaps to the point of coordination. Summarily rejected by the U.S. government and industry, governmental coordination of Japanese industry by the Ministry of International Trade and Industry has been instrumental in Japan's postwar rise to its current position as a high-tech industrial superpower. Studies cited earlier show that all relevant elements of the public and private sectors are now agreed on those areas in which the U.S. must succeed if it is to remain a world-class competitor nation. The U.S. must put aside adversarial relationships among government, industry, and workers.

The Future

When one looks to the East these days, Japan is not the only competitive nation on the landscape. The newly-industrializing countries (NICs) of East Asia (including Singapore, Hong Kong, Taiwan, and South Korea), while not known primarily for technological development, have exhibited proficiency in building components, peripherals, and systems of increasing complexity. It is not inconceivable that these countries will produce high-performance systems in the future. The NICs, however, unlike Japan, are not members of the Coordinating Committee of Export Controls. Worldwide availability of supercomputers from these countries could have a substantial impact on U.S. national security. If and when this time comes, the U.S. government should not be unprepared to address this matter.

In short, and in conclusion, both industry and government have a large stake in the continued health, if not dominance, of America's high-technology sectors, including the supercomputer industry. Both also have important roles to play to insure this continued health. Further, the industry and government need not work toward this goal in isolation from one another.

References and Bibliography

M. Borrus, "Chips of State," *Issues in Science and Technology* **7** (1), 40–48 (1990).

D. H. Brandin and M. A. Harrison, *The Technology War*, John Wiley and Sons, New York (1987).

Federal Coordinating Committee on Science, Engineering, and Technology, "West European High Performance Computer Suppliers," FCCSET memorandum, Washington, DC (1991).

C. H. Ferguson, "Computers and the Coming of the U.S. Keiretsu," *Harvard Business Review* **90** (4), 55–70 (1990).

V. M. Glushkov, V. A. Myasnikov, M. B. Ignat'yev, and V. Torgashev, "Recursive Machines and Computing Technology," in *Information Processing 74: Proceedings of IFIP Congress 74*, J. L. Rosenfeld, Ed., North-Holland Press, Amsterdam, pp. 65–71 (1974).

M. A. Harrison, E. F. Hayes, J. D. Meindl, J. H. Morris, D. P. Siewiorek, and R. M. White, *Advanced Computing in Japan*, Japanese Technology Evaluation Center panel report, M. A. Harrison, Chm., Loyola College, Baltimore, Maryland (1990).

E. F. Hayes, "Advanced Scientific Computing in Japan," in *Far East Scientific Information Bulletin NAVSO P-3580, Vol. 15, No. 3*, Office of Naval Research, Far East APO, San Francisco, pp. 109–117 (July–Sept. 1990).

S. Jarp, "A Review of Japan and Japanese High-End Computers," in *Far East Scientific Information Bulletin NAVSO P-3580, Vol. 16, No. 2*, Office of Naval Research, Far East APO, San Francisco, pp. 59–79 (April–June 1991).

A. K. McAdams, T. Vietorisz, W. L. Dougan, and J. T. Lombardi, "Economic Benefits and Public Support of a National Education and Research Network," *EDUCOM Bulletin* **23** (2–3), 63–71 (1988).

National Research Council, *Global Trends in Computer Technology and Their Impact on Export Control*, S. E. Goodman, Chm., National Academy Press, Washington, DC (1988).

C. V. Prestowitz, Jr., "Life after GATT: More Trade Is Better Than Free Trade," *Technology Review* **94** (3), 22–29 (1990).

C. V. Prestowitz, Jr., *Trading Places: How We Allowed Japan to Take the Lead*, Basic Books, New York (1988).

P. Wolcott and S. E. Goodman, "High-Speed Computers of the Soviet Union," *Computer* **21** (9), 32–41 (1988).

P. Wolcott and S. E. Goodman, "Soviet High-Speed Computers: The New Generation," in *Proceedings, Supercomputing '90*, IEEE Computer Society Press, Los Alamitos, California, pp. 930–939 (1990).

P. Wolcott, "Soviet and Eastern European Computer Performance: Results of Benchmark Tests," Mosaic technical report 1991-003-I, Tucson, Arizona (1991).

Economics, Revelation, Reality, and Computers

Herbert E. Striner

Herbert E. Striner is an economist with a bachelor's and master's degree from Rutgers University and a Ph.D. from Syracuse University, where he was a Maxwell Fellow. Since 1962, he has specialized in manpower and productivity problems and has served as a consultant on productivity to such corporations as IBM, PPG, and Saks Fifth Avenue, as well as to the U.S., Australian, Canadian, and Italian governments.

During a distinguished teaching career at American University, Washington, DC, Dr. Striner served as Dean of the College of Business and as Professor, a post from which he retired in 1989. He has also served at Johns Hopkins University, the Brookings Institution, the Stanford Research Institute, and the W. E. Upjohn Institute for Employment Research. In government, he has worked with NSF and the Department of the Interior. He has appeared frequently as a guest speaker before major business, governmental, and professional organizations and on such television programs as the McNeil-Lehrer Report and various NBC "White Paper" programs.

Dr. Striner has published five books and over 70 articles, including Regaining the Lead: Policies for Economic Growth *(1984). This book focuses on a rethinking of U.S. economic policy as it relates to productivity and economic growth.*

The economic policies of the current administration will lead, undoubtedly, to the loss of U.S. world leadership in the supercomputer industry. By the very early 1990s, unless there is a fundamental rethinking of our economic philosophy, the U.S. will cease to be a major competitor in the field of supercomputers. The implications for our industrial competitiveness will be tragic. We may certainly continue to produce Nobel Laureates. However, the ability to translate their gifted insights into technology and commerce will be held hostage by an archaic economic ideology.

Erwin Schrodinger (1952), a pioneer in quantum mechanics and Nobel Laureate in 1933, put it this way:

> There is a tendency to forget that all science is bound up with human culture in general, and scientific findings, even those which at the moment appear most advanced and esoteric and difficult to grasp, are meaningless outside their cultural context.

Another Nobelist, the physical chemist Ilya Prigogine, in *Order out of Chaos* (Prigogine and Stengers 1984), observed:

> Many results . . . for example those on oscillating chemical reactions, could have been discovered years ago, but the study of these nonequilibrium problems was repressed in the cultural and ideological context of those times.

Since the early 1980s, when I was a consultant on productivity with IBM, I gradually came to recognize that the problem of competitiveness is not an economic problem, it is a values problem. By that I mean the production decision starts with a set of perceptions, or values, reflecting what the decision maker thinks is right. In management there is a broad array of theories on how to manage an organization. But the resistance on the part of managers to move to a demonstrably superior system is almost always the result of personal values. This is equally true in the sciences. Werner Heisenberg (1974) commented on this:

> Once one has experienced the desperation with which clever and conciliatory men of science react to the demand for change in the thought patterns, one can only be amazed that such revolutions in science have actually been possible at all.

Decisioning is anchored in a hierarchy of values that leads to assumptions and finally to policy. In the course of research in which I have been involved—cutting across diverse disciplines—I have observed that failure or success in policies and programs is rarely traced back to the personal hierarchy of values that is the initial motive force of any decision. Our values significantly determine what we see as—and believe to be—reality.

Any perception of reality must be closely related to what actually exists, or we will be tilting at windmills. We are in trouble if what we perceive is only what we want to perceive and is far removed from a truer reality. This problem is not an easy one to deal with. In his book *Vital Lies, Simple Truths: The Psychology of Self-Deception*, Daniel Goleman (1985) put it well: "We do not see what we prefer not to, and do not see that we do not see."

Since the publication of Adam Smith's monumental work, *An Inquiry into the Nature and Causes of the Wealth of Nations*, in 1776, economists have been divided—often bitterly—over Smith's perception of the effectiveness of the so-called "free-market system." Parenthetically, even the word "free" is used effectively to imply that all other possible options are "unfree"—a perfect example of the fallacy of the excluded middle.

It is now critical that we examine the basic values and assumptions surrounding this long-standing debate—"critical" because if we are wrong in assuming the superior efficiency of the free-market system as compared with any other system, we may be, as they say in industry, "betting the store."

The U.S. economy is the largest free-market system in the world, so it is helpful to look at a few facts. Or, to put it differently, to look at a long-term reality somewhat differently from the one usually described by free-market enthusiasts:

- In 1970, among the top 20 banks in the world, based on deposits, six were from the United States. Four of them were among the top 10. By 1989, there were *no* U.S. banks among the top 20.
- In 1970, there were no Japanese banks among the top 10, and only two among the top 20—specifically, in 19th and 20th places. By 1989, all of the top 10 banks in the world were Japanese. Fourteen of the top 20 were Japanese. In 1989, the highest ranking United States banks were in 27th and 44th places. They are Citibank and Bank of America, respectively.
- In 1970, 99 per cent of all telephone sets sold in the U.S. were manufactured by U.S. companies. By 1988, this share was 25 per cent.
- In 1970, 89 per cent of all semiconductors sold in the U.S. were produced by U.S. companies. By 1988, this share was 64 per cent.
- In 1970, almost 100 per cent of all machine tools sold in the U.S. were produced by U.S. companies. By 1988, this share was 35 per cent.
- In 1970, 90 per cent of all color television sets sold in the U.S. were produced by U.S. companies. By 1988 this share was 10 per cent.

- In 1970, 40 per cent of all audio tape recorders sold in the U.S. were produced by U.S. companies. By 1988 this share was one per cent.

Let's look at this long-run period from a slightly different vantage point, that of productivity gain. This is *the* fundamental indicator of competitive ability in a national economy. Looking at productivity figures for industrialized nations (Department of Labor 1990), I have divided seven major competitors into Groups A and B and compared their rates of productivity gain from 1973 through 1989. Group A consists of West Germany, France, Japan, and Italy. Group B consists of Canada, the U.S., and Great Britain. The basis for these groupings is the underlying economic philosophy vis-à-vis government and the private sector:
- Group A: West Germany, France, Japan, and Italy view active government involvement in the economy as either important or essential for achieving economic goals.
- Group B: the U.S., Great Britain, and Canada view government as a problem, best prevented from any involvement in the market system.

How do these groups compare in their 17-year records for average annual gain in output per employed worker? *Group A's gain was 2.43 per cent; Group B's gain was 1.20 per cent.* Group A's productivity gain was twice that of Group B's from 1973 through 1989. The difference is devastating for competitiveness, as the record shows. Interestingly, even if we do not include Japan's productivity gain, the remaining Group A countries exceeded the productivity gain of Group B countries by 86 per cent during the period in question (see Figure 1).

But few in our country would question the value of government investment in a *limited* number of cases. It was the Morrill Act of 1862, the so-called land-grant act, that in combination with state agricultural extension services played a key role, and still does, in producing the highest-productivity agricultural system in the world. It was federal government policies and funds during and immediately following World War II that produced the modern computer. The private aviation industry is the result of federal support for R&D that began in 1915 with the creation of the National Advisory Committee for Aeronautics. Federal expenditures for weather, safety, airport programs, and mail subsidies have been major factors in the growth of our national air transport system. But these, and other existing examples, are viewed as "legitimate" exceptions. The rule is that government is, for the most part, an interloper. The less government, the better is the current, prevailing ideology of the present administration in Washington. If this model remains the rule, however, there is no way for the United States to regain the competitive lead in the world economy.

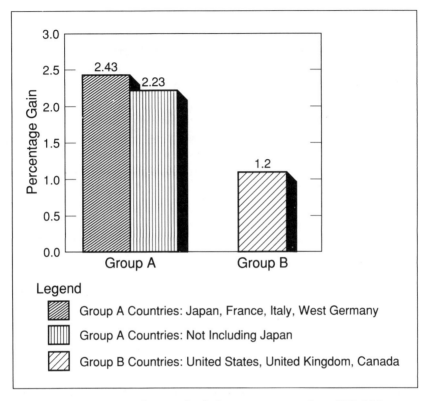

Figure 1. Average annual rates of gain in output per worker, 1973–89 (source: Department of Labor).

Two factors account for what has been happening to cause the erosion of U.S. competitiveness.

First, the computer radically alters the time span in which decisions can be formulated, tested, and finalized. Decisions are, in reality, strategies. Supercomputers have completely changed the decisioning, or strategy-formulation, process. The country capable of introducing computers into the industrial process in the most effective manner gains in design and production capabilities. The ability to shorten design and production cycles is the competitive edge. Supercomputers that can simulate complex relationships are now the core of this process. I speak not only of complex design and production systems but also of marketing, communications, and financial systems. The link between computers and production has revolutionized the economics of long production runs, which has always underlain our concept of mass production and decreasing unit costs. The game now is shorter design and production cycles, flexible manufacturing systems, and high-quality products at both the low- and high-price ends of the production line. The supercomputer is the vital ingredient in this new process. But between 1980 and 1990, the U.S.'s share of installed supercomputer systems in the world has gone from 81 per cent to 50 per cent. Japan's has gone from eight per cent to 28 per cent.

Development and production S-curves (growth curves) exist but in a time period so truncated that the time segment for decision making has become more critical than ever. And catching up becomes more difficult than ever. The time necessary for so-called long-run, natural forces of the marketplace to play out are foreshortened by a new model I will describe shortly.

This state of affairs is even acknowledged in U.S. government reports. Quoting from an International Trade Administration report (1990), we are told:

> With adequate funding available due to the strong and visible presence of the government and the *keiretsu* along with a profitable domestic market, Japan reinvested in the production of all semiconductor devices. Japanese producers' aggressive capital and R&D investment policy enabled them to accelerate new product introductions; in fact, the Japanese were the first to market 64K and 256K DRAMS [dynamic random-access memory chips]. Early market entry is critical to memory device producers, since relative cost advantages accrue to the firm reaching a given production volume first. U.S. firms, faced with waning demand and Japanese pricing pressure, were forced to curtail investment.

The *second* factor has to do with the availability of investment capital. To exploit gains in decision time, massive amounts of capital, available both quickly and reliably over time, are essential. Normal market forces usually look for short-term returns on investment. Investment in processes and products with an eight- or 10-year payout compete poorly when there are good alternatives with a two-to-four-year payout. The level of funds and length of time needed to support a technological development of large magnitude, like supercomputer systems, is such that the free-market-system model simply can not compete. Why is this so? Because nations that are out-competing the U.S. have developed a more effective decisioning process based on a more realistic values system.

Even when U.S. companies undertake to join forces to meet the competition, unless government is ready to share responsibility for providing key aid—usually in necessary investment support—

private-sector initiative is insufficient. No better example of this exists than that of U.S. Memories.

In June 1989, seven U.S. electronics firms announced the formation of a DRAM production consortium that was named U.S. Memories, Inc. The objective was to increase our competitiveness in this area in competition with Japanese producers. However, in January 1990, this effort was canceled after an effort to raise required capital funding failed. Such consortia efforts in Japan, where a national goal of competitiveness is jointly arrived at by industry and government, are adequately funded on the basis of commitments undertaken by the Ministry of International Trade and Industry (MITI) and the Ministry of Finance.

In the race to recover after World War II, the devastated economies of Japan, West Germany, and France developed what I call the key-player model. The model is based on the following:
- a clear perception of an economic goal,
- determining the key ingredients and players necessary for achieving the goal, and
- the use of the key players in such a way as to balance economic and social goals rather than sacrifice any one significant goal in the process of achieving others.

This is not easily done but apparently has been done by some nations—the ones that are succeeding on the world scene. The model is not perfect, but it meets the test of being effective. It is this model that has replaced the free-market-system model that served the industrial world well since the Industrial Revolution of the 19th century.

A values system, or hierarchy of values, that eliminates one or more key players at the outset in effect predetermines that a large systems goal cannot be met. We in the U.S. have chosen to meet the competition with one key player absent, and the results have been apparent for all to see. We start from a false ideological premise. We tend to assume that unless our markets are free of the involvement of government, both efficiency in resource use and individual freedom are threatened. But such capitalist countries as West Germany, Japan, France, and Sweden testify to the fact that public and private sectors can have close, catalytic economic relationships while remaining healthy democracies. Their ability to compete effectively is hardly in question. Anecdotal evidence leads me to suggest that their production/design growth curves are radically different, as shown in Figure 2. The Nobelist Arno Penzias (1989) commented on this truncation in his book *Ideas and Information*, but only in the context of the increasing efficiency of modern laboratories. I am enlarging the concept to include the effects of a systems-decisioning model, the key-player model.

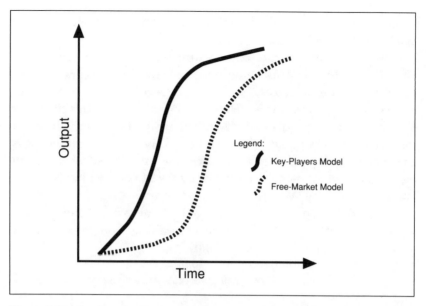

Figure 2. Design/production curves for two growth models.

Using Japan as an example, the key-player model has worked in such different cases as the development of flexible machining systems, very-large-scale integrated circuits, biotechnology, and, of course, supercomputers. The supercomputer industry of today in Japan is the product of the long-term investments started by the government in the late 1950s. Beginning in 1957, MITI supported the planned growth by heavy subsidies, cooperative R&D, protectionist regulations, and an ingenious rental strategy. As was pointed out by Marie Anchordoguy (1990), IBM gained the right to produce in Japan only after agreeing to give the infant Japanese computer companies patent rights at low rates. By 1961—led by MITI—seven computer companies formed the Japan Electronic Computer Company (JECC). This company, with about $2 billion from the government, bought computers from Japanese companies and rented them to Japanese industry at rates below those of IBM in Japan. The interest rates were also well below market rates. Between 1961 and 1981, these policies built the Japanese computer industry. JECC paid computer producers before delivery, with no interest on the loan. Since JECC set computer prices and prohibited discounts, it guaranteed that competition could only be based on quality of both manufacturing and technology. During the same 20 years, the government invested about $6 billion in R&D, new equipment, and operating capital. This is how Japan "grew" its computer industry. Those who currently laud the state of competition in Japan's computer industry are only looking at the peak of the iceberg. They have displayed their ignorance of the key role of government for almost 25 years.

Only government has sufficient long-term capital to supply certain needs of a modern industrial power. Only a key-player model can develop the intricate web of mechanisms calculated to support long-term industrial objectives. Increasingly, U.S. industry is locating facilities and off-shore production at foreign sites where other national policies recognize this fact of modern industrial life.

Public policies calculated to support industrial productivity and technology are not always the obvious ones, such as targeting R&D for special tax treatment. Radically shortening depreciation rates on equipment, supporting the capital needs of such ventures as U.S. Memories, Inc., increasing funds for scholarships and facilities at universities, as well as the direct investment in new areas where the alternative is to lose a vital competitive position, are a few additional means of achieving technological and productivity gains.

In Session 13, there are discussions of the industrial impact of the High Performance Computing Initiative, a project with which both Los Alamos National Laboratory and I have been involved. The discussions will deal with the likely benefits of a five-year increase of $1.9 billion for funding a high-performance computer system. One conclusion of the proposal for this project is that if we do not provide for this funding increase, U.S. market forces are not about to meet the competition that has developed in Japan as a result of past Japanese government support for their computer industry. The supporting evidence is clear. Between 1980 and 1990, the U.S. share of installed supercomputer systems in the world has gone from 81 per cent to 50 per cent, while Japan has gone from eight per cent to 28 per cent—a drop of 38 per cent matched by an increase of 350 per cent.

While the market for supercomputers is relatively small compared to the total computer industry market, less than one per cent of total revenues, this fact can be dangerously misleading. It's about as helpful as assessing the value of the brain based on its per cent of body weight. Only the supercomputer can do the increasingly complex simulations being called for by industry as design and production cycles shorten. And in financial and banking industries, only the supercomputer is capable of tracking and simulating world flows of credit, equities, and foreign exchange.

The problem confronting us is the conflict between economic reality and a values system that wants to believe in a form of revelation—the free-market-system model. This is nowhere better illustrated than in the position taken by the current Chairman of the President's Council of Economic Advisors, Dr. Michael Boskin (1990). He states that he sees the signs of industrial decline as merely parts of a necessary process of free-market competition, which although "uncomfortable . . . makes us perform better." It was this same presidential advisor who observed: "It doesn't matter if we export computer chips or potato chips." This keen analysis is at odds with the Office of Technology Assessment. Their report *Making Things Better* (1990) states, "American manufacturing has never been in more trouble than it is now." And the reason is that the free-market-system model can not compete with the key-player model.

The key-player model exists in capitalistic countries. Basically, it is a far more effective form of capitalism than that which exists in the United States. I have described the evaluation of this newer model (1984), characterizing it as "Shared Capitalism," a system based on a cooperative values system rather than an adversarial one.

Values systems are almost always resistant to change. Placed in terms of Thomas Kuhn's insightful 1970 work, *The Structure of Scientific Revolutions*, normal science, that is, the piecemeal efforts to deal with anomalies, is not capable of major breakthroughs in thinking. Only paradigm changes can do this. Computers and new governance arrangements in other countries have produced a paradigm change for modeling industrial competitiveness. This new key-player model is more effective than the free-market model. But our values system prevents us from meeting the challenge. Clinging to an outmoded decisioning model is rapidly moving us into the ranks of second-rate economic powers. In every competitive system, the rule of success is this: *change in order to cope, or lose out.*

The great economist, Lord Keynes, in answering a critic's jibe when Keynes changed a position he had once taken, retorted, "When I see the facts have changed, I change my opinion. What do you do, Sir?" I have also found the following to be helpful: *you can't be what you want to be by being what you are; change is required.*

If each of us will urge, when the opportunity arises, that our nation use all of its key players in the game of global competition, we just may begin to turn the present situation around. This nation has been at its best when it has junked old solutions that were not working and struck out in new directions. That is the real pioneering spirit. I hope we still have it.

References

M. Anchordoguy, "How Japan Built a Computer Industry," *Harvard Business Review* **68** (4), 65 (1990).

M. Boskin, quoted in *New Technology Week*, p. 12 (July 23, 1990).

Department of Labor, "Comparative Real GDP per Employed Person, 14 Countries, 1950–1989," unpublished BLS data (May 3, 1990).

D. Goleman, *Vital Lies, Simple Truths: The Psychology of Self-Deception*, Simon & Schuster, New York, p. 234 (1985).

W. Heisenberg, *Across the Frontiers*, Harper & Row, New York, p. 162 (1974).

International Trade Administration, "The Competitive Status of the U.S. Electronics Sector, from Materials to Systems," U.S. Department of Commerce report, Washington, DC, p. 100 (April 1990).

Office of Technology Assessment, *Making Things Better: Competing in Manufacturing*, U. S. Government Printing Office, Washington, DC, p. 3 (February 1990).

A. Penzias, *Ideas and Information*, W. W. Norton & Co., New York, p. 186 (1989).

I. Prigogine and I. Stengers, *Order out of Chaos*, Bantam Books, New York, pp. 19–20 (1984).

E. Schrodinger, "Are There Quantum Jumps?" *The British Journal for the Philosophy of Science* **3**, 109–110 (1952).

H. Striner, *Regaining th e Lead: Policies for Economic Growth*, Praeger Publishers, New York, pp. 97–125 (1984).

12

Experience and Lessons Learned

Panelists in this session discussed recent case histories of supercomputing developments and companies: technology problems, financial problems, and what went right or wrong and why. Presentations concerned recent work at Engineering Technology Associates Systems, Bolt Beranek & Newman Inc., FPS Computing, the National Security Agency's Project THOTH, and Princeton University's NSF center.

Session Chair

Lincoln Faurer, Corporation for Open Systems

Supercomputing since 1983

Lincoln Faurer

Lincoln D. Faurer signed on with the Corporation for Open Systems (COS) International on April 7, 1986, as its first employee. Mr. Faurer is a voting ex officio member of the COS Board of Directors and Executive Committee. As Chair of the COS Strategy Forum, Mr. Faurer coordinates and recommends the overall technical direction of COS, and in his role as President and Chief Executive Officer, he oversees the day-to-day business affairs of the company.

Mr. Faurer came to COS after retiring from a distinguished 35-year Air Force career, where he achieved the rank of Lieutenant General. In 1986, President Reagan awarded the National Security Award to Lt. Gen. Faurer for "exemplary performance of duty and distinguished service" as the Director of the National Security Agency from April 1, 1981, through March 31, 1985. Mr. Faurer's prior positions in the Air Force included Deputy Chairman, NATO Military Committee, and Director of Intelligence, Headquarters, U.S. European Command.

Mr. Faurer is a graduate of the United States Military Academy in West Point, New York, the Rensselaer Polytechnic Institute in Troy, New York, and George Washington University in Washington, DC.

I suspect that we all pleasure ourselves occasionally by looking back on events with which we proudly identify, and one such event for me was the establishment of the National Security Agency's (NSA's) Supercomputer Research Center. It was an idea spawned by the predecessor to this meeting in 1983, the first Frontiers of Supercomputing conference. A number of us from NSA left that meeting a little bit taken aback by what we perceived as a lack of a sense of urgency on the part of the government people who were in attendance at that session. Sitting where we did at NSA, we thought there was a lot more urgency to the supercomputer field and therefore set about to create a proposal that culminated in the establishment of a Supercomputer Research Center, which we ended up having to fund alone.

Therefore, it was really a pleasure when I was asked if I would play the role of presiding over one of the sessions at this conference. This session is designed to treat successes and failures, and the latter, I do believe, outnumber the former by a wide margin. But even the so-called failures can spawn successes. In any event, it is the things that are learned that matter, and they matter if we are willing to use what we learn and change.

The objectives of the 1983 conference were to bring together the best in government and industry to understand the directions that technology and requirements were taking. Yet, it was a different industry at that time—small but totally dominant, with a still reasonably healthy U.S. microelectronics industry to support it. That, certainly, now has changed. More foreign competition, tougher technological demands to satisfy, weaker U.S. support industry, microelectronics, storage, etc., affect us now.

So in 1983 at NSA, in addition to starting the Supercomputer Research Center, we made an announcement of intent to buy a heterogeneous element processor from Denelcor. It was their first advance sale. NSA took delivery and struggled for almost a year to get the machine up—our first UNIX four-processor system. However, it did not become operational, and Denelcor subsequently went under. The point is that we took a chance on a new architecture and lost out in the process but learned an important lesson: do not let a firm, even a new one, try to complete the development process in an operational setting.

We could not foresee all the changes of the past seven years at the 1983 conference—the changes in Cray Research, Inc., the loss of Engineering Technology Associates Systems, the loss of Evans & Sutherland supercomputer division, the loss of Denelcor, etc. The major difference is that in 1983, the government market underpinned the industry, certainly to a different extent than today. As of 1990, this is no longer true. The government continues to set the highest demands and is still probably the technological leader, but its market does not and cannot sustain the industry. The supercomputer industry is turning increasingly to an industrial market, an academic market, and foreign markets. Strong pressures from Japan are certainly being encountered in the international marketplace.

One wonders, where will we be seven years from now? What will have occurred that we did not expect? I certainly hope that seven years from now we are not as bad off as some of the current discussion would suggest we could be if we do not do anything right.

Looking back over the past seven years, a number of important developments should not be overlooked.

First, there is a strong trend toward open systems. It is within a niche of that open-systems world that I reside at the Corporation for Open Systems. Second, the evolution of UNIX as a cornerstone of high-performance computer operating systems has lots of pluses and a few minuses. Third, growth of low-end systems, coupled with high-performance workstations, often now with special accelerator boards, has led to truly distributed high-performance computing in many environments. Fourth is the appearance, or imminent appearance, of massively parallel systems with some promise of higher performance at lower cost than traditional serial and vector processing architectures. What remains to be seen is if they can turn the corner from interesting machines to general-purpose systems.

Almost all of this is to the good. Some would argue that high-performance computing does remain the research province of the U.S., by and large. Whether you accept that or not, it is critically important that we dominate the world market with U.S. machines based on U.S. research and composed, at least mostly, of U.S. parts. Aside from the obvious economic reasons for that, it is very important to the nation's security and to the survival of U.S. leadership in areas like aerospace, energy exploration, and genetic research for world scientific prominence.

Lessons Learned

Ben Barker

William B. "Ben" Barker is President of BBN Advanced Computers Inc. and is Senior Vice President of its parent corporation, Bolt Beranek & Newman Inc. (BBN). He joined BBN in 1969 as a design engineer on the Advanced Research Projects Agency ARPANET program and installed the world's first packet switch at the University of California-Los Angeles in October 1969. In 1972 Dr. Barker started work on the architectural design of the Pluribus, the world's first commercial parallel processor, delivered in 1975. In 1979 he started the first of BBN's three product subsidiaries, BBN Communications Corporation, and was its president through 1982. Until April 1990, Dr. Barker served as BBN's Senior Vice President, Business Development, in which role he was responsible for establishing and managing the company's first R&D limited partnerships, formulating initial plans for the company's entry into the Japanese market and for new business ventures, including BBN Advanced Computers.

Dr. Barker holds a B.A. in chemistry and physics and an M.S. and Ph.D. in applied mathematics, all from Harvard University. The subject of his doctoral dissertation was the architectural design of the Pluribus parallel processor.

Abstract

Bolt Beranek & Newman Inc. (BBN) has been involved in parallel computing for nearly 20 years and has developed several parallel-processing systems and used them in a variety of applications. During that time, massively parallel systems built from microprocessors have caught up with conventional supercomputers in performance and are expected to far exceed conventional supercomputers in the coming decade. BBN's experience in building, using, and marketing parallel systems has shown that programmer productivity and delivered, scalable performance are important requirements that must be met before massively parallel systems can achieve broader market acceptance.

Introduction

Parallel processing as a computing technology has been around almost as long as computers. However, it has only been in the last decade that systems based on parallel-processing technology have made it into the mainstream of computing. This paper explores the lessons learned about parallel computing at BBN and, on the basis of these lessons, our view of where parallel processing is headed in the next decade and what will be required to bring massively parallel computing into the mainstream.

BBN has a unique perspective on the trends and history of parallel computation because of its long history in parallel processing, dating back to 1972. BBN has developed several generations of computer systems based on parallel-processing technology and has engaged in advanced research in parallel algorithms and very-large-scale systems (Figure 1). In addition to parallel-processing research, BBN has been shipping commercial parallel processors since 1975 and has installed more than 300 systems. This represents approximately $100 million in investment of private and government funds in BBN parallel-processing technology and products.

BBN has developed extensive experience with parallel-processing systems during this 18-year period. Large numbers of these systems have been used by BBN in communications and simulation applications, most of which are still in operation today. BBN has also used BBN parallel systems in a wide range of research projects, such as speech recognition and artificial intelligence. This extensive experience using our own parallel computers is unique within the industry and has enabled BBN to better understand the needs of parallel computer users.

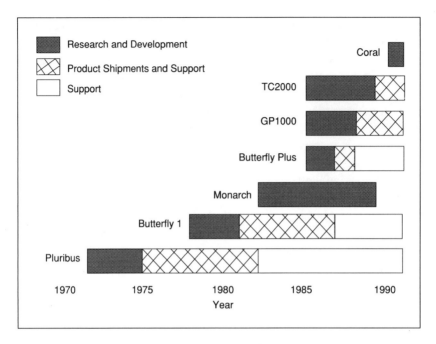

Figure 1. BBN parallel processing systems and projects.

Parallel Processing: 1980 to 2000

In the 1980s, the high-performance computing industry gained experience with parallel processing on a small scale. Vendors such as Sequent Computer Systems and Alliant Computer Systems developed systems with up to tens of processors, and the major vendors, including IBM, Cray Research, and Digital Equipment Corporation, all began marketing systems with four to eight processors. Parallel processing on this scale has now become commonplace in the industry, with even high-end workstations and PC servers employing multiple CPUs.

A key development that helped bring about this acceptance was the symmetric multiprocessing (SMP) operating system. Typically based on UNIX®, but in some cases on proprietary operating systems, SMP operating systems made it much easier to use multiprocessor computers. All of these systems support shared memory, which is needed to develop the parallel operating system kernels used in SMP systems.

However, all of these systems have bus-based or crossbar architectures, limiting the scalability of the systems. The bus in a bus-based architecture has a fixed bandwidth, limited by the technology used and by the physical dimensions of the bus. This fixed bandwidth becomes a bottleneck as more processors are added because of the increase in contention for the bus. Crossbar architectures provide scalable bandwidth, but the cost of crossbars increases as the square of the number of ports, rendering them economically infeasible for more than a few dozen processors.

In the 1990s, massively parallel computers based on scalable interconnects will become a mainstream technology, just as small-scale parallel processing did in the 1980s. The driving force is the economics involved in increasing the performance of the most powerful computers. It is becoming increasingly expensive in both dollars and time to develop succeeding generations of traditional ultra-high clock rate supercomputers and mainframes. Massively parallel systems will be the only affordable way to achieve the performance goals of the 1990s. This shift is made possible by three technologies discussed in the following sections:
- high-performance RISC microprocessors,
- advanced software, and
- versatile, scalable system interconnects.

The Attack of the Killer Micros

One of the key drivers in the high-performance computing industry is the disparity between the price/performance and overall performance gains of microprocessors versus conventional mainframes and vector supercomputers. As Figure 2 illustrates, the gains in microprocessor performance are far more rapid than those for supercomputers, with no end in sight for this trend. When looking at curves such as these, it seems obvious that high-performance microprocessors and parallel systems built from these microprocessors will come to dominate the high-end computing market; this is the "attack of the killer micros."

This changeover is only just now occurring. As Figure 3 illustrates, parallel systems are now capable of higher performance and better price/performance than traditional supercomputers. This transition occurred with the advent of RISC microprocessors, which provided sufficient floating point performance to enable parallel systems to rival supercomputers. This performance and price/performance gap will continue to widen in favor of parallel micro-based systems as microprocessor gains continue to outstrip those of supercomputers.

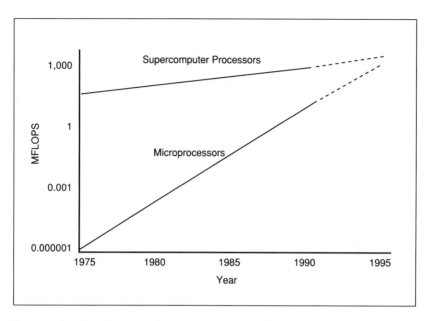

Figure 2. Absolute performance gains of microprocessors versus supercomputers.

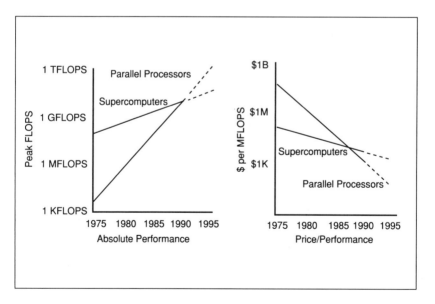

Figure 3. Improvements in parallel processors versus supercomputers.

Programmer Productivity on Massively Parallel Systems

High performance and attractive price/performance are not enough to bring massively parallel systems into the computing mainstream. It is well known that only 10–20 per cent of a computer center's budget goes to paying for computer hardware. The largest portion goes to paying for people to write software and to support the computers. Large gains in price/performance can be quickly erased if the system is difficult to use. In order to be accepted by a larger number of customers, massively parallel systems must provide ease-of-use and programmer productivity that is more like current mainstream high-performance systems.

The conventional wisdom in the 1980s was that parallel systems are difficult to use because it is hard to parallelize code. However, many problems are naturally parallel and readily map to parallel architectures. For these problems, the past has been spent trying to develop serial algorithms that solve these problems on single-CPU systems. Trying to take this serial code and parallelize it is clearly not the most productive approach. A more productive way is to directly map the parallel problem onto a parallel system.

Also, most computer systems today are parallel systems. Minicomputers, workstations, minisupers, supercomputers, even mainframes all have more than a single CPU. Clearly, parallelism itself isn't the only problem, since such systems from major computer vendors are now considered mainstream computers.

Yet there is a programmer productivity gap on most massively parallel systems, as illustrated in Figure 4. While the productivity on

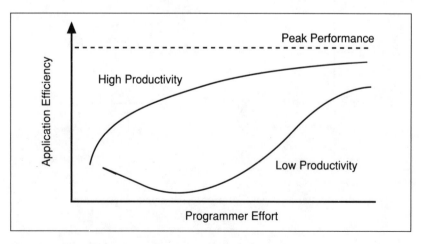

Figure 4. The programmer-productivity gap.

small-scale parallel systems now mirrors the traditionally higher productivity of uniprocessor systems, the productivity on these massively parallel systems is still very low. Given that there are plenty of parallel problems and that parallel processing has reached the mainstream, what is still holding massively parallel systems back? The answer lies in their software development environment.

Front End/Back End versus Native UNIX

One key differentiator between most massively parallel systems and the successful mainstream parallel systems is the relationship of the development environment to the computer. In most massively parallel systems, the computer is an attached processor, or back end, to a front-end workstation, minicomputer, or personal computer (Figure 5). All software development and user interaction is done on the front end, whereas program execution runs on the back-end parallel system. BBN's early parallel processors, such as the Butterfly® I and Butterfly Plus systems, required such front ends. As we learned, there are several problems with this architecture, including
- Bottlenecks: the link between the front end and the back end is a potential bottleneck. It is frequently a local area network, such as Ethernet, with a very limited bandwidth compared with the requirements of high-end supercomputer applications.

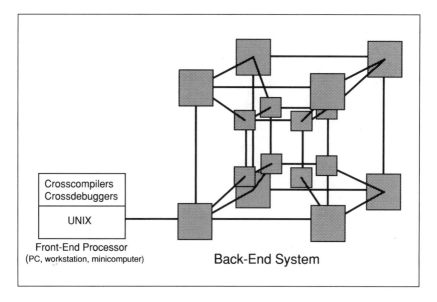

Figure 5. A front-end/back-end system.

- Debugging and tuning difficulties: because the software-development tools are separate from the parallel back end, it can be difficult to debug and tune programs. The tools on the front end cannot directly examine the memory of the back end and must rely on the back-end processors for information. If a program crashes some or all of the parallel nodes' kernels, the debugger may not be able to provide sufficient information.
- Slow development cycle: because development is done on a separate computer, the power of the parallel supercomputer is not available to run the development tools, such as the compiler. Also, executable program images must be downloaded into the back end, adding a step to the development cycle and further slowing down productivity.
- Multiple environments: while the front end typically runs UNIX, the back-end processors run a proprietary kernel. This requires the developer to learn two different environments.
- Limited kernel: the proprietary kernel that runs on the back-end processors does not provide all of the facilities that users expect on modern computers. These kernels provide little protection between tasks, no virtual memory, and few operating-system services.

Contrast this with modern supercomputers, mainframes, minicomputers, and workstations. All have native development environments, typically based on UNIX. This greatly simplifies development because the tools run on the same machine and under the same operating system as the executable programs. The full services of UNIX that are available to the programmer are also available to the executable program, including virtual memory, memory protection, and other system services. Since these systems are all shared-memory machines, powerful tools can be built for debugging and analyzing program performance, with limited intrusion into the programs operation.

Recent BBN parallel computers, such as the GP1000™ and TC2000™ systems, are complete, stand-alone UNIX systems and do not require a front end. The Mach 1000™ and nX™ operating systems that run on these computers contain a highly symmetric multiprocessing kernel that provides all of the facilities that users expect, including load balancing, parallel pipes and shell commands, etc. Since these operating systems present a standard UNIX interface and are compliant with the POSIX 1003.1 standard, users familiar with UNIX can begin using the system immediately. In fact, there are typically many users logged into and using a TC2000 system only hours after it is installed. This is in contrast to our earlier front-end/back-end systems, where users spent days or weeks studying manuals before running their first programs.

Single User versus Multiple Users

A second difference between mainstream computers and most massively parallel systems is the number of simultaneous users or applications that can be supported. Front-end/back-end massively parallel systems typically allow only a few users to be using the back end at one time. This style of resource scheduling is characterized by batch operation or "sign-up sheets." This is an adequate model for systems that will be dedicated to a single application but is a step backward in productivity for multiuser environments when compared with mainstream computers that support timesharing operating systems. As has been known for many years, timesharing provides a means to more highly utilize a computer system. Raw peak MFLOPS (i.e., millions of floating-point operations per second) are not as important as the number of actual FLOPS that are used in real programs; unused FLOPS are wasted FLOPS. The real measure of system effectiveness is the number of solutions per year that the user base can achieve.

Early in BBN's use of the Butterfly I computer, we realized that flexible multiuser access was required in order to get the most productivity out of the system. The ability to cluster together an arbitrary number of processors was added to the Chrysalis™ operating system (and later carried forward into Mach 1000 and nX), providing a simple but powerful "space-sharing" mechanism to allow multiple users to share a system. However, in order to eliminate the front end and move general computing and software development activities onto the system, real timesharing capabilities were needed to enable processors to be used by multiple users. The Mach 1000 and nX operating systems provide true time sharing.

Interconnect Performance, System Versatility, and Delivered Performance

Related to the number of users that can use a system at the same time is the number of different kinds of problems that a system can solve. The more flexible a system is in terms of programming paradigms that it supports, the more solutions per year can be delivered. As we learned while adapting a wide range of applications to the early Butterfly systems, it is much more productive to program using a paradigm that is natural to the problem at hand than to attempt to force-fit the code to a machine-dependent paradigm. Specialized architectures do have a place running certain applications in which the specialized system's architecture provides very high performance and the same code will be run a large number of times. However, many problems are not well suited to these systems.

Current mainstream systems provide a very flexible programming environment in which to develop algorithms. Based on shared-memory architectures, these systems have demonstrated their applicability in a wide range of applications, from scientific problems to commercial applications. BBN's experience with our parallel systems indicates that shared-memory architectures are the best way to provide a multiparadigm environment comparable to mainstream systems. For example, the TC2000 uses the Butterfly® switch (BBN 1990) to provide a large, globally addressable memory space that is shared by the processors yet is physically distributed: "distributed shared memory." This provides the convenience of the shared-memory model for those applications to which it is suited while providing the scalable bandwidth of distributed memory. The TC2000's Butterfly switch also makes it an ideal system for programming with the message-passing paradigm, providing low message-transfer latencies.

Another key difference between mainstream systems and primitive massively parallel computers is the system's delivered performance on applications that tend to randomly access large amounts of memory. According to John Gustafson of NASA/Ames Laboratory, "Supercomputers will be rated by dollars per megabyte per second more than dollars per megaflop . . . by savvy buyers." A system's ability to randomly access memory is called its random-access bandwidth, or RAB. High RAB is needed for such applications as data classification and retrieval, real-time programming, sparse matrix algorithms, adaptive grid problems, and combinational optimization (Celmaster 1990).

High-RAB systems can deliver high performance on a wider range of problems than can systems with low RAB and can provide the programmer with more options for developing algorithms. This is one of the strengths of the traditional supercomputers and mainframes and is a key reason why most massively parallel systems do not run certain parallel applications as well as one would expect. The TC2000 is capable of RAB that is comparable to, and indeed higher than, that of mainstream systems. Table 1 compares the TC2000 with several other systems on the portable random-access benchmark (Celmaster undated). For even higher RAB, the Monarch project (Rettberg et al. 1990) at BBN explored advanced switching techniques and designed a very-large-scale MIMD computer with the potential for several orders of magnitude more RAB than modern supercomputers. Lastly, BBN's experience using early

Table 1. Comparison of Random-Access Bandwidth

System	Number of Processors	RAB (kilo random-access words per second)
TC2000	1	258
	40	9,447
	128	23,587
i860 Touchstone	1	~2.5
	128	~300
IRIS 4D/240	1	349
	4	779
CRAY Y-MP/832	1	28,700

Butterfly systems in real-time simulation and communications applications indicated that special capabilities were required for these areas. The real-time model places very demanding constraints on system latencies and performance and requires software and hardware beyond what is provided by typical non-real-time systems. These capabilities include a low-overhead, real-time executive, low-latency access to shared memory, hardware support such as timers and globally synchronized, real-time clocks, and support for the Ada programming language.

Challenges and Directions for the Future

The challenge facing the vendors of massively parallel processors in the 1990s is to develop systems that provide high levels of performance without sacrificing programmer productivity. When comparing the next generation of parallel systems, it is the interconnect and memory architecture and the software that will distinguish one system from another. All of these vendors will have access to the same microprocessors, the same semiconductor technology, the same memory chips, and comparable packaging technologies. The ability to build scalable, massively parallel systems that are readily programmable will determine the successful systems in the future.

Most vendors have realized this and are working to enhance their products accordingly, as shown in Figure 6. The traditional bus-based

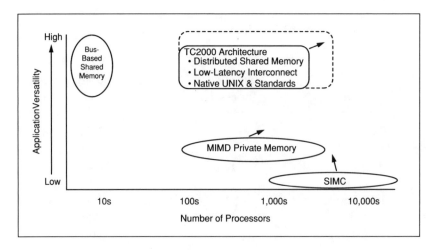

Figure 6. Directions in architecture.

and crossbar architecture systems have always held the lead in application versatility and programmer productivity but do not scale to massively parallel levels. Many of these vendors, such as Cray, have announced plans to develop systems that scale beyond their current tens of processors. At the same time, vendors of data-parallel and private-memory-MIMD systems are working to make their systems more versatile by improving interconnect latency, adding global routing or simulated shared memory, and adding more UNIX facilities to their node kernels. The direction in which all of this development is moving is toward a massively parallel UNIX system with low-latency distributed shared memory.

As in previous transitions in the computer industry, the older technology will not disappear but will continue to coexist with the new technology. In particular, massively parallel systems will coexist with conventional supercomputers, as illustrated in Figure 7. In this "model supercomputer center," a variety of resources are interconnected via a high-speed network or switch and are available to users. The traditional vector supercomputer will provide compute services for those problems that are vectorizable and primarily serial and will continue to run some older codes. The special-purpose application accelerators provide very high performance on select problems that are executed with sufficient frequency to justify the development cost of the application and the cost of the hardware. The general-purpose parallel system will off-load the vector supercomputer of nonvector codes and will provide a production environment for most new parallel applications. It will also serve as a testbed for parallel algorithm research and development.

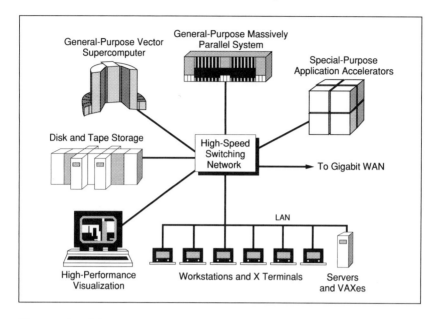

Figure 7. Model supercomputer center.

Summary

In the 1980s, parallel processing moved into the mainstream of computing technologies. The rapid increases in "killer micro" performance will enable massively parallel systems to meet the needs of high-performance users in the 1990s. However, in order to become a mainstream technology, massively parallel systems must close the programmer productivity gap that exists between them and small-scale parallel systems. The keys to closing this gap are standard languages with parallel extensions, native operating systems (such as UNIX), a powerful software development tool set, and an architecture that supports multiple programming paradigms.

Appendix : BBN Parallel-Processing Systems

A summary of the BBN parallel-processing systems appears in Table 2. The Pluribus was BBN's first parallel-processing system. Developed in the early 1970s, with initial customer shipments in 1975, it consisted of up to 14 Lockheed Sue minicomputers interconnected via a bus-based distributed crossbar switch and supported shared global memory. It was used primarily in communications applications, many of which are still operational today.

Table 2. BBN Parallel Processing Systems

Pluribus	Butterfly I	Butterfly Plus	GP1000	TC2000	Coral
Parallel Hardware	Massively Parallel Hardware	Performance Improvement	Mach 1000 pSOS	10X Performance	5-10X CPU Performance
Shared Memory	Shared Memory	More Memory	TotalView	nX, pSOS+m	4X Package Density
Bus and Crossbar	Butterfly Switch		Parallel Fortran	Ada, C++	More Memory
				VME	
	Chrysalis			Xtra	Compilers
	Uniform System			More Memory	HIPPI, FDDI
					VME64
					Tools and Libraries

The initial member of the Butterfly family of systems, the Butterfly I, was developed beginning in 1977. An outgrowth of the Voice Funnel program, a packetized voice satellite communications system funded by the Defense Advanced Research Projects Agency (DARPA), the Butterfly I computer was designed to scale to 256 Motorola, Inc., M68000 processors (a system of this size was built in 1985) but without giving up the advantages of shared memory. The key to achieving this scalability was a multistage interconnection network called the Butterfly switch. BBN developed the proprietary Chrysalis operating system, the Gist™ performance analyzer, and the Uniform System™ parallel programming library for this system. Butterfly I machines were used in a wide variety of research projects and also are used as Internet gateways when running communications code developed at BBN.

In the early 1980s, DARPA also funded BBN to explore very-large-scale parallel-processing systems. The Monarch project explored the design of a 65,536-processor shared-memory MIMD system using a multistage interconnection network similar to the Butterfly switch. The high-speed switch was implemented and tested using very-large-scale integration based on complementary metal oxide semiconductor technology, and a system simulator was constructed to explore the performance of the system on real problems. Some of the concepts and technologies have already been incorporated into Butterfly products, and more will be used in future generations.

The Butterfly Plus system was developed to provide improved processor performance over the Butterfly I system by incorporating Motorola's MC68020 processor and the MC68881 (later, the MC68882) floating-point coprocessor. Since this system used the same Butterfly switch, Butterfly I systems could be easily upgraded to Butterfly Plus performance.

The Butterfly Plus processor boards also included more memory and a memory-management unit, which were key to the development of the Butterfly GP1000 system. The GP1000 used the same processors as the Butterfly Plus but ran the Mach 1000 operating system, the world's first massively parallel implementation of UNIX. Mach 1000 was based on the Mach operating system developed at Carnegie Mellon University but has been extended and enhanced by BBN. The TotalView™ debugger was another significant development that was first released on the GP1000.

The TC2000 system, BBN's newest and most powerful computer, was designed to provide an order of magnitude greater performance than previous Butterfly systems. The world's first massively parallel RISC system, the TC2000 employs the Motorola M88000 microprocessor and a new generation Butterfly switch that has ten times the capacity of the previous generation. The TC2000 runs the nX operating system, which was derived from the GP1000's Mach 1000 operating system. The TC2000 also runs pSOS^{+m} ™, a real-time executive.

The goal of the Coral project is to develop BBN's next-generation parallel system for initial delivery in 1992. The Coral system is targeted at providing up to 200 GFLOPS peak performance using 2000 processors while retaining the shared-memory architecture and advanced software environment of the TC2000 system, with which Coral will be software compatible.

References

BBN, "Inside the TC2000," BBN Advanced Computers Inc. report, Cambridge, Massachusetts (February 1990).

W. Celmaster, "Random-Access Bandwidth and Grid-Based Algorithms on Massively Parallel Computers," BBN Advanced Computers Inc. report, Cambridge, Massachusetts (September 5, 1990).

W. Celmaster, "Random-Access Bandwidth Requirements of Point Parallelism in Grid-Based Problems," submitted to the *5th SIAM Conference on Parallel Processing for Scientific Computing*, Society for Industrial and Applied Mathematics, Philadelphia, Pennsylvania.

R. D. Rettberg, W. R. Crowther, P. P. Carvey, and R. S. Tomlinson, "Monarch Parallel Processor Hardware Design," *Computer* **23** (4), 18–30 (1990).

The John von Neumann Computer Center: An Analysis

Al Brenner

Alfred E. Brenner is Director of Applications Research at the Supercomputing Research Center, a division of the Institute for Defense Analysis, in Bowie, Maryland. He was the first president of the Consortium for Scientific Computing, the corporate parent of the John von Neumann Computer Center. Previously, he was head of the department of Computing at Fermi National Accelerator Laboratory. He has a bachelor's degree in physics and a Ph.D. in experimental high-energy physics from MIT.

Introduction

I have been asked to discuss and analyze the factors involved in the demise of the NSF Office of Advanced Scientific Computing (OASC) in Princeton, New Jersey—the John von Neumann Center (JVNC). My goal is to see if we can extract the factors that contributed to the failure to see whether the experience can be used to avoid such failures in the future. Analysis is much easier in hindsight than before the fact, so I will try to be as objective as I can in my analysis.

The "Pre-Lax Report" Period

During the 1970s, almost all of the supercomputers installed were found in government installations and were not generally accessible to the university research community. For those researchers who could not gain access to these supercomputers, this was a frustrating period. A few found it was relatively easy to obtain time on supercomputers in Europe, especially in England and West Germany.

By the end of the decade, a number of studies, proposals, and other attempts were done to generate funds to make available large-scale computational facilities for some of the university research community. All of this was happening during a period when U.S. policy was tightening rather than relaxing the mechanisms for acquiring large-scale computing facilities.

The Lax Report

The weight of argument in the reports from these studies and proposals moved NSF to appoint Peter Lax, of New York University, an NSF Board Member, as chairman of a committee to organize a Panel on Large-Scale Computing in Science and Engineering. The panel was sponsored jointly by NSF and the Department of Defense in cooperation with the Department of Energy and NASA. The end product of this activity was the "Report of the Panel on Large-Scale Computing in Science and Engineering," usually referred to as the Lax Report, dated December 26, 1982.

The recommendations of the panel were straightforward and succinct. The overall recommendation was for the establishment of a national program to support the expanded use of high-performance computers. Four components to the program were
- increased access to supercomputing facilities for scientific and engineering research;
- increased research in computational mathematics, software, and algorithms;
- training of personnel in high-performance computing; and
- research and development for the implementation of new supercomputer systems.

The panel indicated that insufficient funds were being expended at the time and suggested an interagency and interdisciplinary national program.

Establishment of the Centers

In 1984, once the NSF acquired additional funding from Congress for the program, NSF called for proposals to establish national supercomputer centers. Over 20 proposals were received, and these were evaluated in an extension of the usual NSF peer-review process. In February 1985, NSF selected four of the proposals and announced awards to establish four national supercomputer centers. A fifth center was added in early 1986.

The five centers are organizationally quite different. The National Center for Supercomputing Applications at the University of Illinois-Urbana/Champaign and the Cornell Theory Center are formally operated by the universities in which those centers are located. The JVNC is managed by a nonprofit organization, the Consortium for Scientific Computing, Inc. (CSC), established solely to operate this center. The San Diego Supercomputer Center is operated by the for-profit General Atomics Corporation and is located on the campus of the University of California at San Diego. Finally, the Pittsburgh Supercomputing Center is run jointly by the University of Pittsburgh, Carnegie Mellon University, and Westinghouse Electric Corporation. NSF established the OASC that reported directly to the Director of NSF as the NSF program office through which to fund these centers.

While the selected centers were being established (these centers were called Phase 2 centers), NSF supported an extant group of supercomputing facilities (Phase 1 centers) to start supplying cycles to the research community at the earliest possible time. Phase 1 centers included Purdue University and Colorado State University, both with installed CYBER 205 computers; and the University of Minnesota, Boeing Computer Services, and Digital Productions, Inc., all with CRAY X-MP equipment. It is interesting to note that all these centers, which had been established independent of the OASC initiative, were phased out once the Phase 2 centers were in operation. All Phase 1 centers are now defunct as service centers for the community, or they are at least transformed rather dramatically into quite different entities. Indeed, NSF "used" these facilities, supported them for a couple of years, and then set them loose to "dry up."

From the very beginning, it was evident there were insufficient funds to run all Phase 2 centers at adequate levels. In almost all cases, the centers from the beginning have been working within very tight budgets, which has resulted in difficult decisions to be made by management and a less aggressive program than the user community demands. However, with a scarce and expensive resource such as supercomputers, such limitations are not unreasonable. During the second round of funding for an additional five-year period, the NSF has concluded that the JVNC should be closed. The closing of that center will alleviate some of the fiscal pressure on the remaining four centers. Let us now focus on the JVNC story.

The John von Neumann Center

The Proposal

When the call for proposals went out in 1984 for the establishment of the national supercomputer centers, a small number of active and involved computational scientists and engineers, some very closely involved with the NSF process in establishing these centers, analyzed the situation very carefully and generated a strategy that had a very high probability of placing their proposal in the winning set. One decision was to involve a modest number of prestigious universities in a consortium such that the combined prominence of the universities represented would easily outweigh almost any competition. Thus, the consortium included Brown University, Harvard University, the Institute for Advanced Study, MIT, New York University, the University of Pennsylvania, Pennsylvania State University, Princeton University, Rochester Institute, Rutgers University, and the Universities of Arizona and Colorado. (After the establishment of the JVNC, Columbia University joined the consortium.) This was a powerful roster of universities indeed.

A second important strategy was to propose a machine likely to be different from most of the other proposals. At the time, leaving aside IBM and Japan, Inc., the only two true participants were Cray Research and Engineering Technology Associates Systems (ETA). The CRAY X-MP was a mature and functioning system guaranteed to be able to supply the necessary resources for any center. The ETA-10, a machine under development at the time, had much potential and was being designed and manufactured by an experienced team that had spun off from Control Data Corporation (CDC). The ETA-10, if delivered with the capabilities promised, would exceed the performance of the Cray Research offerings at the time. A proposal based on the ETA-10 was likely to be a unique proposal.

These two strategic decisions were the crucial ones. Also, there were other factors that made the proposal yet more attractive. The most important of these was the aggressive networking stance of the proposal in using high-performance communications links to connect the consortium-member universities to the center.

Also, the plan envisioned a two-stage physical plant, starting with temporary quarters to house the center at the earliest possible date, followed by a permanent building to be occupied later. Another feature was to contract out the actual operations functions to one of the firms experienced in the operation of supercomputing centers at other laboratories.

Finally, the proposal was nicely complemented with a long list of proposed computational problems submitted by faculty members of the 12 founding institutions. Although these additional attributes of the proposal were not unique, they certainly enhanced the strong position of a consortium of prestigious universities operating a powerful supercomputer supplied by a new corporation supported by one of the most prominent of the old-time supercomputer firms. It should surprise no one that on the basis of peer reviews, NSF found the JVNC proposal to be an attractive one.

I would like now to explore the primary entities involved in the establishment, operation, funding, and oversight of the JVNC.

Consortium for Scientific Computing

The CSC is a nonprofit corporation formed by the 12 universities of the consortium for the sole purpose of running the JVNC. Initially, each university was to be represented within the consortium by the technical representative who had been the primary developer of the proposal submitted to NSF. Early in the incorporation process, representation on the consortium was expanded to include two individuals from each university—one technical faculty and one university administrator. The consortium Board of Directors elected an Executive Committee from its own membership. This committee of seven members, as in normal corporate situations, wielded the actual power of the consortium. The most important function of the CSC included two activities: (1) the appointment of a Chief Operating Officer (the President) and (2) the establishment of policies guiding the activities of the center. As we analyze what went wrong with the JVNC, we will see that the consortium, in particular the Executive Committee, did not restrict itself to these functions but ranged broadly over many activities, to the detriment of the JVNC.

The Universities

The universities were the stable corporate entities upon which the consortium's credibility was based. Once the universities agreed to go forth with the proposal and the establishment of the consortium, they played a very small role.

The proposal called for the universities to share in the support of the centers. Typically, the sharing was done "in kind" and not in actual dollars, and the universities were involved in establishing the bartering chips that were required.

The State of New Jersey

The State of New Jersey supported the consortium enthusiastically. It provided the only, truly substantial, expendable dollar funding to the JVNC above the base NSF funding. These funds were funneled through the New Jersey State Commission for Science and Technology. The state was represented on the consortium board by one nonvoting member.

The NSF

NSF had moved forward on the basis of the proposals of the Lax Report and, with only modest previous experience with such large and complex organizations, established the five centers. The OASC reported directly to the Director of NSF to manage the cooperative agreements with the centers. Most of the senior people in this small office were tapped from other directorates within NSF to take on difficult responsibilities, and these people often had little or no prior experience with supercomputers.

ETA

In August 1983, ETA had been spun off from CDC to develop and market the ETA-10, a natural follow-on of the modestly successful CYBER 205 line of computers. The reason for the establishment of ETA was to insulate from the rest of CDC the ETA development team and its very large demands for finances. This was both to allow ETA to do its job and to protect CDC from an arbitrary drain of resources.

The ETA machine was a natural extension of the CYBER 205 architecture. The primary architect was the same individual, and much of the development team was the same team, that had been involved in the development of the CYBER 205.

Zero One

The JVNC contracted the actual daily operations of its center to an experienced facilitator. The important advantage to this approach was the ability to move forward as quickly as possible by using the resources of an already extant entity with an existing infrastructure and experience.

Zero One, originally Technology Development Corporation, was awarded the contract because it had experience in operating supercomputing facilities at NASA Ames, and they appeared to have an adequate, if not large, personnel base. As it turned out, apart from a small number of people, all of the personnel assigned to the JVNC were newly hired.

JVNC

During the first half of 1985, the consortium moved quickly and initiated the efforts to establish the JVNC. One of the first efforts was to find a building. Once all the factors were understood, rather than the proposed two-phase building approach, it was decided to move forward with a permanent building as quickly as possible and to use temporary quarters to house personnel, but not equipment, while the building was being readied.

The site chosen for the JVNC was in the Forrestral Research Center off Route 1, a short distance from Princeton University. The building shell was in place at the time of the commitment by the consortium and it was only the interior "customer modification" that was required. Starting on July 1, 1986, the building functioned quite well for the purposes of the JVNC.

A small number of key personnel were hired. Contracts were written with the primary vendors. The Cooperative Agreement to define the funding profile and the division of responsibility between the consortium and the NSF was also drawn up.

What Went Wrong?

The Analysis

The startup process at JVNC was not very different from the processes at the other NSF-funded supercomputing centers. Why are they still functioning today while the JVNC is closed? Many factors contributed to the lack of strength of the JVNC. As with any other human endeavor, if one does not push in all dimensions to make it right, the sum of a large number of relatively minor problems might mean failure, whereas a bit more strength or possibly better luck might make for a winner.

I will first address the minor issues that, I believe, without more detailed knowledge, may sometimes be thought of as being more important than they actually were. I will then address what I believe were the real problems.

Location

Certainly, the location of the JVNC building was not conducive to a successful intellectual enterprise. Today, with most computer accesses occurring over communications links, it is difficult to promote an intellectually vibrant community at the hardware site. If the hardware is close by, on the same campus or in the same building where many of the user participants reside, there is a much better chance of generating the collegial spirit and intellectual atmosphere for the center and its support personnel. The JVNC, in a commercial industrial park sufficiently far from even its closest university customers, found itself essentially in isolation.

Furthermore, because of the meager funding that allowed no in-house research-oriented staff, an almost totally vacuous intellectual atmosphere existed, with the exception of visitors from the user community and the occasional invited speaker. For those centers on campuses or for those centers able to generate some internal research, the intellectual atmosphere was certainly much healthier and more supportive than that at the JVNC.

Corporate Problems

Some of the problems the JVNC experienced were really problems that emanated from the two primary companies that the JVNC was working with: ETA and Zero One. The Zero One problem was basically one of relying too heavily on a corporate entity that actually had very little flex in its capabilities. At the beginning, it would have been helpful if Zero One had been able to better use its talent elsewhere to get the JVNC started, but it was not capable of doing that, with one or two exceptions. The expertise it had, although adequate, was not strong, so the relationship JVNC had with Zero One was not particularly effective in establishing the JVNC. Toward the end of June 1989, JVNC terminated its relationship with Zero One and took on the responsibility of operating the center by itself. Consequently, the Zero One involvement was not an important factor in the long-term JVNC complications.

The problems experienced in regard to ETA were much more fundamental to the demise of JVNC. I believe there were two issues that had a direct bearing on the status of the JVNC. The first was compounded by the inexperience of many of the board members. When the ETA-10 was first announced, the clock cycle time was advertised as five nanoseconds. By the time contractual arrangements had been completed, it was clear the five-nanosecond time was not attainable and that something more like seven or eight nanoseconds was the best goal to be achieved. As we know, the earliest machines were delivered with cycle times twice those numbers. The rancor and associated interactions concerning each of the entities' understanding of the clock period early in the relationship took what could have been a cooperative interaction and essentially poisoned it. Both organizations were at fault. ETA advertised more than they could deliver, and the consortium did not accommodate the facts.

Another area where ETA failed was in its inability to understand the importance of software to the success of the machine. Although the ETA hardware was first-rate in its implementation, the decision to make the ETA-10 compatible with the CYBER 205 had serious consequences. The primary operating-system efforts were to replicate the functionality of the CYBER 205 VSOS; any extensions would be shells around that native system. That decision and a less-than-modern approach to the implementation of the approach bogged down the whole software effort. One example was the high-performance linkages; these were old, modified programs that gave rise to totally unacceptable communications performance. As the pressures mounted for a modern operating system, in particular UNIX, the efforts fibrillated, no doubt consuming major resources, and never attained maturity. The delays imposed by these decisions certainly were not helpful to ETA or to the survival of the JVNC.

NSF, Funding, and Funding Leverage

We now come to an important complication, not unique to the JVNC but common to all of the NSF centers. To be as aggressive as possible, NSF extended itself as far as the funding level for the OASC would allow and encouraged cost-sharing arrangements to leverage the funding. This collateral funding, which came from universities, states, and corporate associates interested in participating in the centers' activities, was encouraged, expected, and counted upon for adequate funding for the centers.

As the cooperative agreements were constructed in early 1985, the funding profiles for the five-year agreements were laid out for each individual center's needs. The attempt to meet that profile was a painful experience for the JVNC management, and I believe the same could be said for the other centers as well. For the JVNC, much of the support in kind from universities was paper; indeed, in some cases, it was closer to being a reverse contribution.

As the delivery of the primary computer equipment to JVNC was delayed while some of the other centers were moving forward more effectively, the cooperative agreements were modified by NSF to accommodate these changes and stay within the actual funding profile at NSF. Without a modern functioning machine, the JVNC found it particularly difficult to attract corporate support. The other NSF centers, where state-of-the-art supercomputer systems were operational, were in much better positions to woo industrial partners, and they were more successful. Over the five-year life of the JVNC, only about $300,000 in corporate support was obtained; that was less than 10 per cent of the proposed amount and less than three-quarters of one per cent of the actual NSF contribution.

One corporate entity, ETA, contributed a large amount to the JVNC. Because the delivery of the ETA-10 was so late, the payment for the system was repeatedly delayed. The revenue that ETA expected from delivery of the ETA-10 never came. Thus, in a sense, the hardware that was delivered to the JVNC—two CYBER 205 systems and the ETA-10—represented a very large ETA corporate contribution to the JVNC. The originally proposed ETA contribution, in discounts on the ETA-10, personnel support, and other unbilled services, was $9.6 million, which was more than 10 per cent of the proposed level of the NSF contribution.

A year after the original four centers were started, the fiscal stress in the program was quite apparent. Nevertheless, NSF chose to start the fifth center, thereby spreading its resources yet thinner. It is true that the NSF budgets were then growing, and it may have seemed to the NSF that it was a good idea to establish one more center. In retrospect, the funding level was inadequate for a new center. Even today, the funding levels of all the centers remain inadequate to support dynamic, powerful centers able to maintain strong, state-of-the-art technology.

Governance

I now come to what I believe to be the most serious single aspect that contributed to the demise of the JVNC: governance. The governance, as I perceive it, was defective in three separate domains, each defective in its own right but all contributing to the primary failure, which was the governance of the CSC. The three domains I refer to are the universities, NSF, and the consortium itself.

Part of the problem was that the expectations of almost all of the players far exceeded the possible realities. With the exception of the Director of NSF, there was hardly a person directly or indirectly involved in the governance of the JVNC who had any experience as an operator of such complex facilities as the supercomputing centers represented. Almost all of the technical expertise was as end users. This was true for the NSF OASC and for the technical representatives on the Board of Directors of the consortium. The expertise, hard work, maturation, and planning needed for multi-million-dollar computer acquisitions were unknown to this group. Their expectations both in time and in performance levels attainable at the start-up time of the center were totally unrealistic.

At one point during the course of the first year, when difficulties with ETA meeting its commitments became apparent, the consortium

Experience and Lessons Learned   479

negotiated the acquisition of state-of-the-art equipment from an alternate vendor. To move along expeditiously, the plan included acquiring a succession of two similar but incompatible supercomputing systems from that vendor, bringing them up, networking them, educating the users, and bringing them down in sequence—all over a nine-month period! This was to be done in parallel with the running of the CYBER 205, which was then to be the ETA interim system—all of this with the minuscule staff at JVNC. At a meeting where these plans were enunciated to NSF, the Director of NSF very vocally expressed his consternation of and disbelief in the viability of the proposal. The OASC staff, the actual line managers of the centers, had no sense of the difficulty of the process being proposed.

At a meeting of the board of the consortium, the board was frustrated by the denial of this alternate approach that had by then been promulgated by NSF. A senior member of the OASC, who had participated in the board meeting but had not understood the nuances of the problem, when given the opportunity to make clear the issues involved, failed to do so, thereby allowing to stand misconceptions that were to continue to plague the JVNC. I believe that incident, which was one of many, typified a failure in governance on the part of NSF's management of the JVNC Cooperative Agreement.

With respect to the consortium itself, the Executive Committee, which consisted of the small group of people who had initiated the JVNC proposal, insisted on managing the activities as they did their own individual research grants. On a number of occasions, the board was admonished by the nontechnical board members to allow the president to manage the center. At no point did that happen during the formation of the JVNC.

These are my perceptions of the first year of operation of the JVNC. I do not have first-hand information about the situation during the remaining years of the JVNC. However, leaving aside the temporary management provided by a senior Princeton University administrator on a number of occasions, the succession of three additional presidents of the consortium over the next three years surely supports the premise that the problems were not fixed.

Since NSF was not able to do its job adequately in its oversight of the consortium, where were the university presidents during this time? The universities were out of the picture because they had delegated their authority to their representatives on the board. In one instance, the president of Princeton University did force a change in the leadership of the Board of Directors to try to fix the problem. Unfortunately, that action was not coupled to a simultaneous change of governance that was really needed to fix the problem. One simple fix would have been to rotate the cast of characters through the system at a fairly rapid clip, thereby disengaging the inside group that had initiated the JVNC.

Although the other centers had to deal with the same NSF management during the early days, their governance typically was in better hands. Therefore, they were in a better position to accommodate the less-than-expert management within the NSF. Fortunately, by the middle of the second year, the NSF had improved its position. A "rotator" with much experience in operating such centers was assigned to the OASC. Once there was a person with the appropriate technical knowledge in place at the OASC, the relationship between the centers and the NSF improved enormously.

Conclusions

I have tried to expose the problems that contributed to the demise of the JVNC. In such a complex and expensive enterprise, not everything will go right. Certainly many difficult factors were common to all five centers. It was the concatenation of the common factors with the ones unique to the JVNC that caused its demise and allowed the other centers to survive. Of course, once the JVNC was removed, the funding profile for the other centers must have improved.

In summary, I believe the most serious problem was the governance arrangements that controlled the management of the JVNC. Here seeds of failure were sown at the inception of the JVNC and were not weeded out. A second difficulty was the lack of adequate funding. I believe the second factor continues to affect the other centers and is a potential problem for all of them in terms of staying healthy, acquiring new machines, and maintaining challenging environments.

I have tried as gently as possible to expose the organizations that were deficient and that must bear some responsibilities for the failure of the JVNC. I hope, when new activities in this vein are initiated in the future, these lessons will be remembered and the same paths will not be traveled once again.

Project THOTH: An NSA Adventure in Supercomputing, 1984–88

Larry Tarbell

> *Lawrence C. Tarbell, Jr., is Chief of the Office of Computer and Processing Technology in the Research Group at the National Security Agency (NSA). His office is charged with developing new hardware and software technology in areas such as parallel processing, optical processing, recording, natural language processing, neural nets, networks, workstations, and database-management systems and then transferring that technology into operational use. For the last five years, he has had a special interest in bringing parallel processing technology into NSA for research and eventually operational use. He served as project manager for Project THOTH at NSA for over two years. He is the chief NSA technical advisor for the Supercomputing Research Center.*
>
> *Mr. Tarbell holds a B.S. in electrical engineering from Louisiana State University and an M.S. in the same field from the University of Maryland. He has been in the Research Group at the NSA for 25 years.*

I will discuss an adventure that the National Security Agency (NSA) undertook from 1984 to 1988, called Project THOTH. The reason the project was named THOTH was because the person who had the ability to choose the name liked Egyptian gods. There is no other meaning to the name.

NSA's goal for Project THOTH was to design and build a high-speed, high-capacity, easy-to-use, almost-general-purpose supercomputer. Our performance goal, which we were not really sure we could meet but really wanted to shoot for, was to have a system that ended up being 1000 times the performance of a Cray Research CRAY-1S on the kinds of problems we had in mind. The implication of that, which I do not think we really realized at the time, was that we were trying to skip a generation or so in the development of supercomputers. We really hoped that, at the end, THOTH would be a general-purpose, commercial system.

The high-level requirements we set out to achieve were performance in the range of 50 to 100×10^9 floating-point operations per second (GFLOPS), with lots of memory, a functionally complete instruction set (because we needed to work with integers, with bits, and with floating-point numbers), and state-of-the-art software. We were asking an awful lot! In addition, we wanted to put this machine into the operational environment we already had at NSA so that workstations and networks could have access to this system.

This was a joint project between the research group that I was in and the operations group that had the problems that needed to be solved. The operations organization wanted a very powerful system that would let them do research in the kinds of operational problems they had. They understood those problems and they understood the algorithms they needed to use for it. We supposedly understood something about architecture and software and, more importantly, about how to deal with contracting, which turned out to be a big problem.

After a lot of thought, we came up with the project structure. To characterize the kinds of computations we wanted to do, the operations group developed 24 sample problems. We realized our samples were not complete, but they were representative. In addition, we asked ourselves to imagine that there were 24 or 48 or 100 other kinds of problems that needed the same kind of computation.

These problems were not programs; they were just mathematical and word statements because we did not want to bias anybody by using old code. We were going to use these problems to try to evaluate the performance of the architectures we hoped would come out of THOTH. So they were as much a predictor of success as they were a way to characterize what we needed to do.

We ended up deciding to approach the project in three phases. The first phase was architectural studies, where we tried to look at a broad range of options at a really high level and then tried to choose several possibilities to go into a detailed design. For the second phase, we hoped at the end to be able to choose one of the detailed designs to actually build something. The third phase was the implementation of THOTH. We probably deluded ourselves a bit by saying that what we were trying to do was not to develop a new supercomputer; we were just trying to hasten the advent of something that somebody was already working on. In the end, that was not true.

We chose contractors using three criteria. First, they had to be in the business of delivering complete systems with hardware and software. Second, they had to have an active and ongoing R&D program in parallel processing and supercomputing. Third, we wanted them to have at least one existing architecture already under development because we thought that, if they were already involved in architecture development, we would have a much better chance of success.

We had envisioned six such companies participating in Phase 1. We hoped to pick the two best from Phase 1 to go into Phase 2. Then we hoped at the end of Phase 2 to have two good alternatives to choose from, to pick the better of those, and to go with it. We also had some independent consulting contractors working with us who were trying to make sure that we had our feet on the ground.

Figure 1 shows a time line of what actually happened. The concept for this project started in 1984. It was not until late 1985 that we actually had a statement of work that would be presented to contractors. Following the usual government procedure, it was not until mid-1986 that we

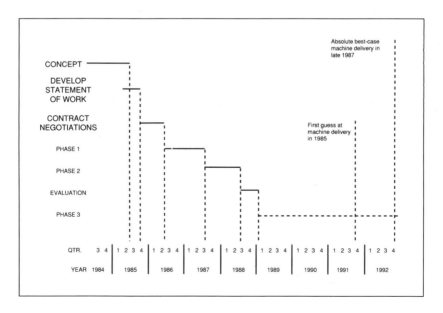

Figure 1. THOTH time line.

actually had some contracts under way. Thus, we spent a lot of time getting started, and while we were getting started, technology was moving. We were trying to keep up with the technology, but that was a hard thing to do.

At the beginning of 1985, we had as our first guess that we would get a machine in late 1991. By the time we got around to beginning the second phase, it was clear that the absolute best case for a machine delivery would be late in 1992, if then.

We ended up with nine companies invited to participate in Phase 1, seven of which chose to participate with us in a nine-month, low-cost effort. We had three parts of Phase 1. The first part, the definition study, was for the companies to show us that they understood what we were after. They were supposed to analyze the problems, to show us possibility in the problems for parallelism, and to talk about how their potential architecture might work on those problems. In the architectural study, we actually asked them to formally assess their favorite architecture as to how it performed against the THOTH problems that we had given them. In the third task, we actually paid them to produce a formal proposal for a high-level hardware/software system that would be designed in the second phase if we had enough belief in what they were trying to do. This took a lot more work on our part than we had initially envisioned. It essentially burned up the research organization involved in this; we did nothing else but this. A similar thing happened with the operations group and with some of the people that were supporting us.

We had some problems in Phase 1. I had not expected much personnel change in a nine-month phase, but in one company we had a corporate shakeup. They ended up with all new people and a totally new technical approach, and they waited until they were 40 per cent into the project to do this. That essentially sunk them. In several companies we had major personnel changes in midstream, and in nine months there was not time to catch up from changes in project managers.

Several companies had a lack of coordination because they had software people on one side of the country and hardware people on another side of the country, and once again that presented some problems. Even though most of them told us that they really understood our requirements, it turned out that they really had not fully understood them. I suppose we should not have been surprised. In one case, the division that was doing THOTH was sold. The project was then transferred from the division that was sold to another part of the company that actually built computers. The late submission of reports did not surprise us at all, but in nine months this caused us some problems. Also, there was one company that just could not pick out which way they wanted to go. They had two possible ways, and they never chose.

The first result in Phase 1 was that two companies dropped out before Phase 1 was over. We got five proposals at the end of Phase 1, but then one company subsequently removed their proposal, leaving us with only four. We ended up choosing three of those four, which surprised us. But when we sat down and looked at what was presented to us, it looked like there were three approaches that had some viability to them. Strangely enough, we had enough money to fund all three, so we did.

Then we moved into Phase 2, detailed design. Our goal was to produce competing detailed system designs from which to build THOTH, and we had three companies in the competition to produce a detailed design. That was what we thought would happen. They were to take the high-level design they had presented us at the end of Phase 1, refine it, and freeze it. We went through some detailed specifications, preliminary design reviews, and critical design reviews. Near the end of this one-year phase, they had to tell us that the performance they predicted in Phase 1 was still pretty much on target, and we had to believe them. This was a one-year, medium-cost effort, and once again, it cost us lots more in people than we had expected that it would.

Phase 2 had some problems. Within one company, about three months into Phase 2, the THOTH team was reorganized and moved from one part of the company to the other. This might not have been so bad except that for three months they did not have any manager, which led to total chaos. One company spent too much time thinking about hardware at the expense of software, which was exactly the reverse of what they had done in Phase 1, and that really shocked us. That, nevertheless, is what happened. Another company teamed, which was an interesting arrangement, with another software firm. Midway through, at a design review, the software firm stood up and quit right there. Another company had problems with their upper management. The lower-level people actually running the project had a pretty good sense of what to do, but upper management kept saying no, that is not the right way to do it—do it our way. They had no concept of what was going on, so that did not work out very well either.

There was a lack of attention to packaging and cooling, there were high compiler risks and so forth, and the real killer was that the cost of this system began to look a lot more expensive than we had thought. We had sort of thought, since we were going to be just accelerating somebody else's development, that $30 or $40 million would be enough. That ended up not being the case at all.

Several results came from Phase 2. One company withdrew halfway through. After the costs began to look prohibitive, we finally announced to the two remaining contractors that the funding cap was $40 million, which they agreed to compete for. We received two proposals. One was acceptable, although just barely. In the end, NSA's upper management canceled Project THOTH. We did not go to Phase 3.

Why did we stop? After Phase 2, we really only had one competitor that we could even think about going with. We really had hoped to have two. The one competitor we had looked risky. Perhaps we should have taken that risk, but at that time budgets were starting to get tighter, and upper management just was not willing to accept a lot of risk. Had we gone to Phase 3, it would have doubled the number of people that NSA would have had to put into this—at a time when our personnel staff was being reduced.

Also, the cost of building the system was too high. Even though the contractor said they would build it for our figure, we knew they would not.

There was more technical risk than we wanted to have. The real killer was that when we were done, we would have had a one-of-a-kind machine that would be very difficult to program and that would not have met the goals of being a system for researchers to use. It might have done really well for a production problem where you could code up one problem and then run it forever, but it would not be suited as a mathematical research machine.

So, in the end, those are the reasons we canceled. If only one or two of those problems had come up, we might have gone ahead, but the confluence of all those problems sunk the whole project.

As you can imagine from an experience like this, you learn some things. While we were working on THOTH, the Supercomputing Research Center (SRC) had a paper project under way, called Horizon, that began to wind down just about the time that THOTH was falling apart. So in April 1989, after the THOTH project was canceled in December, we had a workshop to try to talk about what had we learned from both of these projects together. Some of our discoveries were as follows:

- We did get some algorithmic insights from the Phase 1 work, and even though we never built a THOTH, some of the things we learned in THOTH have gone back into some of the production problems that were the source of THOTH and have improved things somewhat.
- We believe that the contractors became more aware of NSA computing needs. Even those who dropped out began to add things we wanted to their architectures or software that they had not done before, and we would like to believe that it was because of being in THOTH.

- We pushed the architectures and technology too hard. It turns out that parallel software is generally weak, and the cost was much higher than either we or the contractors had estimated at the time.
- We learned that the companies we were dealing with could not do it on their own. They really needed to team and team early on to have a good chance to pull something like this off.
- We learned that the competition got in our way. We did this as a competitive thing all along because the government is supposed to promote competition. We could see problems with the three contractors in Phase 2 that, had we perhaps been able to say a word or two, might have gone away. But we were constrained from doing that. So we were in a situation where we could ask questions of all three contractors, and if they chose to find in those questions the thing we were trying to suggest, that was fine. And if not, we couldn't tell them.

Hindsight is always 20/20. If we look backward, we realize the project took too long to get started. Never mind the fact that maybe we should not have started, although I think we should have, and I think we did some good. However, it took too long to get moving.

While we were involved in the project, especially in software, the technology explosion overtook us. When we started this system, we were not talking about a UNIX-like operating system, C, and Fortran compilers. By the end, we were. So the target we were shooting for kept moving, and that did not help us.

NSA also was biased as we began the project. We had built our own machines in the past. We even built our own operating system and invented our own language. We thought that since we had done all that before, we could do it again. That really hid from us how much trouble THOTH would be to do.

Our standards for credibility ended up being too low. I don't know how we could have changed that, but at the end we could see that we had believed people more than we should have. Our requirements were too general at first and became more specific as we went along. Thus, the contractors, unfortunately, had a moving target.

We also learned from these particular systems that your favorite language may not run well, if at all, and that the single most important thing that could have been built from this project would have been a compiler—not hardware, not an interconnection network, but a good compiler. We also learned that parallel architectures can get really weird if you let some people just go at it.

In the end, success might have been achieved if we had had Company A design the system, had Company B build it, and had Company C provide the software because in teaming all three companies we would have had strength.

There is a legacy to THOTH, however. It has caused us to work more with SRC and elsewhere. We got more involved with the Defense Advanced Research Projects Agency during this undertaking, and we have continued that relationship to our advantage. The project also resulted in involvement with another company that came along as a limited partner who had a system already well under way. There was another related follow-on project that came along after THOTH was canceled, but it relied too much on what we had done with THOTH, and it also fizzled out a little later on. Last, we now believe that we have better cooperation among NSA, SRC, and industry, and we hope that we will keep increasing that cooperation.

The Demise of ETA Systems

Lloyd Thorndyke

Lloyd M. Thorndyke is currently the CEO and chairman of DataMax, Inc., a startup company offering disk arrays. Before joining DataMax, he helped to found and was President and CEO of Engineering Technology Associates Systems, the supercomputer subsidiary of Control Data Corporation. At Control Data he held executive and technical management positions in computer and peripheral operations. He received the 1988 Chairman's Award from the American Association of Engineering Societies for his contributions to the engineering professions.

In the Beginning

Engineering Technology Associates Systems, or just plain ETA, was organized in the summer of 1983, and as some of you remember, its founding was announced here at the first Frontiers of Supercomputing conference in August 1983. At the start, 127 of the 275 people in the Control Data Corporation (CDC) CYBER 205 research and applications group were transferred to form the nucleus of ETA. This was the first mistake—we started with too many people.

The original business plan called for moving the entire supercomputer business, including the CYBER 205 and its installed base of 40 systems, to ETA. That never happened, and the result was fragmentation of the product line strategies and a split in the management of the CYBER 200 and ETA product lines. As a consequence it left the CDC CYBER 205 product line without dedicated management and direction and undermined any upgrade strategy to the ETA-10. Another serious consequence was the lack of a migration path for CYBER 200 users to move to the ETA-10.

ETA was founded with the intention of eventually becoming an independent enterprise. Initially, we had our own sales and marketing groups because the CYBER 205 was planned as part of ETA. Because the CYBER 205 was retained by CDC, the sales and marketing organizations of the two companies were confused. It seemed that CDC repeatedly reorganized to find the formula for success. The marketing management at CDC took responsibility when ETA was successful and returned responsibility to us when things did not go well. Without question, the failure to consolidate the CYBER 200 product line and marketing and sales management at ETA was a major contributing factor to the ETA failure.

There was an initial major assumption that the U.S. government would support the entry of a new supercomputer company through a combination of R&D funding and orders for early systems. The blueprint of the 1960s was going to be used over again. We read the tea leaves wrong. No such support ever came forth, and we did not secure orders from the traditional leading-edge labs. Furthermore, we did not receive R&D support from the funding agencies for our chip, board, and manufacturing technologies. We had four meetings with U.S. government agencies, and they shot the horse four times: the only good result was that they missed me each time.

The lack of U.S. government support was critical to our financial image. The lack of early software and systems technical help also contributed to delays in maturing our system. Other vendors did and still do receive such support, but such was not case with ETA. Our planning anticipated that the U.S. government would help a small startup. That proved to be a serious error.

Control Data played the role of the venture capitalist at the start and owned 90 per cent of the stock, with the balance held by the principals. The CDC long-range intent was to dilute to 40 per cent ownership through a public offering or corporate partnering as soon as possible. The failure to consummate the corporate partner, although we had a willing candidate, was a major setback in the financial area.

The first systems shipment was made to Florida State University (FSU) in December of 1986—three years and four months from the start. From that standpoint, we feel we reduced the development schedule of a complex supercomputer by almost 50 per cent.

At the time of the dynamiting of ETA on April 17, 1989, there were six liquid nitrogen-cooled systems installed. Contrary to the bad PR you might have heard, the system at FSU was a four-processor G system operating at seven nanoseconds. In fact, for a year after the closing, the system ran for a year at high levels of performance and quality, as FSU faculty will attest.

We had about 25 air-cooled systems installed (you may know them as Pipers) at customer sites. Internally, there were a total of 25 processors, both liquid- and air-cooled, dedicated to software development. Those are impressive numbers if one considers the costs of carrying the inventory and operating costs.

Hardware

From a technology viewpoint, I believe the ETA-10 was an outstanding hardware breakthrough and a first-rate manufacturing effort. We used very dense complementary metal oxide semiconductor (CMOS) circuits, reducing the size of the supercomputer processor to a single 16- by 22-inch board. I'm sure many of you have seen that processor. The CMOS chips reduced the power consumption of the processor to about 400 watts—that's watts, not 400 kilowatts. The use of CMOS chips operating in liquid nitrogen instead of ambient air resulted in doubling the speed of the CMOS. As a result of the two cooling methods and the configuration span from a single air-cooled processor to an eight-processor, liquid-cooled machine, we achieved a 27-to-one performance range. That range was able to use the same software and training for the diagnostics, operating-system software, and manufacturing checkout. We had broad commonality on a product line and inventory from top to bottom. We paid for the design only once, not many times. Other companies have proposed such a strategy—we executed it.

The liquid-nitrogen cryogenic cooling was a critical part of our design. I would suggest liquid nitrogen cooling as a technology other people should seriously consider. For example, a 20-watt computer will boil off one gallon of liquid nitrogen in an eight-hour period. Liquid nitrogen can be bought in bulk at a price cheaper than milk—it is as low as 25 cents a gallon in large quantities. This equals eight hours of operation for $0.40, assuming $0.40 per gallon. We get about 90 cubic feet of -200°C nitrogen gas. This gas can also help cool the rest of your computer room, greatly reducing the cooling requirements.

The criticism that liquid nitrogen resulted in a long mean time to repair was erroneous because at the time of the ETA closure, we could replace a processor in a matter of hours. The combination of CMOS and liquid-nitrogen cooling coupled with the configuration range provided a broad product family. These were good decisions—not everything we did was wrong.

The ETA-10 manufacturing process was internally developed and represented a significant advance in the state of the art. The perfect processor board yield at the end was 65 per cent for a board that was 16 by 22 inches with 44 layers and a 50-ohm controlled impedance. Another 30 per cent were usable with surface ECO wires. The remaining five per cent were scrap. This automated line produced enough boards to build two computers a day with just a few people involved.

For board assembly, we designed and built a pick-and-place robot to set the CMOS chips onto the processor board, an operation it could perform in less than four hours. The checkout of the computer took a few more hours. We really did have a system designed for volume manufacturing.

Initially, the semiconductor vendor was critical to us because it was the only such vendor in the United States that would even consider our advanced CMOS technology. In retrospect, our technology requirements and schedule were beyond the capabilities of the vendor to develop and deliver. Also, this vendor was not a merchant semiconductor supplier and did not have the infrastructure or outside market to support the effort. We were expected to place enough orders and supply enough funding to keep them interested in our effort. Our mistake was teaming with a nonmerchant vendor needing our resources to stay in the commercial semiconductor business.

We believed that we should work with U.S. semiconductor vendors because of the critical health of the U.S. semiconductor business. I would hasten to point out that the Japanese were very willing to supply us with the technology, both logic and memory that met or exceeded what we needed. Still, we stayed with the U.S. suppliers longer than good judgment warranted because we thought there was value to having a U.S.-made supercomputer with domestic semiconductors. We believed that our government encouraged such thinking, but ETA paid the price. In essence, we were trying to sell a computer with 100 per cent U.S. logic and memory components against a computer with 90 per cent Japanese logic and memory components, but we could not get any orders. I found the government's encouragement to us to use only U.S. semiconductor components and the subsequent action of buying competitive computers with the majority of their semiconductor content produced in Japan inconsistent and confusing.

Very clearly, the use of Japanese components does not affect the salability of the system in the U.S.—that message should be made clear to everyone. This error is not necessarily ETA's alone, but if the U.S.

government wants healthy U.S. semiconductor companies, then it must create mechanisms to encourage products with high U.S. semiconductor content only and to support R&D to keep domestic suppliers up to the state of the art.

Software

It is difficult to say much good about the early ETA software and its underlying strategy, although it was settling down at the end. A major mistake was the early decision to develop a new operating system, as against porting the CYBER 205 VSOS operating system. Since the CYBER 205 remained at CDC, we did not have product responsibility or direction, and the new operating system seemed the best way at the time.

In hindsight there has been severe criticism for not porting UNIX to the ETA-10 at the beginning—that is, start with UNIX, only. But in 1983 it was not that clear. I now hear comments from people saying, "If ETA would have started with UNIX, I would have bought." It was only two years later that they said, "Well, you should have done UNIX." However, we did not get UNIX design help, advice, or early orders for a UNIX system.

After we completed a native UNIX system and debugged the early problems, the UNIX system stabilized and ran well on the air-cooled systems, and as a result, several additional units were ordered. While the ETA UNIX lacked many features needed for supercomputer operation, users knew that these options were coming, but we were late to market. In hindsight, we should have ported VSOS and then worked only on UNIX.

Industry Observations

To be successful in the commercial supercomputer world, one must have any array of application packages. While we recognized this early on, as a new entrant to the business, we faced a classical problem that was talked about by other presenters: you can't catch up if you can't catch up.

We were not able to stimulate the applications vendors' interest because we didn't have a user base. Simultaneously, it's hard to build a user base without application packages. This vicious circle has to be broken because all the companies proposing new architectures are in the same boat. Somehow, we must figure out how to get out of it, or precious few new applications will be offered, except by the wealthiest of companies.

We need to differentiate the true supercomputer from the current situation, where everyone has a supercomputer of some type. The PR people have confiscated the supercomputer name, and we must find a new name. Therefore, I propose that the three or four companies in this business should identify their products as superprocessor systems. It may not sound sexy, but it does the job. We can then define a supercomputer system as being composed of one or more superprocessors.

The supercomputer pursuit is equivalent to a religious crusade. One must have the religion to pursue the superprocessors because of the required dedication and great, but unknown, risks. In the past, CDC and some of you here pioneered the supercomputer. Mr. Price had the religion, but CDC hired computer executives who did not, and in fact, they seemed to be supercomputer atheists. It was a major error by CDC to hire two and three levels of executives with little or no experience in high-performance or supercomputer development, marketing, or sales and place them in the computer division. Tony Vacca, ETA's long-time technologist, now at Cray Research, Inc. (see Session 2), observed that supercomputer design and sales are the Super Bowl of effort and are not won by rookies. It seems CDC has proved that point.

Today we all use emitter-coupled-logic (ECL), bipolar, memory chips, and cooling technologies in product design because of performance and cost advantages. Please remember that these technologies were advanced by supercomputer developers and indirectly paid for by supercomputer users.

If Seymour Cray had received a few cents for every ECL chip and bipolar chip and licensing money for cooling technology, he wouldn't need any venture capital today to continue his thrust. However, that is not the case. I believe that Seymour is a national treasure, but he may become an endangered species if the claims I have heard at this conference about massively parallel systems are true. However, remember that claims alone do not create an endangered species.

I have learned a few things in my 25 years in the supercomputer business. One is that the high-performance computers pioneer costly technology and bear the brunt of the startup costs. The customers must pay a high price partly because of these heavy front-end costs. Followers use this developed technology without the heavy front-end costs and then argue supercomputers are too costly without considering that the technology is low-cost because supercomputers footed the early bills.

Somehow, some way, we in the U.S. must find a way to help pay the cost of starting up a very expensive, low-volume gallium arsenide facility so that all of us can reap the performance and cost benefits of the technology. Like silicon, the use will develop when most companies can afford to use it. That occurs only after someone has paid to put the technology in production, absorbed the high learning-curve costs, proved the performance, and demonstrated the packaging. Today we are asking one company to support those efforts. Unfortunately, we hear complaints that supercomputers with new technology cost too much. We should all be encouraging Seymour's effort, not predicting doom, and we should be prepared to share in the expenses.

The Japanese supercomputer companies are vertically integrated—an organizational structure that has worked well for them. Except for IBM and AT&T, the U.S. companies practice vertical cooperation. However, vertical cooperation must change so that semiconductor vendors will underwrite a larger part of the development costs. The user cannot continue to absorb huge losses while the vendor is making a profit and still expecting the relationship to flourish. This is not vertical cooperation; it is simply a buyer-seller relationship. To me, vertical cooperation means that the semiconductor vendors and the application vendors underwrite their costs for part of the interest in the products. That is true cooperation, and the U.S. must evolve to this or ignore costly technology developments and get out of the market.

I have been told frequently by the Japanese that they push the supercomputer because it drives their semiconductor technology to new components leading to new products that they know will be salable in the marketplace. In their case, vertical integration is a market-planning asset. I maintain that vertical cooperation can have similar results.

I believe that we have seen the gradual emergence of parallelism in the supercomputers offered by Cray and the Japanese—I define those architectures as Practical Parallelism. During the past two days, we have heard about the great expectations for massively parallel processors and the forecasted demise of the Cray dynasty. I refer to these efforts as Research Parallelism, and I want to add that Research Parallelism will become a practicality not when industry starts to buy them but when the Japanese start to sell them. The Japanese are attracted to profitable markets. Massively parallel systems will achieve the status of Practical Parallelism when the Japanese enter that market—that will be the sign that users have adopted the architecture, and the market is profitable.

I would like to close with a view of the industry. I lived through the late 1960s and early 1970s, when the U.S. university community was mesmerized by the Digital Equipment Corporation VAX and held with the belief that VAXs could do everything and there was no need for supercomputers. A few prophets like Larry Smarr, at the University of Illinois (Session 10), kept saying that supercomputers were needed in universities. That they were right is clearly demonstrated by the large number of supercomputers installed in universities today.

Now I hear that same tune again. We are becoming mesmerized with superperformance workstations: they can do everything, and there is again no need for supercomputers. When will we learn that supercomputers are essential for leading-edge work? It is not whether we need supercomputers or super-performance workstations but that we need both working in unison. The supercomputer will explore new ideas, new applications, and new approaches. Therefore, I believe very strongly that it is both and not one or the other. The supercomputer has a place in our industry, so let's start to hear harmonious words of support in place of the theme of supercomputers being too costly and obsolete while massively parallel systems are perfect, cheap, and the only approach.

FPS Computing: A History of Firsts

Howard Thrailkill

> *Howard A. Thrailkill is President and CEO of FPS Computing. He received his bachelor of science degree in electrical engineering from the Georgia Institute of Technology and his master of science degree in the same field from the Florida Institute of Technology. He has received several patents for work with electronic text editing and computerized newspaper composition systems. For the past 20 years, he has held successively more responsible management positions with a variety of high-technology and computer companies, including General Manager of two divisions of Harris Corporation, President of Four-Phase Systems, Inc., Corporate Vice President of Motorola, Inc., and President and CEO of Saxpy Computer Corporation.*

I suspect that a significant number of you in this audience were introduced to high-performance computing on an FPS-attached processor of some type. That technology established our company's reputation as a pioneer and an innovator, and it had a profound effect on the evolution of supercomputing technology. For the purposes of this session, I have been asked to discuss a pioneering product whose failure interrupted a long and successful period of growth for our company.

Pioneers take risks. Such was the case with our T-Series massively parallel computer, which was announced in 1985 with considerable fanfare and customer interest. It was the industry's first massively parallel machine that promised hypercube scalability and peak power in the range of multiple GFLOPS (i.e., 10^9 floating-point operations per second). Regrettably, it was not a successful product.

Before presenting a T-Series "postmortem," I would like to retrace some history. FPS Computing, formerly known as Floating Point Systems, is a 20-year-old firm with a strong tradition of innovation. In 1973, we introduced the first floating-point processor for minicomputers and, in 1976, the first array processor—the FPS 120B. That was followed in 1981 by the first 64-bit minisupercomputer, the FPS 164, and by the FPS 264 in 1985; both enjoyed widespread acceptance. Buoyed by those successes, FPS then failed to perceive a fundamental shift in the direction of minisupercomputer technology: the shift toward the UNIX software environment and architectures with tightly coupled vector and scalar processors.

Other companies correctly recognized that shift and introduced competitive machines, which stalled our rapid growth. We responded with heavy investment in our T-Series machine, a radically new product promising extraordinary peak performance. It never lived up to its promise, and it is interesting to understand why.

First, a small company like FPS lacked the resources to absorb comfortably a mistake costing a few tens of millions of dollars. We simply overreached our resources.

Second, we failed to recognize that our software system, based on Occam, could never compete with the widespread acceptance of UNIX. I had not yet joined FPS at the time, and I recall my reading the T-Series announcement and becoming concerned that I knew so little about this new software environment. Clearly, I was not alone. In retrospect, this miscalculation crippled the product from the outset.

Third, the machine exhibited unbalanced performance even when its software shortcomings were overlooked. Highly parallel portions of an application code could be hand tuned and impressive speeds achieved. However, portions of the code that did not yield to such treatment ran very slowly. Too often we were limited to the speed of a single processor, and performance also suffered from the inefficiencies of message passing in our distributed-memory subsystem. The complexity of I/O with our hypercube architecture exacerbated the problem.

Fourth, we went to market with very little applications software. This deficiency was never overcome because our tedious programming environment did not encourage porting of any significant existing codes.

Finally, we entered a market that was much too small to justify our product development expenditures. While some talented researchers achieved impressive performance in a small number of special-purpose applications, users wanted broader capabilities even in networked environments.

I have been asked to relate the lessons we learned from this experience. Clearly, no $100 million company like FPS wanted to face the serious consequences of a market miscalculation like we experienced. We had counted on the T-Series as a primary source of revenue upon which we would maintain and build our position in the industry.

Among my first duties upon joining FPS was to assess the prospects for the T-Series to fulfill the needs of our customers and our company. My first conversations with customers were not at all encouraging. They wanted standards-compliant systems that would integrate smoothly into their heterogeneous computing environment. They wanted balanced performance. While they valued the promise of parallel processing, they were reluctant to undertake the complexity of parallel programming, although a few users did justify the effort simply to avail themselves of the machine's raw speed. They also wanted a comprehensive library of third-party application codes.

Unfortunately, we saw no way to meet the needs of a large enough number of customers with the T-Series. Thus, we suspended further investment in the T-Series perhaps 45 days after I joined FPS.

Having made that decision, a senior representative from FPS was dispatched to meet individually with every T-Series customer and reconcile our commitments to them. It was a step that cost us a lot of money. With our commitment to continuing as a long-term participant in the high-performance computing business, we believed that nothing less than the highest standards of business ethics and fairness to our customers would be acceptable. That painful process was completed before we turned our attention to redirecting our product strategy.

We then took steps to move back into the mainstream of high-performance computing with our announcement in late 1988 of a midrange UNIX supercomputer from FPS—the Model 500. An improved version, the Model 500EA, followed in 1989.

In the summer of 1990, FPS emerged again as an industry innovator with our announcement of Scalable Processor ARChitecture (SPARC) technology, licensed from Sun Microsystems, Inc., as the foundation for future generation supercomputers from FPS. With our concurrent announcement of a highly parallel Matrix Coprocessor, we also introduced the notion of integrated heterogeneous supercomputing. Integrated heterogeneous supercomputing features modular integration of multiple scalar, vector, and matrix processors within a single, standards-compliant software environment. In our implementation, that software environment is to be an extended version of SunOS. We also announced an alliance with Canon Sales in Japan, a major market for our products.

FPS has made a major commitment to industry standards for high-performance computing—SPARC, SunOS (UNIX), network file system, high-performance parallel interface, and others. We believe we have a strong partner, Sun, in our corner now as we move forward.

We are reassured as we visit customers and observe SPARC technology we have licensed from Sun enjoying such widespread acceptance. Our joint sales and marketing agreement with Sun has also served us well since its announcement in June 1990. We support their sales activities and they support ours.

We will be announcing shortly further steps toward standardization in the software environment we make available to customers. Our compiler will soon have a full set of Cray Research and Digital Fortran extensions, along with a set of ease-of-use tools.

Our customers seem to be receptive to our product strategy, as dozens of machines are now installed in major universities, research laboratories, and industrial sites around the world. Interestingly, about one-third of them are in Japan.

We believe we have defined a new direction for high-speed computing—integrated heterogeneous supercomputing. Our current implementation is shown in Figure 1.

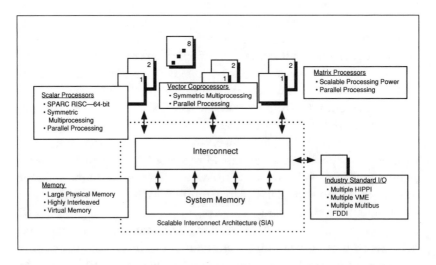

Figure 1. FPS 500 series integrated heterogeneous supercomputing.

As you may observe from the figure, up to eight high-speed SPARC RISC processors may be plugged in independently to run as a shared-memory multiprocessor. Conveniently, all of our installed customers can be upgraded to SPARC and run SunOS software in this modular fashion.

This high-speed SPARC RISC multiprocessor architecture can be augmented by modular addition of multiple vector and matrix coprocessors. The system configuration can be tailored to the application at hand. As technology advances, new processors can modularly replace current versions, essentially obsoleting the industry's tradition of "forklift" replacement of the entire supercomputer when upgrades are undertaken.

While vector processing from FPS and others is a familiar technology, matrix coprocessing may be a new concept to you. We consider this technology as applied parallel processing. FPS matrix coprocessors, currently implemented with up to 84 parallel processing elements, attack the locality of computation present in the compute-intensive portions of many application codes that address big problems. Linear solvers, eigenvalue problems, fast Fourier transforms, and convolutions all involve many computations per data element, i.e., high locality. These portions of code run very fast on our machine, achieving sustained performance in the one- to three-GFLOPS range, and this performance can be achieved in the familiar SunOS software environment.

We believe we have now overcome the setbacks that the ill-fated T-Series dealt to our company. We have returned to our roots in high-performance computing and are now moving forward and building upon our notion of integrated heterogeneous supercomputing as implemented in our current supercomputer product line. Innovation is once again driving FPS, but we are now much more sensitive to customer demands.

13

Industry Perspective: Policy and Economics for High-Performance Computing

Panelists in this session presented a general discussion of where the U.S. high-performance computing industry is and how and why we got there. Topics included government helps and hindrances, competitive issues, financing and venture capital problems, and future needs.

Session Chair

Robert White, Department of Commerce

Why Supercomputing Matters: An Analysis of the Economic Impact of the Proposed Federal High Performance Computing Initiative

George Lindamood

George E. Lindamood is Vice President and Director of High-Performance Computing at Gartner Group, Inc., Stamford, Connecticut. He received his B.S., magna cum laude, in mathematics and physics from Wittenberg University and his M.A. in mathematics from the University of Maryland. He has more than 30 years of experience in the computer field, spanning academia, government, and industry, in activities ranging from research and development to international trade negotiations.

Introduction

On September 8, 1989, the Office of Science and Technology Policy (OSTP) published a report proposing a five-year, $1.9 billion federal High Performance Computing Initiative (HPCI). The goals of this program are to

- maintain and extend U.S. leadership in high-performance computing and encourage U.S. sources of production;
- encourage innovation in high-performance computing by increasing its diffusion and assimilation into the U.S. science and engineering communities; and
- support U.S. economic competitiveness and productivity through greater utilization of networked high-performance computing in analysis, design, and manufacturing.

In response to a Congressional request, OSTP and the Department of Energy, acting through Los Alamos National Laboratory, engaged Gartner Group, Inc., to develop a quantitative assessment of the likely economic impact of the proposed HPCI program over the coming decade. This study is proceeding in two phases.

In Phase I, which was completed in July 1990, two alternative scenarios (A and B), both depicting supercomputing through the year 2000, were developed. One scenario assumes full funding for the proposed HPCI program that would commence in FY 1992. The other scenario assumes "business as usual," that is, no additional federal funding above what is expected for HPCI-related activities now under way.

Phase II, which is the more important phase, was completed in September 1990. In Phase II, the two scenarios are extended to encompass the impact of the HPCI program, first upon selected industrial segments that are the major users of supercomputers and then upon the U.S. economy as a whole.

I will summarize the results of Phase I and describe the methodology employed in Phase II.

Phase I Methodology

During Phase I, two scenarios were developed. Scenario A assumes that current levels of HPCI funding will remain constant. Scenario B assumes full HPCI support, thereby changing the rate and direction of supercomputer development and utilization.

Scenario A

Our projection of the future of supercomputing is rooted in our understanding of the past, not only of supercomputing but also of other elements of the information industry. Over the last three years, we have developed a quantitative model that characterizes the information industry in terms of MIPS (millions of instructions per second), systems, and dollars for various classes of systems—mainframes, minicomputers, personal computers, etc., as well as the components of these systems, such as CPUs, peripherals, and software. Both the methodology and the results of this model have been applied in the development of the two alternative scenarios for the coming decade in supercomputing.

Basically, the model assumes that technology is the driver of demand because it is the principal determiner of both the overall performance and the price/performance of various types of information systems. Hence, future projections are based on anticipated technological advances, interpreted through our understanding of the effects of similar advances in the past and the changing competitive conditions in the industry. Industry revenues are derived from these projections of price/performance, MIPS, and systems shipments, using average system price and average MIPS per system as "reasonability" checks. Historically, the model also reflects macroeconomic cycles that have affected overall demand, but there has been no attempt to incorporate macroeconomic forecasts into the projections.

Assumption 1. In modeling the supercomputer industry, we have assumed that supercomputer systems can be aggregated into three classes:

- U.S.-made vector supercomputers, such as those from Cray Research, Inc., Cray Computer Corporation, and (in the past) Control Data Corporation and Engineering Technology Associates Systems;
- Japanese-made vector supercomputers, such as those marketed by Nippon Electric Corporation, Hitachi, and Fujitsu; and
- parallel supercomputers, such as those made by Intel and Thinking Machines Corporation.

We assume that the price, performance, and price/performance characteristics of systems in each of these classes should be sufficiently uniform that we do not have to go into the details of individual vendors and models (although these are present in the "supporting layers" of our analysis). We do not anticipate any future European participation in vector supercomputers, but we assume no restrictions on the nationality of future vendors of the third class of systems.

Assumption 2. For our base scenario, we assume that the signs of maturity that have been observed in the market for vector supercomputers since 1988 will become even more evident in the 1990s after the current generation of Japanese supercomputers and the next generation of U.S. vector supercomputers—the C90 from Cray Research and the CRAY-3 (and -4?) from Cray Computer—have had their day.

Assumption 3. For parallel systems, however, we assume that the recent successes in certain applications will expand to other areas once the technical difficulties with programming and algorithms are overcome. When that happens, use of parallel systems will increase significantly, somewhat displacing vector systems—at least as the platform of choice for new applications—because of superior overall performance and price/performance. Growth rate percentages for millions of floating-point operations per second (MFLOPS) until the year 2000, installed, are shown in Table 1.

Table 1. Compound Annual Growth Rate for Installed MFLOPS

	1980–84	1985–89	1990–94	1995–99
U.S. Vector Supercomputers	64%	45%	34%	23%
Japanese Vector Supercomputers		86%	63%	39%
Parallel Supercomputers		80%	66%	71%

Assumption 4. We also assume that the price/performance of U.S.-made vector supercomputers will continue to improve, or decrease, at historical rates (about 15 per cent per year) and that the price/performance of Japanese-made vector systems will gradually moderate from 30+ per cent per year levels to 15 per cent per year by the year 2000. For parallel systems, we assume an accelerating improvement, to 20 per cent per year by the year 2000, in price/performance as a result of increasing R&D in this area.

Assumption 5. Despite the decrease in price per MFLOPS, average prices for supercomputer systems have actually increased a few percentage points per year historically. The reason, of course, is that average system size has grown significantly, especially because of expanded use of multiprocessing. We assume that these trends will continue for vector systems, albeit at a slowed rate of increase after 1995, because of anticipated difficulties in scaling up these systems to ever-higher levels of parallelism. For parallel systems, technological advances should lead to accelerated growth rates in processing power, resulting in systems capable of one-TFLOPS sustained performance by the year 2000. Growth rate percentages for average MFLOPS per system are shown in Table 2.

Assumption 6. Finally, we assume that retirement rates for all classes of supercomputer systems will follow historical patterns exhibited by U.S.-made vector systems.

Table 2. Compound Annual Growth Rate for Average MFLOPS/System

	1980–84	1985–89	1990–94	1995–99
U.S. Vector Supercomputers	21%	23%	27%	22%
Japanese Vector Supercomputers		48%	30%	24%
Parallel Supercomputers			60%	42%

These assumptions are sufficient to generate a projection of supercomputer demand for the next 10 years:
- The number of installed systems will more than triple by the year 2000. Table 3 shows how this installed base will be divided, as compared with today. The number of supercomputers installed in Japan will exceed the number installed in the U.S. after 1996.
- Installed supercomputer power (measured in peak MFLOPS) will increase more than 125-fold over the next decade, from almost 1.4 million MFLOPS in 1990 to over 175 million MFLOPS in 2000. (However, this is substantially less than the growth rate in the 1980s—from about 4000 MFLOPS in 1980 to 340 times that amount in 1990.) Of the MFLOPS installed in 2000, 90 per cent will be parallel supercomputers, two per cent will be U.S.-made systems, and eight per cent will be Japanese-made systems.
- As shown in Table 4, the "average" vector supercomputer will increase about 10 times in processing power, whereas the "average" parallel system will increase about 60 times over the decade. Average supercomputer price/performance will improve by a factor of 25,

Table 3. Growth in Supercomputer Demand, 1990–2000

	1990	2000
Source		
U.S. Vector Supercomputers	347 (57%)	640 (34%)
Japanese Vector Supercomputers	183 (30%)	669 (36%)
Parallel Supercomputers	81 (13%)	552 (30%)
User		
Government	174 (28%)	463 (25%)
Academia	130 (21%)	402 (22%)
Industry	250 (41%)	833 (45%)
In-House	57 (9%)	163 (9%)
Installation Site		
U.S.	301 (49%)	683 (37%)
Europe	115 (19%)	345 (19%)
Japan	174 (28%)	768 (41%)
Other	21 (3%)	65 (3%)
Total Installations	611	1861

Table 4. Supercomputer Power, Scenario A

	U.S. Vector Supercomputers	Japanese Vector Supercomputers	Parallel Supercomputers
Average System Price (Millions)	$24.8	$16.4	$35.3
Average System Power (Peak GFLOPS)	12.0	38.5	630,000
Price per MFLOPS	$2000	$425	$56

mostly as a result of increased usage of parallel systems that have more than 10 times better price/performance than vector systems.
- Annual revenues for vector supercomputers will peak at just under $3 billion in 1998. Revenues for parallel systems will continue to grow, surpassing those for vector systems by 1999 and exceeding $3.1 billion by 2000.

Scenario B

For this scenario, we assume that the federal HPCI program will change the direction of high-performance computing (HPC) development and utilization and the rate of HPC development and utilization.

Assumption 1. As in Scenario A, supercomputers are grouped into three classes in Scenario B:
- U.S.-made vector supercomputers,
- Japanese-made vector supercomputers, and
- parallel supercomputers.

Assumptions 2 and 3. We assume that demand for supercomputer systems of both the vector and parallel varieties will be increased by the HPCI program components concerned with the evaluation of early systems and high-performance computing research centers. All funding for early evaluation ($137 million over five years) will go toward the purchase of parallel supercomputers, whereas funding for research centers ($201 million over five years) will be used for U.S.-made vector and parallel supercomputers, tending more to the latter over time. We also assume that federal funding in these areas will precipitate increased state government expenditures, as well, although at lower levels. Although all of these systems would be installed in academic and government facilities (primarily the former), we also postulate in Scenario B that the technology transfer components of HPCI would succeed in stimulating industrial demand for supercomputer systems. Here, the emphasis will be more on U.S.-made vector systems in the near term, although parallel systems will also gain popularity in the industrial sector in the late 1990s as a result of academic and government laboratory developmental efforts supported by HPCI.

Assumption 4. This increased demand and intensified development will also affect the price/performance of supercomputer systems. For U.S.-made vector systems, we conservatively assume that price/performance will improve one percentage point faster than the rates used in Scenario A. For parallel supercomputers, we assume that price/performance improvement will gradually approach levels typical of microprocessor chips and RISC technology (that is, 30+ per cent per year) by the year 2000.

Assumption 5. The increased R&D stimulated by HPCI should also result in significantly more powerful parallel supercomputers, namely, a TFLOPS system by about 1996. However, we do not assume any change in processing power for vector supercomputers, as compared with Scenario A, because we expect that HPCI will have little effect on hardware development for such systems. (This is distinct, however, from R&D into the use of and algorithms for vector systems, which definitely will be addressed by HPCI.)

Assumption 6. We assume that retirement rates for supercomputer systems of all types will be the same as in Scenario A.

As before, these assumptions are sufficient to generate a projection of supercomputer demand for the next 10 years:

- The number of installed supercomputers will approach 2200 systems by the year 2000. Table 5 shows how this installed base will be divided, as compared with Scenario A.
- Particularly noteworthy is the difference between these two scenarios in terms of U.S. standing relative to Japan. In Scenario A, Japan takes the lead in installed supercomputers, but in Scenario B, the U.S. retains the lead.
- Installed supercomputer power (measured in peak MFLOPS) will be increased by a factor of more then 300, to over 440 million MFLOPS, by the year 2000 (which is slightly less than the rate of growth in the 1980s). Of the MFLOPS installed in 2000, 96 per cent will be parallel supercomputers, one per cent will be U.S.-made vector supercomputers, and three per cent will be Japanese-made vector supercomputers.

Table 5. Supercomputer Installations in the Year 2000, by Scenario

	Scenario A	Scenario B
Source		
U.S. Vector Supercomputers	640 (34%)	754 (35%)
Japanese Vector Supercomputers	669 (36%)	669 (31%)
Parallel Supercomputers	552 (30%)	750 (34%)
User		
Government	463 (25%)	488 (22%)
Academia	402 (22%)	518 (24%)
Industry	833 (45%)	984 (45%)
In-House	163 (9%)	183 (8%)
Installation Site		
U.S.	683 (37%)	995 (46%)
Europe	345 (19%)	345 (16%)
Japan	768 (41%)	768 (35%)
Other	65 (3%)	65 (3%)
Total Installations	1861	2173

- As shown in Table 6, the "average" vector supercomputer will increase about 10 times in processing power, whereas the "average" parallel system will increase nearly 125-fold over the decade. Average supercomputer price/performance will improve by a factor of 55.
- Annual revenues for vector supercomputers will peak at just over $3 billion in 1998. Revenues for parallel systems will continue to grow, surpassing those for vector systems by 1997 and exceeding $5 billion in 2000.

The differences between Scenarios A and B, as seen by the supercomputer industry, are as follows:
- 17 per cent more systems installed;
- almost three times as many peak MFLOPS shipped and two and one-half times as many MFLOPS installed in 2000;
- 39 per cent greater revenues in the year 2000—an $8 billion industry (Scenario B) as opposed to a $5 billion industry (Scenario A); and
- $10.4 billion more supercomputer revenues for the 1990–2000 decade.

In addition to these differences for supercomputers, HPCI would cause commensurate increases in revenues and usage for minisupercomputers, high-performance workstations, networks,

Table 6. Supercomputer Power, Scenario B

	U.S. Vector Supercomputers	Japanese Vector Supercomputers	Parallel Supercomputers
Average System Price (Millions)	$22.1	$16.4	$37.9
Average System Power (Peak GFLOPS)	12.0	38.5	1,300,000
Price per MFLOPS	$1840	$425	$29

software, systems integration and management, etc. However, the largest payoff is expected to come from enhanced applications of high-performance computing.

Phase II Methodology

To estimate the overall economic benefit of HPCI, we have sought the counsel of major supercomputer users in five industrial sectors representing a variety of experience and sophistication:
- aerospace,
- chemicals,
- electronics,
- oil and gas exploration and production, and
- pharmaceuticals.

Our assumption is that supercomputers find their primary usage in R&D, as an adjunct to and partial replacement for laboratory or field experimentation and testing—for example, simulating the collision of a vehicle into a barrier instead of actually crashing thousands of new cars into brick walls. Hence, high-performance computing enables companies to bring more and better new products to market and bring new products to market faster.

In other words, high-performance computing improves R&D productivity. Even if there is no other benefit, the use of HPC, which in turn affects overall company productivity in direct proportion to the share of expenditures for R&D, provides a way to determine a conservative estimate of productivity improvement, as shown by the following steps:
- Scenarios A and B are presented to company R&D managers, who are then asked to give estimates, based on their expertise and experience, of the change in R&D productivity over the coming decade.

- For both scenarios, these estimates are translated into overall productivity projections, using information from the company's annual report, for the ratio of R&D spending to total spending.
- Productivity projections for several companies in the same industrial sector are combined, with weightings based on relative revenues, to obtain overall Scenario A and B projections for the five industrial sectors identified above. At this point, projections for other industrial sectors may be made on the basis of whatever insights and confidence have been gained in this process.

These productivity projections are interesting in and of themselves, but we do not intend to stop there. Rather, we plan to use them to drive an input/output econometric model that will then predict the 10-year change in gross national product (GNP) under Scenarios A and B. By subtracting the GNP prediction for Scenario A from that for Scenario B, we expect to obtain a single number, or a range, that represents the potential 10-year payoff from investing $1.9 billion of the taxpayers' money in the federal HPCI program.

Government as Buyer and Leader

Neil Davenport

Neil Davenport is the former President and CEO of Cray Computer Corporation. For more complete biographical information, see his presentation in Session 3.

The market for very high-performance supercomputers remains relatively small. Arguably, the market worldwide in 1990 was not much more than $1 billion. In the 1990s, the development of a machine to satisfy this market—certainly to get a viable market share—requires the development of components, as well as of the machine itself. The marketplace presented to component manufacturers by suppliers of supercomputers is simply not large enough to attract investment necessary for the production of very fast next-generation logic and memory parts. High performance means high development costs and high price. This is a far cry from the days of development of the CRAY-1, when standard logic and memory components were put together in innovative packaging to produce the world's fastest computer.

The market for very large machines is small. It would clearly be helpful if there were no inhibitions to market growth. An easier climate for export of such technology would help the manufacturers. This is a small aspect of a general preference for free and open competition, which would give better value to the buyer.

Government remains the largest customer throughout the world, without whom there would probably not be a supercomputer industry. It is very important that government continue to buy and use supercomputers and in so doing, direct the efforts of the manufacturers of supercomputers. The world market is so small that it clearly cannot sustain a large number of competitors, given the high cost of entry and maintenance. The essential element for success in the supercomputing business is that there be a reasonable size of market that is looking for increased performance and increased value. In this way, the survival of the fittest can be assured, if not the survival of all.

Concerns about Policies and Economics for High-Performance Computing

Steven J. Wallach

> *Steven J. Wallach is Senior Vice President of Technology of CONVEX Computer Corporation. For more complete biographical information, see his presentation in Session 3.*

First, I would like to "take a snapshot" of the state of the supercomputing industry. I think today we certainly have leadership roles because the U.S. is the world leader in supercomputing. I think one of the key areas that people do not often talk enough about is in the area of application leadership. Even when you go to Japan to use a Japanese supercomputer, more than likely the application was developed in the United States, not in Japan. I do not think that I have heard this mentioned in other presentations, but this is a very, very important point. He who has the applications ultimately wins in this business. Also, we are establishing worldwide standards. If the Japanese or Europeans build a new supercomputer, they tend to follow what we are doing, as opposed to trying to establish new standards.

Those are some positive points. What are some of the negatives? Most of the semiconductor technology of today's supercomputers is based on Japanese technology. That is a problem because it is something that we do not necessarily have under our control.

What scares me the most is that for all U.S. supercomputing companies, other than IBM, supercomputers are their only business. They cannot afford to fund efforts for market share over a three-to-five-year period. For the Japanese companies—Hitachi, Fujitsu, and Nippon Electric

Corporation—the supercomputer business is a very small percentage of their overall business, and they are multibillion-dollar-a-year companies. If they chose to sell every machine at cost for the next five years, you would not even see a dent in the profit-and-loss statements of the Japanese companies. Personally, this is what scares me more than anything in competing against the Japanese.

In contrast, how does the U.S. work? We have venture capital. Some people call it "vulture" capital. When a product is very successful and makes a lot of money for its creators and inventors, that one success tends to bring about many, many "clones" that want to cash in on the market. We can go back in the late 1960s, when the minicomputer market started and we had Digital Equipment Corporation and 15 other companies; yet the only companies that really grew out of that boom that are still around are Data General and Prime. Almost everyone else went out of business.

The problem is that five companies cannot each have 40 per cent of the market, so there is a shakeout. This happened in the minicomputer business, the tandem business, and the workstation business; it certainly happened in the midrange supercomputer business.

Now let us take a look at government policy. Typically, the revenue of most companies today is approximately 50 per cent U.S. and 50 per cent international. This is true for almost every major U.S. manufacturer. At CONVEX Computer Corporation, we are actually 45 per cent U.S. (and 35 per cent Europe and five per cent other), but it is always surprising that 15 per cent of our revenue is in Japan. We have not found any barriers to selling our machines in Japan; some of our largest customers are in Japan.

In five years, when you buy a U.S.-made high-definition television (HDTV), it probably will have been simulated on a CONVEX in Japan. We have over 100 installations and literally zero barriers. The only barrier that we have come across was at a prestigious Japanese university that said, "If you want to sell a machine to us, that's great; we'll buy it. But when we have a 20-year relationship with a Japanese company, we typically pay cost. If you want to sell us your machine at cost, we will consider it." Now, if that is a barrier, then so be it. But personally, I say we have had no barriers whatsoever.

U.S. consumption, from CONVEX's viewpoint, is anywhere from 30 per cent to 50 per cent and is affected by the U.S. government directly or indirectly—directly when the Department of Defense (DoD) buys a machine and indirectly when an aerospace contractor buys a machine based on a government grant. From an international viewpoint, our export policy, of course, is controlled. The policy is affected by U.S. export laws like those promulgated in accordance with the Coordinating

Industry Perspective: Policy and Economics for High-Performance Computing

Committee on Export Controls (COCOM), especially with respect to non-COCOM countries, such as Korea, Taiwan, and Israel. The key is that we now have competition from countries that are not under our control (such as Germany, Britain, and France), where there are new developments in supercomputers. If we were to try to export one of these machines, the export would be precluded. So I think we are losing control because of our export policies.

In the current state of government policy, government spending impacts revenues and growth. For companies like CONVEX, effects from government money tend to be the early adopters (universities, national laboratories, etc.). These institutions buy the first machines and take the risk because the risk is on government money. Sometimes proving something does not work is as significant a contribution as proving something does work because if you can prove it does not work, then someone else does not have to go down that path.

The other thing we find that helps us is long-term contracts. That is, buyers will commit to a three- or four-year contract with the government helping, via the Defense Advanced Research Project Agency (for example, through the Thinking Machines Corporation Connection Machine and the Touchstone project) and the NSF centers. The NSF centers have absolutely helped a company like CONVEX because they educated the world in the use of supercomputers.

One of the reasons we do very well in Japan is because Japanese business managers ask their engineers why they are not using a supercomputer, not the other way around. So we are received with open arms, as opposed to reluctance.

So where are we going? The term I hear today more and more is COTS, commercial off-the-shelf, especially in DoD procurements. Also, I think we have totally underestimated Taiwan, Korea, Hong Kong, and Singapore. Realistically, we have to worry about Korea and Taiwan. I have traveled extensively in these countries, and I would worry more about them than I would about the SUPRENUM and similar efforts.

The thing that worries me is that we Americans compete with each other "to the death" among our companies. Can the U.S. survivors have enough left to survive the foreign competition?

Another concern I have is about the third-party software suppliers. Will these suppliers begin to reduce the number of different platforms they support?

My last concern is whether the U.S. capital investment environment can be changed to be more competitive. In Japan, if a company has money in the bank, it invests some extra money to diversify its base. In Japan, the price of stock goes up when a company explains how it is investing money for long-term benefit; because a lot of the stock is owned in these banking groups, earnings might be depressed for two years until the investment shows a profit. By contrast, in U.S. companies, we live quarter to quarter. If you blow one quarter, your epitaph is being written.

So what am I encouraging? I think we have to have changes in the financial infrastructure. Over half the market is outside the U.S., and we have no control—U.S. dumping laws mean nothing if the Japanese want to acquire our market share in Germany. So I think somehow we have to address that issue. My experience with Japanese companies is that in Germany they will bid one deutsche mark if they have to; in Holland, one guilder; but they will never lose a deal based on price. It can be a $20-million machine, but if they want to make that sale, they will not lose it on price. So, we must deal with the fact that U.S. dumping laws affect less than 50 per cent of the market.

One last thing in terms of export control: I personally think we should export our technology as fast as we can and make everyone dependent on us so that the other countries do not have a chance to build it up. One of the reasons we do not have a consumer electronics industry now is that the Japanese put the U.S. consumer electronics industry out of business. Now they are in control of us because we cannot build anything. The same thing, potentially, is true with HDTV. We should export it; let others be totally dependent on us, and then we will actually have more control because other countries will have to come to us.

High-Performance Computing in the 1990s

Sheryl L. Handler

After receiving her Ph.D. from Harvard University, Sheryl L. Handler founded PACE/CRUX, a domestic and international economic development consulting firm. Clients ranged from biotechnology and telecommunications companies to the World Bank, the U.S. State Department, and numerous other agencies and companies. She was President of PACE/CRUX for 12 years.

In June 1983, Dr. Handler founded Thinking Machines Corporation and within three years introduced the first massively parallel high-performance computer system, the Connection Machine supercomputer. The Connection Machine was the pioneer in a new generation of advanced computing and has become the fastest and most cost-effective computer for large, data-intensive problems on the market. Thinking Machines is now the second largest supercomputer manufacturer in America.

Supercomputing has come to symbolize the leading edge of computer technology. With the recognition of its importance, supercomputing has been put on the national agenda in Japan and Europe. Those countries are actively vying to become the best. But a national goal in supercomputing has not yet been articulated for America, which means that all the necessary players—the designers, government laboratories, software developers, students, and corporations—will not be inspired to direct their energies toward a big and common goal. This is potentially dangerous.

Supercomputing represents the ability to dream and to execute with precision. What is a country without dreams? What is a country without the ability to execute its own ideas better than anyone else?

What drew me into the supercomputing industry was an awe, an almost kid-like fascination with it. Supercomputing is a tool that allows you to

> contemplate huge and complex topics
> or zero in on the smallest details
> while adjusting the meter of time
> or the dimensions of space.
> Or with equal ease, to build up big things
> or to take them apart.

To me, there is poetry in our business and a big business in this poetry. In addition, supercomputing is now getting sexy. I recently saw a film generated on the Connection Machine* system that was really sensual. The color, shapes, and movement had their own sense of life.

It is very important to this country to take the steps to be the leader in this field, both in the present and in the future. How can we do this? In short, we must be bold: set our sights high and be determined and resourceful in how we get there. Big steps must be taken. But the question is, by whom?

Some look at our industry and economic structure and say that big steps are not possible. Innovation requires new companies, which require venture capitalists, which require quick paybacks, which only leaves time for small, incremental improvements, etc. In fact, big steps *can* be taken, and taken successfully, if one has the will and the determination.

Sometimes these big steps can be taken by a single organization. At other times, it requires collective action. I would like to look at an example of each.

It is generally agreed that the development of the Connection Machine supercomputer was a big step. Some argued at the time that it was a step in the wrong direction, but all agreed that it was a bold and decisive step. How did such a product come about? It came about because the will to take a big step came before the product itself. We looked around us in the early 1980s and saw a lot of confusion and halfway steps. We didn't know what the answer was, but we were sure it wasn't the temporizing that we saw around us.

---

*Connection Machine is a registered trademark of Thinking Machines Corporation.

So we organized to get back to basics. We gathered the brightest people we could find and gave them only one request: find the right big step that needed to be taken. We needed people whose accomplishments were substantial, so their egos weren't dependent on everything being done their way.

In addition to Danny Hillis, we were fortunate to have two other prominent computer architects who had their own designs. Eventually even their enthusiasm for the Connection Machine became manifest, and then we knew we were onto something good. I thought of this initial phase of the company as building a team of dreamers.

Then we added another dimension to the company—we built a team of doers. And they were as good at "doing" things as the theorists were at dreaming. This team was headed by Dick Clayton, who had vast experience at Digital Equipment Corporation as Vice President of Engineering. His responsibilities ranged from building computers to running product lines. When he arrived at Thinking Machines, we put a sign on his door: "Vice President of Reality." And he was.

So we had the dreamers and the doers. Then there was a third phase—coupling the company to the customer in a fundamental way.

We built a world-class scientific team that was necessary to develop the new technology. But as you know, many companies keep a tight reign on R&D expenses. We viewed R&D as the fuel for growth, not just a necessary expense that had to be controlled.

We had a powerful opportunity here. Our scientific team became a natural bridge to link this new technology to potential customers. These scientists and engineers who had developed the Connection Machine supercomputer were eager to be close to customers to understand this technology from a different perspective and, therefore, more fully. Our early customers had a strong intuition that the Connection Machine system was right for them. But the ability to work hand-in-hand with the very people who had developed the technology enabled our users to get a jump on applications. As a result, our customers were able to buy more than just the product: they were buying the company. The strategy of closely coupling our scientists and our customers has become deeply embedded in the corporate structure. In fact, it is so important to us that we staff our customer support group with applications specialists who have advanced degrees.

So the creation of the Connection Machine supercomputer is an example of a big step that was taken by a single organization. In the years since, massively parallel supercomputers have become part of everyone's supercomputing plans. A heterogeneous environment, with vector supercomputers, massively parallel supercomputers, and workstations, is becoming the norm at the biggest, most aggressive centers.

And now another big step needs to be taken collectively by many of the players in the computer industry. Right now, there is no good way for a scientist to write a program that runs unchanged on all of these platforms. We have not institutionalized truly scalable languages, languages that allow code to move gracefully up and down the computing hierarchy. And the next generation of software is being held up as a result. (If you don't believe that this is a problem, let me ask you the following question: how many of you would be willing to install a meter on your supercomputers that displays the year in which the currently running code was originally written?)

How long can we wait until we give scientists and programmers a stable target environment that takes advantage of the very best hardware that the 1990s have to offer? We already know that such languages are possible.

Fortran 90 is an example. It is known to run efficiently on massively parallel supercomputers such as the Connection Machine computer. It is a close derivative of the Control Data Fortran that is known to run efficiently on vector supercomputers. It is known to run efficiently on coarse-grain parallel architectures, such as Alliant. And while it has no inherent advantages on serial workstations, it has no particular disadvantages, either.

Is Fortran 90 the right scalable language for the 1990s? We don't know that for sure, either. But it is proof that scalable languages are there to be had. Languages that operate efficiently across the range of hardware will be those that will be most used in the 1990s. It is hard for this step to come solely from the vendors. Computer manufacturers don't run mixed shops. My company does not run any Crays, and, to the best of my knowledge, John Rollwagen doesn't run any Connection Machine systems. But many shops run both.

So there is a step to be taken. A big step. It will take will. It will take a clear understanding that things need to be better than they are today—and that they won't get better until the step gets taken. That is where we started with the Connection Machine computer, with the clear conviction that things weren't good enough and the determination to take a big step to make them good enough. And as the computer industry matures and problems emerge that affect a wide segment of the industry, we should come together. It works. I recommend it.

A High-Performance Computing Association to Help the Expanding Supercomputing Industry

Richard Bassin

> *Richard Bassin has been a pioneer in the development of relational database-management systems throughout Europe, having introduced this important new technology to a wide array of influential and successful international organizations. In 1988, Mr. Bassin joined nCUBE Corporation, a leading supplier of massively parallel computing systems, where he served as Vice President of Sales until April 1991. Starting in 1983, Mr. Bassin spent five years helping build the European Division of Oracle Corporation, where he was General Manager of National Accounts. During this time, Mr. Bassin also developed and was a featured speaker in an extremely successful series of relational database-management seminars, entitled Fourth Generation Environments for Business and Industry. This series is today a vigorously functioning institution and constitutes the worldwide standard as a forum for the exchange of information on innovations in database management. Before working in the database-management field, Mr. Bassin was a Technical Manager for Computer Sciences Corporation.*

It is evident to me that there are a lot of people fighting for a very small supercomputer marketplace. It is a growing marketplace, but it is still not big enough. The number of vendors represented among the presenters in this session confirms a relatively small marketplace. If we are talking about a billion dollars, it's a relatively small marketplace.

We need to expand that marketplace if we're going to have strength in high-performance computing. I would state that the High Performance Computing Initiative (as opposed to the supercomputing initiative), as the government calls it, is probably a better angle because a lot of people already have a misconception of what supercomputing is.

But we need to expand because people need higher-performance computing. We need to expand it to a greater degree, especially in industry. Both vendors and users will see advantages from this expansion. Vendors will have the financial security to drive the R&D treadmill from which users benefit.

There has been a lot of discussion over the last few days about the foreign threat, be it Japanese, European, from the Pacific rim, or otherwise. Again, if we expand the industry, as Steve Wallach suggests in his presentation, we have to go worldwide. We must not only be concerned about the billion dollars the government has made available to the community, but we must also look at the worldwide market and expand it. And we must expand out within the national market, getting supercomputing into the hands of people who can benefit from it. There's not enough supercomputing or high-performance computing on Wall Street. Financial analysts, for instance, could use a lot of help. Maybe if it were available, they would not make some of their most disastrous miscalculations on what will go up and what will go down.

How do we strengthen that marketplace? How do we expand it? Well, in my view, there's a need for the vendors to get together and do it in concert—in, for example, a high-performance computing association, where the members are from the vendor community, both in hardware and in software. That organization, based in places like supercomputing centers, should represent the whole high-performance computing community and should work to expand the entire industry rather than address the needs of an individual vendor.

All too often, government is influenced by those most visible at the moment. If we had an association that would address the needs of the industry, that would probably be the best clearing-house that the government could have for getting to know what is going on and how the industry is expanding.

It would also provide an ideal clearing-house for users who are confused as to what's better for them and which area of high-performance computing best suits their needs. Today, they're all on their own and make a lot of independent decisions on types of computing, price of computing, and price/performance of computing relative to their needs. Users could get a lot of initial information through an association.

The last thing I would say is that such an association could also propose industry-wide standards. We have a standard called HIPPI (high-performance parallel interface), but unfortunately we don't have a standard that stipulates the protocol for HIPPI yet. A lot of people are going a lot of different ways. If we had an organization where the industry as a whole could get together, we might be able to devise something from which all the users could benefit because all the users would be using the same interface and the same protocol.

I am a firm believer in open systems. Our company is a firm believer in open systems. Open systems benefit the industry *and* the user community, not *just* the user community.

In conclusion I will tell you that we at nCUBE Corporation have discussed the concept of a high-performance computing organization at the executive level, and our view is that we will gladly talk to the other vendors, be they big, small, or new participants in high-performance computing. We have funds to put into an association, and we think we should build such an association for the betterment of the industry.

The New Supercomputer Industry

Justin Rattner

Justin Rattner is founder of and Director of Technology for Intel Supercomputer Systems Division. He is also principal investigator for the Touchstone project, a $27 million research and development program funded jointly by the Defense Advanced Research Projects Agency and Intel to develop a 150-GFLOPS parallel supercomputer.

In 1988, Mr. Rattner was named an Intel Fellow, the company's highest ranking technical position; he is only the fourth Intel Fellow named in the company's 20-year history. In 1989, he was named Scientist of the Year by R&D Magazine and received the Globe Award from the Oregon Center for Advanced Technology Education for his contributions to educational excellence. Mr. Rattner is often called the "Father of Parallel Valley," the concentration of companies near Portland, Oregon, that design and market parallel computers and parallel programming tools.

Mr. Rattner received his B.S. and M.S. degrees in electrical engineering and computer science from Cornell University, Ithaca, New York, in 1970 and 1972, respectively.

An observation was made by Goldman, Sachs & Co. in about 1988 about the changing structure of the computer industry. They talked about "New World" computing companies versus "Old World" computing companies. In my observation of the changing of the guard in high-performance computing, I group companies such as Cray Research, IBM,

Digital Equipment Corporation, Nippon Electric Corporation, and Fujitsu as Old World supercomputing companies, and Intel, Silicon Graphics IRIS, Thinking Machines Corporation, nCUBE Corporation, and Teradata as New World supercomputing companies.

The point of this grouping is that the structure of the industry associated with Old World computing companies is very different from the structure of the industry associated with New World computing companies. I am not saying that the New World companies will put all Old World companies out of business or that there will be an instantaneous transformation of the industry as we forsake all Old World computers for New World computers. What I am trying to emphasize is the fact that the New World industries have fundamentally different characteristics from Old World industries, largely because of the underlying technology.

Of course, the underlying agent of change here is the microprocessor because micro-based machines show more rapid improvement. Intel's forecast for a high-performance microprocessor in the year 2000—in contrast to a high-integration microprocessor in the year 2000—is something on the order of 100 million transistors, about an inch on a side, operating with a four-nanosecond cycle time, averaging 750 million instructions per second, and peaking at a billion floating-point operations per second (GFLOPS), with four processing units per chip.

I think that Intel is every bit as aggressive with superprocessor technology as other people have been with more conventional technologies. For instance, see Figure 1, which is a picture of a multichip module substrate for a third-generation member of the Intel i860 family, surrounded by megabit-cache memory chips. It is not that unusual, except that you are looking at optical waveguides. This is not an illustration of an aluminum interconnect. These are electro-optic polymers that are interconnecting these chips with light waves.

We are also working on a cryogenically cooled chip. In fact, there is a 46 microprocessor operating at Intel at about 50 per cent higher than its commercially available operating frequency, and I think it is running at about -30°C.

Figure 2 shows the impact of this technology. I tried to be as generous as I could be to conventional machines, and then I plotted the various touchstone prototypes and our projections out to the point I ran off the performance graph. We believe we can reach sustained TFLOPS performance by the middle of the decade.

Industry Perspective: Policy and Economics for High-Performance Computing   531

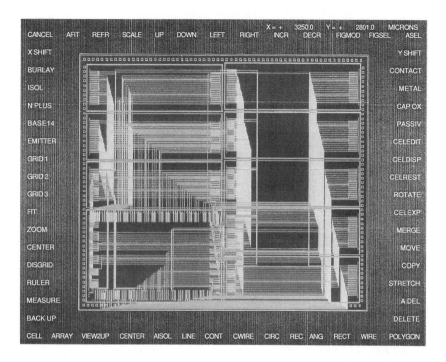

Figure 1. A multichip module substrate for a third-generation Intel i860.

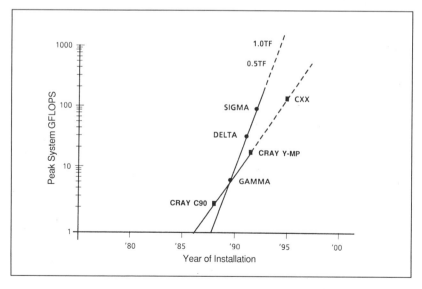

Figure 2. Micro-based machines show more rapid improvement in peak system performance.

The "new" industry technology is what produces a "paradigm shift," if I can borrow from Thomas Kuhn, and leads to tremendous crisis and turmoil as old elements and old paradigms are cast aside and new ones emerge. In short, it has broad industry impact.

Among these impacts are technology bridges—things that would help us get from the old supercomputing world to the new supercomputing world. One of those bridges is some type of unifying model of parallel computation, and I was delighted to see a paper by Les Valiant (1990), of Harvard University, on the subject. He argues that the success of the von Neumann model of computation is attributable to the fact that it is an efficient bridge between software and hardware and that if we are going to have similar success in parallel computing, we need to find models that enable that kind of bridging to occur.

I think new industry machines result in changes in usage models. This is something we have seen. I have talked to several of the supercomputer center directors that are present at this conference. We get somewhat concerned when we are told to expect some 1500 logins for one of these highly parallel machines and all the attendant time-sharing services that are associated with these large user communities. I do not think these architectures now and for some time will be conducive to those kinds of conventional models of usage, and I think we need to consider new models of usage during this period of crisis and turmoil that accompanies the paradigm shift.

Issues associated with product life cycles affect the investment strategies that are made in developing these machines. The underlying technology is changing very rapidly. It puts tremendous pressure on us to match our product life cycles to those of the microprocessors. It is unusual in the Old World of supercomputers to do that.

Similarly, we have cost and pricing effects. I cannot tell you how many times people have said that the logic in one of our nodes is equivalent to what they have in their workstations. They ask, "Why are you charging me two or three times what someone would charge to get that logic in the workstation?" These are some of the issues we face.

The industry infrastructure is going to be changing. We are witnessing the emergence of a whole new industry of software companies that see the advent of highly parallel machines as leveling the playing field—giving them an opportunity to create new software that is highly competitive in terms of compatibility with old software while avoiding the problem of having 20-year-old algorithms at the heart of these codes. Many other industry infrastructure changes can be anticipated, such as the one we see beginning to take place in the software area.

Procurement policies for overall supercomputers are based on the product life cycle, cost, and pricing structures of Old World supercomputers. And we see that as creating a lot of turmoil during the paradigm shift. When we sit down with representatives of the various government agencies, they tend to see these things in three- or four- or five-year cycles, and when we talk about new machines every 18 to 24 months, it's clear that procurement policies just don't exist to deal with machines that are advancing at that rapid rate.

Finally, export policies have to change in response to this. In a *New York Times* article entitled "Export Restrictions Fail to Halt Spread of Supercomputers," the reporter said that one thing creating this problem with export restrictions was that among the relatively powerful chips that are popular with computer makers abroad is the Intel i860 microprocessor, which is expected to reach 100 million floating-point operations per second sometime in late 1990 or early 1991. This is just an example of the kind of crisis that the new computer industry will continue to create, I think, for the balance of this decade, until the paradigm shift is complete.

Reference

L. Valiant, "A Bridging Model for Parallel Computation," *Communications of the ACM* **33** (8), 103–111 (1990).

The View from DEC

Sam Fuller

Samuel H. Fuller, Vice President of Research, Digital Equipment Corporation, is responsible for the company's corporate research programs, such as work carried out by Digital's research groups in Maynard and Cambridge, Massachusetts, Palo Alto, California, and Paris, France. He also coordinates joint research with universities and with the Microelectronics and Computer Technology Corporation (MCC). Dr. Fuller joined Digital in 1978 as Engineering Manager for the VAX Architecture Group. He has been instrumental in initiating work in local area networks, high-performance workstations, applications of expert systems, and new computer architectures.

Before coming to Digital, Dr. Fuller was Associate Professor of Computer Science and Electrical Engineering at Carnegie Mellon University, where he was involved in the performance evaluation and design of several experimental multiprocessor computer systems. Dr. Fuller is a member of the boards of directors of MCC, MIPS Corporation, and the National Research Initiatives. He also serves as a member of the advisory councils of Cornell University, Stanford University, and the University of Michigan and is on the advisory board of the National Science Resource Center (Smithsonian Institution-National Academy of Sciences). Dr. Fuller is a Fellow of the Institute of Electrical and Electronics Engineers and a member of the National Academy of Engineering.

I would like to cover several topics. One is that Digital Equipment Corporation (DEC) has been interested for some time in forms of parallel processing and, in fact, massively parallel processing, usually called here SIMD processing.

There are two things that are going on in parallel processing at DEC that I think worthy of note. First, for a number of years, we have had an internal project involving four or five working machines on which we are putting applications while trying to decide whether to bring them to market. The biggest thing holding us back—as stated by other presenters in this session—is that it is a limited market. When there is a question of how many people can be successful in that market, does it really make sense for one more entrant to jump in? I would be interested in discussing with other researchers and business leaders the size of this market and the extent of the opportunities.

The second thing we are doing in massively parallel processing is the data parallel research initiative that we have formed with Thinking Machines Corporation and MasPar. In this effort, we have focused on the principal problem, which is the development of applications. A goal in January of 1989 with Thinking Machines was to more than double the number of engineers and scientists who were writing applications for massively parallel machines.

An interesting aspect I did not perceive when we went into the program was the large number of universities in this country that are interested in doing some work in parallel processing but do not have the government contracts or the grants to buy some of the larger machines we are talking about at this conference. As the smaller massively parallel machines have come forward, over 18 of the MasPar machines with DEC front ends have gone into various universities.

Some people have spoken to me about having supercomputer centers where people are trained in vector processing and having that concept filter down into the smaller machines. Also, as more schools get small massively parallel machines, those students will begin to learn how to develop applications on parallel machines, and then we will begin to see that trend trickle upward, as well.

A very healthy development over the past 12 months is the availability of low- as well as high-priced massively parallel machines. The goal of the DEC-Thinking Machines-MasPar initiative involving universities is no longer to double the number of engineers and scientists. It is now, really, to more than quadruple the number of engineers and scientists that are working on these types of machines, and I think that is quite possible in the year or two ahead.

Industry Perspective: Policy and Economics for High-Performance Computing 537

Our next goal, now that several of these machines are in place, is to begin having a set of workshops and conferences where we publish the results of those applications that have been developed on these machines at universities around the country.

Another significant initiative at DEC is to look at how far we can push single-chip microprocessors. The goal is a two-nanosecond cycle time on a two-way superscalar machine. Our simulations so far indicate that we can achieve capacities on the order of 300 million instructions per second (MIPS). Looking forward and scaling up to our 1993 and 1994 potential, we expect performance peaks to be in the neighborhood of 1000 MIPS.

I hasten to add that in this research program we are doing some of the work with universities, although the bulk of it is being done internally in our own research labs. The idea is to try and show the feasibility of this—to see whether we can make this the basis of our future work. The methodology is to use the fastest technology and the highest level of integration. Attempting to use the fastest technology means using emitter-coupled logic (ECL). We are continuing to work with U.S. vendors, Motorola and National. We've gone through two other vendors over the course of the past 18 months now, and there's no doubt in our minds that while the U.S. vendors are dropping back in some of their commitments to ECL, the Japanese are not. It would have been a lot easier for us to move forward and work with the Japanese. But we made a decision that we wanted to try and work with the U.S. vendor base to develop a set of CAD tools. We're doing custom design in ECL, and the belief is we can get as high a density with the ECL as we can get today with complementary metal oxide semiconductors (CMOS). It's a somewhat different ECL process. I think some people might even argue that it's closer to bipolar CMOS than ECL. But, in fact, all of the transistors in the current effort are ECL.

Today, packaging techniques can let you dissipate 150 to 175 watts per package. But the other part of the project, in addition to the CAD tools, is to develop the cooling technology so that we can do that on a single part.

Another reason it is not appropriate to call this a supercomputer is the large impact on workstations, because you can surround this one ECL part with fairly straightforward CMOS dynamic random-access-memory-chip second-level and third-level caches. So I think we can provide a fairly powerful desktop device in the years ahead.

What we are building is something that can get the central processing unit, the floating-point unit, and the translation unit, as well as instruction and data caches, on a single die. By getting the first-level caches on a single die, we hope to go off-chip every tenth to fifteenth cycle, not every cycle, which allows us to run the processor two to ten times faster than the actual speed on the board. So we just use a phase-lock loop on the chip to run it at a clock rate higher than the rest of the system. It also lets us use a higher performance on the processor but then use lower technology for the boards themselves.

Because this is a research project, not a product development, it seems to me it's useful to discuss whether we meet our goal—whether our U.S. suppliers can supply us the parts in 1992 and 1993. This is clearly an achievable task. It will require some aggressive development of CAD tools, some new packaging technology, and the scaling of the ECL parts. But all of those will be happening, so in terms of looking at one-chip microprocessors, it's clear that this is coming, whether it happens in 1994, 1995, or a year or two later.

The next main topic I wanted to talk about is the question posed for this session by the conference organizers, i.e., where the government might be of help or be of hindrance in the years ahead, and I have three points. One is that I think it would be relatively straightforward for the government to ease up on export controls and allow us to move more effectively into new markets, particularly eastern Europe.

DEC has set up subsidiaries in Hungary and Czechoslovakia. It would like to go elsewhere. But a number of the rules hamper us. Now, the other people have talked about the supercomputing performance rules. Well, because DEC doesn't make supercomputers—we make minicomputers—we've run into other problems. We actually began to develop a fairly good market there. Then, in Afghanistan, we followed the direction of the government and stopped all further communication and stopped our delivery of products to eastern Europe.

As things opened up here this past year, it's turned out that the largest installed base of computers in Hungary is composed of Digital machines. Yes, they are all clones, not built by us, but it's a wonderful opportunity to service and provide new software. Right now we're precluded from doing that because it would violate various patent and other laws, so we're basically going to give that market over to the Japanese, who will go in and upgrade the cloned DEC computers and provide service.

The second point is that the government needs to be more effective in helping collaboration between U.S. industry, universities, and the government laboratories. The best model of that over the past couple decades has been the Defense Advanced Research Projects Agency (DARPA). Digital, in the early years, certainly with timesharing and networking, profited and contributed well in those two areas. We didn't do as well with work-stations, I think. Obviously Sun Microsystems, Inc., and Silicon Graphics Inc. got the large benefit of that. We finally woke up. We're doing better on workstations.

The point is that DARPA has done well, I think, in fostering the right type of collaboration with universities and industries in the years past. We need to do more of that in the years ahead, I think. So I would, number one, encourage that.

I have a final point on government collaboration that I think I've got to get on the table. People have said that their companies are for open systems and that you've got to have more collaboration. DEC also is absolutely committed to open systems. We need more collaboration. But let me caution you. In helping to set up a number of these collaborations—Open Software Foundation, Semiconductor Manufacturing Technology Consortium, and others—the government needs to play a central role if you want that collaboration to be focused on this country.

Unless you have the government involved in helping to set up that forum and providing some of the funding for the riskiest of the research, you will have an international, rather than a national, forum. High-performance computing is the ideal place, I think, for somebody in the government—whether it's the Department of Energy or DARPA or the civilian version of DARPA—to cause that forum to bring the major U.S. players together so we can develop some of the common software that people have talked about at this conference.

Industry Perspective: Remarks on Policy and Economics for High-Performance Computing

David Wehrly

> *David S. Wehrly is President of ForeFronts Computational Technologies, Inc. He received his Ph.D. and joined IBM in 1968. He has held numerous positions with responsibility for IBM's engineering, scientific, and supercomputing product development. Dr. Wehrly pioneered many aspects of technical computing at IBM in such areas as heterogeneous interconnect and computing systems, vector, parallel, and clustered systems, and computational languages and libraries. He was until August 1992 Director of IBM's High-Performance/ Supercomputing Systems and Development Laboratories, with overall, worldwide systems management and development responsibility for supercomputing at IBM.*

I would like to share some of my thoughts on high-performance computing. First, I would like to make it clear that my opinions are my own and may or may not be those of IBM in general.

The progress of high-performance computing, in the seven years since the last Frontiers of Supercomputing conference in 1983, has been significant. The preceding sessions in this conference have done an excellent job of establishing where we are in everything from architecture to algorithms and the technologies required to realize them. There are, however, a few obstacles in the road ahead, and perhaps we are coming to some major crossroads in the way we do business and what the respective roles of government, industry, and academia are.

There is a lot more consensus on where to go than on a plan of how to get there, and we certainly fantasize about more than we can achieve in any given time. However, in a complex situation with no perfect answers—and without a doubt, no apparent free ride—most would agree that the only action that would be completely incorrect is to contemplate no action at all.

Herb Striner (Session 11) was far more articulate and Alan McAdams (same session) was more passionate than I will be, but the message is the same. Leading edge in supercomputing is certainly not for the faint of heart or light of wallet!

The Office of Science and Technology Policy report on the federal High Performance Computing Initiative of 1989 set forth a framework and identified many of the key issues that are inhibiting the advance of U.S. leadership in high-performance computing. There are many observers of the current circumstances that contend that it is a lack of a national industrial policy and extremely complex, bilateral relationships with countries such as Japan, with which we are both simultaneously allies and competitors, that assures our failure.

So, what are the major problems that we face as a nation? We have seen this list before:
- the high cost of capital;
- a focus on short-term revenue optimization;
- inattention to advanced manufacturing, quality, and productivity;
- unfair trade practices; and
- the realization that we are behind in several key and emerging strategic technologies.

Although this is not a total list, it represents some of the major underlying causes of our problems, and many of these problems arise as a result of the uncertain role government has played in the U.S. domestic economy, coupled with sporadic efforts to open the Japanese markets.

When viewed with respect to the U.S. Trade Act of 1988, Super 301, and the Structural Impediment Initiative talks, statements by high-profile leaders such as Michael Boskin, Chairman, Council of Economic Advisors—who said, "potato chips, semiconductor chips, what is the difference? They are all chips."—or Richard Darman, Director, Office of Management and Budget—who said, "Why do we want a semiconductor industry? We don't want some kind of industrial policy in this country. If our guys can't hack it, let them go."—must give one pause to wonder if we are not accelerating the demise of the last pocket of support for American industry and competitiveness and moving one step closer to carrying out what Clyde V. Prestowitz, Jr., a veteran U.S.-Japanese negotiator, characterized as "our own death wish." His observations (see the July 1990 issue of *Business Tokyo*) were made in the context of Craig Fields's departure from the Defense Advanced Research Projects Agency. We must recognize such confusion about our technology policy that is in contrast to the Japanese Ministry of International Trade and Industry (MITI), in which a single body of the Japanese government fine tunes and orchestrates an industrial policy.

I am not advocating that the U.S. go to that extreme. However, some believe the U.S. will become a second- or third-rate industrial power by the year 2000 if we do not change our "ad hoc" approach to technology policy. The competitive status of the U.S. electronic sector of the industry is such that Japan is headed toward replacing the U.S. as the world's number one producer and trader of electronic hardware by mid-1990, if not earlier.

So, what is needed? The U.S. needs a focused industrial policy that is committed to rejuvenating and maintaining the nation's high-tech strength. Such a policy must be focused—a committed government working in close conjunction with American business and academic institutions. Technical and industrial policy must be both tactical and strategic and neither isolated from nor confused by the daily dynamics of politics.

I would summarize the key requirements in the following way:
- We need to recognize economic and technological strength as vital to national security. We must understand fully the strategic linkages between trade, investment, technology, and financial power. The Japanese achieve this through their *keiretsu* (business leagues).
- Institutional reforms are needed to allow greater coordination between government, academia, and business. MITI has both strengths and weaknesses, but first and foremost, it has a vision—the promotion of Japan's national interest.
- The U.S. must strengthen its industrial competitiveness. Measures are necessary to encourage capital formation, increase investment, improve quality, promote exports, enhance education, and stimulate research and development.
- America must adopt a more focused, pragmatic, informed, and sophisticated approach toward Japan on the basis of a clear industrial policy and coherent strategy with well-defined priorities and objectives, without relegating Japan to a position of adversary. We must recognize Japan for what it is—a brilliant negotiator and formidable competitor.

From an economic standpoint, the U.S. has a growing dependence on others for funding our federal budget deficit, providing capital for investment, selling consumer products to the American public, and supplying components for strategic American industrial production. American assets are half of their 1985 value, and Americans pay twice as much for Japanese products and property.

Sometimes we are obsessed with hitting home runs rather than concentrating on the fundamentals that score runs and win games. Our focus is on whose machine is the fastest, whose "grand challenges" are grandest, and who gets the most prestigious award. While all of this is important, we must not lose sight of other, perhaps less glamorous but possibly more important, questions: How do we design higher quality cars? How do we bring medicines to market more quickly? How can we apply information technologies to assist in the education of our children? The answers to these questions will come by focusing on designing high-performance computers for the sake of technology and by focusing on applications.

To achieve these objectives, significant work will be required in the hardware, software, and networking areas. Work that the private sector does best should be left to the private sector, including hardware and software design and manufacture and the operation of the highly complex network required to communicate between systems and to exchange information. Even the Japanese observed in their just-completed supercomputing initiative that industry, through the demands of the market, had far exceeded the goals set by MITI and that the government had, in fact, become an inhibitor in a project that spanned almost a decade. Our government, however, could assist tremendously if it would focus attention on the use of high-performance computing to strengthen both American competitiveness and scientific advances by serving as a catalyst, by using high-performance computing to advance national interests, and by participating with the private sector in funding programs and transferring critical skills and technologies.

We have said a lot about general policy and countering the Japanese competitive challenge, but what about the technology challenge, from architecture to system structure? What are the challenges that we face as an industry?

At the chip level, researchers are seeking new levels of component density and new semiconductor materials and cooling technologies and exploring complex new processing and lithographic techniques, such as soft X-ray, for fabricating new chips. Yet, as the density increases and fundamental device structures become smaller, intrinsic parasitics and interconnect complexities bring diminishing returns and force developers to deal with an ever-increasing set of design tradeoffs and parameters.

This is demanding ever-higher capital investment for research and development of tooling and manufacturing processes.

Ultimately, as we approach one-nanosecond-cycle-time machines and beyond, thermal density and packaging complexities are causing a slowing of the rate of progress in both function and performance for traditional architectures and machine structures. With progress slowing along the traditional path, much research is now being focused on higher, or more massive, degrees of parallelism and new structural approaches to removing existing road blocks to enable the computational capacity required for the challenges of tomorrow. Japan, too, has discovered this potential leverage.

At the system level, the man-machine interface is commanding a lot of development attention in such areas as voice recognition and visualization and the algorithms needed to enable these new technologies. These areas are opening up some exciting possibilities and creating new sets of challenges and demands for wideband networking and mass storage innovations.

Probably the greatest technology challenge in this arena is parallel software enablement. At this juncture, it is this barrier more than anything else that stands in the way of the 100- to 10,000-fold increase in computing capacity required to begin to address the future scientific computational challenges.

In summary, scientific computing and supercomputers are essential to maintaining our nation's leadership in defense technologies, fundamental research and development, and, ultimately, our industrial and economic position in the world. Supercomputers and supercomputing have become indispensable tools and technology drivers, and the U.S. cannot afford to relinquish its lead in this key industry.

I believe the High-Performance Computing Act of 1989 offers us in its objectives an opportunity to stay in the lead, though I do have some concerns involving the standards and intellectual-property aspects of the bill. I encourage a national plan with an application focus and comprehensive network and believe the time for action is now. I am persuaded that work resulting from implementation of this act will encourage the government and the private sector to build advanced systems and applications faster and more efficiently.

14

What Now?

Panelists in this session presented a distillation of issues raised during the conference, especially about the government's role in a national high-performance computing initiative and its implementation. David B. Nelson, of the Department of Energy, summarized the conference, and the panel discussed the role of the government as policy maker and leader.

Session Chair

David B. Nelson, Department of Energy

Conference Summary

David B. Nelson

David B. Nelson is Executive Director of the Office of Energy Research, U.S. Department of Energy (DOE), and concurrently, Director of Scientific Computing. He is also Chairman of the Working Group on High Performance Computing and Communications, an organization of the Federal Coordinating Committee on Science, Engineering, and Technology. His undergraduate studies were completed at Harvard University, where he majored in engineering sciences, and his graduate work was completed at the Courant Institute of Mathematical Sciences at New York University, where he received an M.S. and Ph.D. in mathematics.

Before joining DOE, Dr. Nelson was a research scientist at Oak Ridge National Laboratory, where he worked mainly in theoretical plasma physics as applied to fusion energy and in defense research. He headed the Magnets-Hydrodynamics Theory Group in the Fusion Energy Division.

Introduction

I believe all the discussion at the conference can be organized around the vision of a seamless, heterogeneous, distributed, high-performance computing environment that has emerged during the week and that K. Speierman alluded to in his remarks (see Session 1). The elements in this environment include, first of all, the people—skilled, imaginative users, well trained in a broad spectrum of applications areas. The second ingredient of that environment is low-cost, high-performance, personal workstations and visualization engines. The third element is mass storage and accessible, large knowledge bases. Fourth is heterogeneous high-performance compute engines. Fifth is very fast local, wide-area, and national networks tying all of these elements together. Finally, there is an extensive, friendly, productive, interoperable software environment.

As far as today is concerned, this is clearly a vision. But all of the pieces are present to some extent. In this summary I shall work through each of these elements and summarize those aspects that were raised in the conference, both the pluses and the minuses.

Now, we can't lose sight of what this environment is for. What are we trying to do? The benefits of this environment will be increased economic productivity, improved standard of living, and improved quality of life. This computational environment is an enabling tool that will let us do things that we cannot now do, imagine things that we have not imagined, and create things that have never before existed. This environment will also enable greater national and global security, including better understanding of man's effect on the global environment.

Finally, we should not ignore the intellectual and cultural inspiration that high-performance computing can provide to those striving for enlightenment and understanding. That's a pretty tall order of benefits, but I think it's a realistic one; and during the conference various presenters have discussed aspects of those benefits.

Skilled, Imaginative Users and a Broad Spectrum of Applications

It's estimated that the pool of users trained in high-performance computing has increased a hundredfold since our last meeting in 1983. That's a lot of progress. Also, the use of high-performance computing in government and industry has expanded into many new and important areas since 1983. We were reminded that in 1983, the first high-performance computers for oil-reservoir modeling were just being introduced. We have identified a number of critical grand challenges whose solution will be enabled by near-future advances in high-performance computing.

We see that high-performance computing centers and the educational environment in which they exist are key to user education in computational science and in engineering for industry. You'll notice I've used the word "industry" several times. Unfortunately, the educational pipeline for high-performance computing users is drying up, both through lack of new entrants and through foreign-born people being pulled by advantages and attractions back to their own countries.

One of the points mentioned frequently at this conference, and one point I will be emphasizing, is the importance of broadening the use of this technology into wider industrial applications. That may be one of the most critical challenges ahead of us. Today there are only pockets of high-performance computers in industry.

Finally, the current market for high-performance computing—that user and usage base—is small and increasingly fragmented because of the choices now being made available to potential and actual users of high-performance computing.

Workstations and Visualization Engines

The next element that was discussed was the emergence of low-cost, high-performance personal workstations and visualization engines. This has happened mainly since 1983. Remember that in 1983 most of us were using supercomputers through glass teletypes. There has been quite a change since then.

The rapid growth in microprocessor compatibility has been a key technology driver for this. Obviously, the rapid fall in microprocessor and memory costs has been a key factor in enabling people to buy these.

High-performance workstations allow cooperative computing with compute engines. As was pointed out, they let supercomputers be supercomputers by off-loading smaller jobs. The large and increasing installed base of these workstations, plus the strong productive competition in the industry, is driving hardware and software standards and improvements.

Next, several of the multiprocessor workstations that are appearing now include several processors that allow a low-end entry into parallel processing. It was pointed out that there may be a million potential users of these, as compared with perhaps 10,000 to 100,000 users of the very high-end parallel machines. So this is clearly the broader base and therefore the more likely entry point.

Unfortunately, in my opinion, this very attractive, seductive, standalone environment may deflect users away from high-end machines, and it's possible that we will see a repetition on a higher plane of the VAX syndrome of the 1970s. That syndrome caused users to limit their problems to those that could be run on a Digital Equipment Corporation VAX machine; a similar phenomenon could stunt the growth of high-performance computing in the future.

Mass Storage and Accessible Knowledge Bases

Mass storage and accessible, large knowledge bases were largely ignored in this meeting—and I think regrettably so—though they are very important. There was some discussion of this, but not at the terabyte end.

What was pointed out is that mass-storage technology is advancing slowly compared with our data-accumulation and data-processing capabilities. There's an absence of standards for databases such that interoperability and human interfaces to access databases are hit-and-miss things.

Finally, because of these varied databases, we need but do not have expert systems and other tools to provide interfaces for us. So this area largely remains a part of the vision and not of the accomplishment.

Heterogeneous High-Performance Computer Engines

What I find amazing, personally, is that it appears that performance on the order of 10^{12} floating-point operations per second—a teraflops or a teraops, depending on your culture—is really achievable by 1995 with known technology extrapolation. I can remember when we were first putting together the High Performance Computing Initiative back in the mid-1980s and asking ourselves what a good goal would be. We said that we would paste a really tough one up on the wall and go for a teraflops. Maybe we should have gone for a petaflops. The only way to achieve that goal is by parallel processing. Even today, at the high end, parallel processing is ubiquitous.

There isn't an American-made high-end machine that is not parallel. The emergence of commercially available massively parallel systems based on commodity parts is a key factor in the compute-engine market—another change since 1983. Notice that it is the same commodity parts, roughly, that are driving the workstation evolution as are driving these massively parallel systems.

We are still unsure of the tradeoffs—and there was a lively debate about this at this meeting—between fewer and faster processors versus more and slower processors. Clearly the faster processors are more effective on a per-processor basis. On a system basis, the incentive is less clear. The payoff function is probably application dependent, and we are still searching for it. Fortunately, we have enough commercially available architectures to try out so that this issue is out of the realm of academic discussion and into the realm of practical experiment.

Related to that is an uncertain mapping of the various architectures available to us into the applications domain. A part of this meeting was the discussion of those application domains and what the suitable architectures for them might be. Over the next few years I'm sure we'll get a lot more data on this subject.

It was also brought out that it's important that one develops balanced systems. You have to have appropriate balancing of processor power, memory size, bandwidth, and I/O rates to have a workable system. By and large, it appears that there was consensus in this conference on what that balance should be. So at least we have some fairly good guidelines.

There was some discussion at this conference of new or emerging technologies—gallium arsenide, Josephson junction, and optical—which may allow further speedups. Unfortunately, as was pointed out, gallium arsenide is struggling, Josephson junction is Japanese, and optical is too new to call.

Fast, Local, Wide-Area, and National Networks

Next, let's turn to networking, which is as important as any other element and ties the other elements together.

Some of the good news is that we are obtaining more standards for I/O channels and networks. We have the ability to build on top of these standards to create things rather quickly. As an example of that, I mention the emergence of the Internet and the future National Research and Education Network (NREN) environment, which is based on standards, notably transmission control protocol/Internet protocol, and on open systems and has already proved its worth.

Unfortunately, as we move up to gigabit speeds, which we know we will require for a balanced overall system, we're going to need new hardware and new protocols. Some of the things that we can do today simply break down logically or electrically when we get up to these speeds. Still, there's going to be a heavy push to achieve gigabit speeds.

Another piece of bad news is that today the network services are almost nonexistent. Such simple things as yellow pages and white pages and so on for national networks just don't exist. This is like the Russian telephone system: if you want to call somebody, you call them to find out what their telephone number is because there are no phone books.

Another issue that was raised is how we can extend the national network into a broader user community. How can we move NREN out quickly from the research and education community into broader industrial usage? Making this transition will require dealing with sticky issues such as intellectual property rights, tariffed services, and interfaces.

Software Environment

Archiving an extensive, friendly, productive, interoperable software environment was acknowledged to be the most difficult element to achieve. We do have emerging standards such as UNIX, X Windows, and so on that help us to tie together software products as we develop them.

The large number of workstations, as has previously been the case with personal computers, has been the motivating factor for developing quality, user-friendly interfaces. These are where the new things get tried out. That's partly based on the large, installed base and, therefore, the opportunities for profitable experimentation.

Now, these friendly interfaces can be used and to some extent are used as access to supercomputers, but discussion during this conference showed that we have a way to go on that score. Unfortunately—and this was a main topic during the meeting—we do not have good standards for software portability and interfaces in a heterogeneous computing environment. We're still working with bits and pieces. It was acknowledged here that we have to tie the computing environment together through software if we are to have a productive environment. Finally—this was something that was mentioned over and over again—significant differences exist in architectures, which impede software portability.

Concluding Remarks

First, if we look back to the meeting in 1983, we see that the high-performance computing environment today is much more complex. Before, we could look at individual boxes. Largely because of the reasons that I mentioned, it's now a distributed system. To have an effective environment today, the whole system has to work. All the elements have to be at roughly the same level of performance if there is to be a balanced system. Therefore, the problem has become much more complex and the solution more effective.

To continue high-performance computing advances, it seems clear from the meeting that we need to establish effective mechanisms to coordinate our activities. Any one individual, organization, or company can only work on a piece of the whole environment. To have those pieces come together so that you don't have square plugs for round holes, some coordination is required. How to do that is a sociological, political, and cultural problem. It is at least as difficult, and probably rather more so, than the technical problems.

Next, as I alluded to before, high-performance computing as a business will live or die—and this is a quote from one of the speakers—according to its acceptance by private industry. The market is currently too small, too fragmented, and not growing at a rapid enough rate to remain operable. Increasing the user base is imperative. Each individual and organization should take as a challenge how we can do that. To some extent, we're the apostles. We *believe*, and we know what can be done, but most of the world does not.

Finally, let me look ahead. In my opinion, there's a clear federal role in high-performance computing. This role includes, but is not limited to, (1) education and training, (2) usage of high performance for agency needs, (3) support for R&D, and (4) technology cooperation with industry. This is not transfer; it goes both ways. Let's not be imperialistic.

The federal role in these and other areas will be a strong motivator and enabler to allow us to achieve the vision discussed during this meeting. It was made clear over and over again that the federal presence has been a leader, driver, catalyst, and strong influence on the development of the high-performance computing environment to date. And if this environment is to succeed, the federal presence has to be there in the future.

We cannot forget the challenge from our competitors and the fact that if we do not take on this challenge and succeed with it, there are others who will. The race is won by the fleet, and we need to be fleet.

The High Performance Computing Initiative

Eugene Wong

> *Eugene Wong is Associate Director for Physical Sciences and Engineering in the Office of Science and Technology Policy. Dr. Wong began his research career in 1955. From 1962 to 1969, he served as Assistant Professor and Professor and, from 1985 to 1989, as Chairman in the Electrical Engineering and Computer Sciences Department of the University of California at Berkeley. When he was confirmed by the U.S. Senate on April 4, 1990, he continued serving in a dual capacity as Professor and Departmental Chairman at the University of California. Dr. Wong received his B.S., M.A., and Ph.D. degrees in electrical engineering from Princeton University. His research interests include stochastic processes and database-management systems.*

I would like to devote my presentation to an overview of the High Performance Computing Initiative as I see it. This is a personal view; it's a view that I have acquired over the last six months.

The High Performance Computing Initiative, which many people would like to rename the High Performance Computing and Communications Initiative, is a proposed program of strategic investment in the frontier areas of computing. I think of it as an investment—a long-term investment. If the current proposal is fully funded, we will have doubled the original $500 million appropriation over the next four to five years.

There is a fairly long history behind the proposal. The first time the concept of high-performance computing—high-performance computing as distinct from supercomputing—was mentioned was probably in the White House Science Council Report of 1985. At that time the report recommended that a study be undertaken to initiate a program in this area.

A strategy in research and development for high-performance computing was published in November 1987 under the auspices of the Office of Science and Technology Policy (OSTP) and FCCSET. FCCSET, as some of you know, stands for Federal Coordinating Council on Science, Engineering and Technology, and the actual work of preparing the program and the plan was done by a subcommittee of that council.

In 1989, shortly after Allan Bromley assumed the office of Director of OSTP, a plan was published by OSTP that pretty much spelled out in detail both the budget and the research program. It is still the road map that is being followed by the program today.

I think the goal of the overall program is to preserve U.S. supremacy in this vital area and over the next few years to accelerate the program so as to widen the lead that we have in the area. Secondly, and perhaps equally important, is to effect a timely transfer of both the benefits and the responsibilities of the program to the private sector as quickly as possible.

"High performance" in this context really means advanced, cutting edge, or frontiers. It transcends supercomputing. It means pushing the technology to its limits in speed, capacity, reliability, and usability.

There are four major components in the program:
- systems—both hardware and system software;
- application software—development of environment, algorithms, and tools;
- networking—the establishment and improvement of a major high-speed, digital national network for research and education; and
- human resources and basic research.

Now, what is the basic motivation for the program, aside from the obvious one that everybody wants more money? I think the basic motivation is the following. In information technology, we have probably the fastest-growing, most significant, and most influential technology in our economy. It has been estimated that, taken as a whole, electronics, communications, and computers now impact nearly two-thirds of the gross national product.

Advanced computing has always been a major driver of that technology, even though when it comes to dollars, the business may be only a very small part of it. Somebody mentioned the figure of $1 billion. Well, that's $1 billion out of $500 billion.

It is knowledge- and innovation-intensive. So if we want as a nation to add value, this is where we want to do it. It is also leveraging success.

This is an area where, clearly, we have the lead, we've always had the lead, and we want to maintain that lead.

In my opinion, a successful strategy has to be based on success and not merely on repair of flaws. Clearly, this is our best chance of success.

Why have a federal role? This is the theme of the panel. You'll hear more about this. But to me, first of all, this is a leading-edge business. I think several presenters at this conference have already mentioned that for such a market, the return is inadequate to support the R&D effort needed to accelerate it and to move ahead. That's always the case for a leading-edge market. If we really want to accelerate it rather than let it take its natural time, there has to be a federal role. The returns may be very great in the long term. The problem is that the return accrues to society at large and not necessarily to the people who do the job. I think the public good justifies a federal role here.

Networking is a prominent part of the program, which is an infrastructure issue and therefore falls within the government s purview. The decline in university enrollment in computer science needs to be reversed, and that calls for government effort—leadership.

Finally, and most importantly, wider use and access to the technology and its benefits require a federal role. It is the business of the federal government to promote applications to education and other social use of high-performance computing.

As a strategy, I think it's not enough to keep the leading edge moving forward. It's not enough to supply enough money to do the R&D. The market has to follow quickly. In order for that market to follow quickly, we have to insure that there is access and there is widespread use of the technology.

There are eight participating federal agencies in the program. Four are lead agencies: the Defense Advanced Research Projects Agency, the Department of Energy, NSF, and NASA, all of whom are represented here in this session. The National Institute of Standards and Technology, the National Oceanic and Atmospheric Administration, the National Institutes of Health, and the EPA will also be major participants in the program.

So, where are we in terms of the effort to implement the program? The plan was developed under FCCSET, and I think the group that did the plan did a wonderful job. I'm not going to mention names for fear of leaving people out, but I think most of you know who these people are.

That committee under FCCSET, under the new organization, has been working as a subcommittee of the Committee on Physical Sciences,

Engineering, and Mathematics, under the leadership of Erich Bloch (see Session 1) and Charlie Herzfeld. Over the last few months, we've undertaken a complete review of the program, and we've gotten OMB's agreement to look at the program as an integral program, which is rare. For example, there are only three such programs that OMB has agreed to view as a whole. The other two are global change and education, which are clearly much more politically visible programs. For a technology program to be treated as a national issue is rare, and I think we have succeeded.

We are targeting fiscal year 1992 as the first year of the five-year budget. This is probably not the easiest of years to try to launch a program that requires money, but no year is good. I think it's a test of the vitality of the program to see how far we get with it.

The President's Council of Advisors on Science and Technology (PCAST) has provided input. I'm in the process of assembling an ad hoc PCAST panel, which has already met once. It will provide important private-sector input to the program. In that regard, someone mentioned that we need to make supercomputing, high-performance computing, a national issue. I think it has succeeded.

The support is really universal. You'll hear later that, in fact, committees in the Senate are vying to support a program. I think the reason why it succeeded is that it's both responsive to short-term national needs and visionary, and I think both are required for a program to succeed as a national issue. It's responsive to national needs on many grounds—national security, economy, education, and global change. It speaks to all of these issues in a deep and natural way.

The grand challenges that have been proposed really have the potential of touching every citizen in the country, and I think that's what gives it the importance that it deserves.

The most visionary part of the program is the goal of a national high-speed network in 20 years, and I think most people in Washington are convinced that it's going to come. It will be important, and it will be beneficial. The question is how it is going to come about. This is the program that we will spearhead.

The program has spectacular and ambitious goals, and in fact the progress is no less spectacular. At the time it was conceived as a goal, a machine with a capacity of 10^{12} floating-point operations per second was considered ambitious. But now it doesn't look so ambitious. The program has a universal appeal in its implications for education, global change, social issues, and a wide range of applications. It really has the potential of touching everyone. Thus, it's a program that I'm excited about. I'm fortunate to arrive in Washington just in time to help to get it started, and I'm hopeful that it will get started this year.

Let me move now to the theme of this session-What Now? Given the important positions that the presenters in this session occupy, their thoughts on the theme of the federal role in high-performance computing will be most valuable. In the course of informal exchanges at this conference, I have already heard a large number of suggestions as to what that role might be. Let me list some of these before we hear from our presenters.

What is the appropriate government role? I think the most popular suggestion is that it be a good customer. Above all, it's the customer that pays the bills. It should also be a supporter of innovation in a variety of ways — through purchase, through support R&D, and through encouragement. It needs to be a promoter of technology, not only in development but also in its application. It should be a wise regulator that regulates with a deft and light touch. The government must be a producer of public good in the area of national security, in education, and in myriad other public sectors. It should also be an investor, a patient investor for the long term, given the current unfriendly economic environment. And last, above all, it should be a leader with vision and with sensitivity to the public need. These are some of the roles that I have heard suggested at this conference.

Government Bodies as Investors

Barry Boehm

> *Barry W. Boehm is currently serving as Director of the Defense Advanced Research Projects Agency's (DARPA's) Software and Intelligent Systems Technology Office, the U.S. Government's largest software research organization. He was previously Director of DARPA's Information Science and Technology Office, which also included DARPA's research programs in high-performance computing and communications.*

I'd like to begin by thanking the participants at this conference for reorienting my thinking about high-performance computing (HPC). Two years ago, to the extent that I thought of the HPC community at all, I tended to think of it as sort of an interesting, exotic, lost tribe of nanosecond worshippers.

Today, thanks to a number of you, I really do feel that it is one of the most critical technology areas that we have for national defense and economic security. Particularly, I'd like to thank Steve Squires (Session 4 Chair) and a lot of the program managers at the Defense Advanced Research Projects Agency (DARPA); Dave Nelson and the people in the Federal Coordinating Committee on Science, Engineering, and Technology (FCCSET); Gene Wong for being the right person at the right time to move the High Performance Computing Initiative along; and finally all of you who have contributed to the success of this meeting. You've really given me a lot better perspective on what the community is like and what its needs and concerns are, and you've been a very stimulating group of people to interact with.

In relation to Gene's list, set forth in the foregoing presentation, I am going to talk about government bodies as investors, give DARPA as an example, and then point to a particular HPC opportunity that I think we have as investors in this initiative in the area of software assets.

If you look at the government as an investor, it doesn't look that much different than Cray Research, Inc., or Thinking Machines Corporation or IBM or Boeing or the like. It has a limited supply of funds, it wants to get long-range benefits, and it tries to come up with a good investment strategy to do that.

Now, the benefits tend to be different. In DARPA's case, it's effective national defense capabilities; for a commercial company, it's total corporate value in the stock market, future profit flows, or the like.

The way we try to do this at DARPA is in a very interactive way that involves working a lot with the Department of Defense (DoD) users and operators and aerospace industry, trying to figure out the most important things that DoD is going to need in the future, playing those off against what technology is likely to supply, and evaluating these in terms of what their relative cost-benefit relationships are. Out of all of that comes an R&D investment strategy. And I think this is the right way to look at the government as an investor.

The resulting DARPA investment strategy tends to include things like HPC capabilities, not buggy whips and vacuum tubes. But that does not mean we're doing this to create industrial policy. We're doing this to get the best defense capability for the country that we can.

The particular way we do this within DARPA is that we have a set of investment criteria that we've come up with and use for each new proposed program that comes along. The criteria are a little bit different if you're doing basic research than if you're doing technology applications, but these tend to be common to pretty much everything that we do.

First, there needs to be a significant long-range DoD benefit, generally involving a paradigm shift. There needs to be minimal DoD ownership costs. Particularly with the defense budget going down, it's important that these things not be a millstone around your neck. The incentive to create things that are commercializable, so that the support costs are amortized across a bigger base, is very important.

Zero DARPA ownership costs: we do best when we get in, hand something off to somebody else, and get on to the next opportunity that's there. That doesn't mean that there's no risk in the activity. Also, if Cray is already doing it well, if IBM is doing it, if the aerospace industry is doing it, then there's no reason for DARPA to start up something redundant.

A good many of DARPA's research and development criteria, such as good people, good new ideas, critical mass, and the like, are self-explanatory. And if you look at a lot of the things that DARPA has done in the past, like ARPANET, interactive graphics, and Berkeley UNIX, you see that the projects tend to fit these criteria reasonably well.

So let me talk about one particular investment opportunity that I think we all have, which came up often during the conference here. This is the HPC software problem. I'm exaggerating it a bit for effect here, but we have on the order of 4000 live HPC projects at any one time, and I would say there's at least 4000, or maybe 8000 or 12,000 ad hoc debuggers that people build to get their work done. And then Project 4001 comes along and says, "How come there's no debugger? I'd better build one." I think we can do better than that. I think there's a tremendous amount of capability that we can accumulate and capitalize on and invest in.

There are a lot of software libraries, both in terms of technology and experience. NASA has Cosmic, NSF has the supercomputing centers, DoD has the Army Rapid repository, DARPA is building a Stars software repository capability, and so on. There's networking technology, access-control technology, file and database technology, and the like, which could support aggregating these libraries.

The hardware vendors have user communities that can accumulate software assets. The third-party software vendor capabilities are really waiting for somebody to aggregate the market so that it looks big enough that they can enter.

Application communities build a lot of potentially reasonable software. The research community builds a tremendous amount just in the process of creating both machines, like Intel's iWarp, and systems, like Nectar at Carnegie Mellon University; and the applications that the research people do, as Gregory McRae of Carnegie Mellon will attest, create a lot of good software.

So what kind of a capability could we produce? Let's look at it from a user's standpoint.

The users ought to have at their workstations a capability to mouse and window their way around a national distributed set of assets and not have to worry about where the assets are located or where the menus are located. All of that should be transparent so that users can get access to things that the FCCSET HPC process invests in directly, get access to various user groups, and get access to software vendor libraries.

There are certain things that they, and nobody else, can get access to. If they're with the Boeing Company, they can get the Boeing airframe software, and if they're in some DoD group that's working low observables, they can get access to that. But not everybody can get access to that.

As you select one of these categories of software you're interested in, you tier down the menu and decide that you want a debugging tool, and then you go and look at what's available, what it runs on, what kind of capabilities it has, etc.

Behind that display are a lot of nontrivial but, I believe, workable issues. Stimulating high-quality software asset creation is a nontrivial job, as anybody who has tried to do it knows. I've tried it, and it's a challenge.

An equally hard challenge is screening out low-quality assets—sort of a software-pollution-control problem. Another issue is intellectual property rights and licensing issues. How do people make money by putting their stuff on this network and letting people use it?

Yet another issue is warranties. What if the software crashes right in the middle of some life- or company- or national-critical activity?

Access-control policies are additional challenges. Who is going to access the various valuable assets? What does it mean to be "a member of the DoD community," "an American company," or things like that? How do you devolve control to your graduate students or subcontractors?

Distributed-asset management is, again, a nontrivial job. You can go down a list of additional factors. Dave Nelson mentioned such things as interface standards during his presentation in this session, so I won't cover that ground again except to reinforce his point that these are very important to good software reuse. But I think that all of these issues are workable and that the asset base benefits are really worth the effort. Right now one of the big entry barriers for people using HPC is an insufficient software asset base. If we lower the entry barriers, then we also get into a virtuous circle, rather than a vicious circle, in that we increase the supply of asset producers and pump up the system.

The real user's concern is reducing the calendar time to solution. Having the software asset base available will decrease the calendar time to solution, as well as increase application productivity, quality, and performance.

A downstream research challenge is the analog to spreadsheets and fourth-generation languages for high-performance applications. These systems would allow you to say, "I want to solve this particular structural dynamic problem," and the system goes off and figures out what kind of mesh sizes you need, what kind of integration routine you should use, etc. Then it would proceed to run your application and interactively present you with the results.

We at DARPA have been interacting quite a bit with the various people in the Department of Energy, NASA, and NSF and are going to try to come up with a system like that as part of the High Performance Computing Initiative. I would be interested in hearing about the reactions of other investigators to such a research program.

Realizing the Goals of the HPCC Initiative: Changes Needed

Charles Brownstein

Charles N. Brownstein is the Acting Assistant Director of NSF and is a member of the Executive Directorate for Computer and Information Science and Engineering (CISE) at NSF. Dr. Brownstein chairs the interagency Federal Networking Council, which oversees federal management of the U.S. Internet for research and education. He participates in the Federal Coordinating Committee on Science, Engineering and Technology, which coordinates federal activities in high-performance computing, research, and educational networking. He has served as a Regent of the National Library of Medicine and recently participated in the National Critical Technologies panel. Before the creation of CISE in 1986, Dr. Brownstein directed the Division of Information Science and Technology. He also directed research programs on information technology and telecommunications policy at NSF from 1975 to 1983. He came to NSF from Lehigh University, Bethlehem, Pennsylvania, where he taught and conducted research on computer and telecommunications policy and information technology applications.

The President's FY 1992 High Performance Computing and Communications Initiative (HPCC Initiative)—also known as the High Performance Computing Initiative (HPCI)—may signal a new era in U.S. science and technology policy. It is the first major technology initiative of the 1990s.

It integrates technology research with goals of improving scientific research productivity, expanding educational opportunities, and creating new kinds of national "information infrastructure."

In the post-Cold War and post-Gulf War environment for R&D, we're going to need a precompetitive, growth-producing, high-technology, highly leveraged, educationally intensive, strategic program. The HPCC Initiative will be a prime example. This paper explores the changes that had to occur to get the initiative proposed and the changes which are needed to realize its goals. It is focused on the issue of the human resource base for computing research.

The HPCC program, as proposed, will combine the skills, resources, and missions of key agencies of the government with those of industry and leading research laboratories. The agency roles are spelled out in the report "Grand Challenges: High Performance Computing and Communications," which accompanied the President's budget request to Congress.

On its surface, the HPCC Initiative seeks to create faster scientific computers and a more supple high-performance computer communications network. But the deep goals deal more with creating actual resources for research and national economic competitiveness and security. One essential part is pure investment: education and human resources are an absolutely critical part of both the HPCC Initiative and the future of our national industrial position in computing.

HPCC became a budget proposal for FY 1992 because the new leaders of the President's Office of Science & Technology Policy (OSTP) had the vision to create an innovative federal R&D effort. The President's Science Advisor, Dr. Allan Bromley, came in, signed on, and picked up an activity that had been under consideration for five years. The plan represents the efforts of many, many years of work from people in government, industry, and user communities. Dr. Bromley's action elevated the HPCC program to a matter of national priority. Gene Wong (see his presentation earlier in this session) joined OSTP and helped refine the program, translating between the language of the administration and the language of the scientific community.

Industry participation is a central feature of HPCC. In the period of planning the program, industry acknowledged the fact that there's a national crisis in computing R&D, education, and human resources. Research has traditionally been supported in the U.S. as a partnership among the government, universities, laboratories, and industry. Today, it's possible to get money to do something in education from almost all of the major companies. That's a good sign.

Computing as an R&D field has some obvious human-resource deficiencies now and will have some tremendous deficiencies in the labor pool of the future. There are fewer kids entering college today with an intent in majoring in natural sciences and engineering than was the case at any time in the past 25 years. Entry into college with the goal of majoring in natural sciences and engineering is the best predictor of completing an undergraduate degree in the field. Very few people transfer into these fields; a lot of people transfer out. The net flow is out into the business schools and into the humanities.

The composition of the labor pool is changing. The simple fact is that there will be greater proportions of students whose heritages have traditionally not led them into natural sciences and engineering. There are no compelling reasons to suggest that significant numbers of people from that emerging labor pool can't be trained. A cultural shift is needed, and intervention is needed to promote that cultural shift.

The resource base for education is in local and state values and revenues; the ability for the federal government to intervene effectively is really pretty small. Moreover, the K–12 curriculum in natural sciences and engineering, with respect to computational science, is marginal. The problem gets worse the higher up the educational system that you go. Teaching is too often given second-class status in the reward structure at the country's most prestigious universities. So we have a daunting problem, comparable to any of the grand challenges that have been talked about in the HPCC Initiative.

One place to start is with teachers. There's an absolute hunger for using computers in education. Parts of the NSF supercomputer center budgets are devoted to training undergraduate and graduate students, and the response has been encouraging. One recent high school student participant in the summer program at the San Diego Center placed first in the national Westinghouse Science Talent Search.

The Washington bureaucracy understands that dealing with education and human resources is a critical part of the effort. Efforts undertaken by the Federal Coordinating Committee on Science, Engineering, and Technology have produced an HPCC Initiative with a substantial education component and, separately, an Education and Human Resources Initiative capable of working synergistically in areas like networking.

One group that needs to get much more involved is our leading "computing institutions." We have, it seems, many isolated examples of "single-processor" activities in education and human resources. Motives ranging from altruism to public relations drive places like the NSF and

NASA Centers, the Department of Energy (DOE) labs, and other computational centers around the country to involve a broad range of people in educational activities. That includes everything from bringing in a few students to running national programs like Superquest. We need to combine these efforts.

A national Superquest program might be created to educate the country about the importance of high-performance computing. Cornell University and IBM picked up the Superquest idea and used it in 1990. They involved students from all over the country. One of the winners was Thomas Jefferson High School, in Virginia; Governor John Sununu, whose son attends school there, spoke at the ceremonies announcing the victory.

We need to run interagency programs that put together a national Superquest program with the assets of DOE, NASA, NSF, and the state centers and use it to pull a lot of kids into the process of scientific computing. They'll never escape once they get in, and they will pull a lot of people such as their parents and teachers into the process.

We also need to find ways to involve local officials. School systems in the country have daunting problems. They have been asked to do a lot of things apart from education. Science education is just a small part of the job they do. They need a lot of help in learning how to use computing assets. I believe that concerned officials are out there and will become involved. If we're going to do anything significant, ever, in the schools at the precollege level, these officials have to be the people who are convinced. The individual teacher is too much at the whim of the tax base and what happens to residential real-estate taxes. The local officials have to be pulled in and channeled into these programs. The people to do that are the local members of the research community from our universities, laboratories, and businesses.

Getting these people to make an investment in the student base and in the public base of understanding for this technology is very important. I would love to see people like Greg McRae of Carnegie Mellon University spend about six months being an evangelist in the sense of reaching out to the educational community and the public at large.

What will the HPCC Initiative do for education in computational science? The initiative has about 20 per cent of its activity in education and human resources, and to the extent that we can get this program actually supported, there is a commitment, at least within the agencies, to maintain that level. The program will support graduate students on research projects, along with training for postdocs in computational science, curriculum improvement, and teacher training. About 20 per cent is for the National Research and Education Network (NREN). This used to be called NRN. The educational opportunity is absolutely critical, and we are serious about keeping it NREN—with an E.

Over the past few years, the Internet has been extended from about 100 institutions—the elite—to over 600 educational institutions. The number we aim for is 3000. That sounds daunting, except that we've driven the price of an incremental connection to the network, through a structure of regional networks, down from the $60,000 range, to about the $10,000 or $11,000 range. That will come down much further, making it easily available to anyone who wants it as commercial providers enter the market.

That means that small, "have-not" institutions in remote or poor areas of the country will be viable places for people to reach out to national assets of research and education, including the biggest computers at the best-equipped institutions in the country. One finds odd people in odd places using odd machines over the network to win Gordon Bell awards. It's really quite amazing what a transforming effect an accessible net can have.

Much has occurred in networking recently. We've gone from 56 kilobytes to 45 megabits in the NSFnet backbone that is shared with the mission agencies. The regional nets will be upgrading rapidly. The mission agency nets are linked via federal internet exchanges, resources are shared, and the mix is becoming transparent.

A commercial internet exchange (CIX) has been created. Network services are also changing. The resources are there for X-400 and X-500. There are programs under way to get national white pages. Many scientific communities have begun to create their own distributed versions, ranging from bulletin boards to software libraries.

Long-time users have experienced a transformation in reliability, compatibility, reach, and speed. We have, in the wings, commercial providers of services, from electronic publishing to educational software, preparing to enter this network. The U.S. research community has been engaged, and usage is growing rapidly in all of the science and engineering disciplines. Like supercomputing, high-performance networking is turning out to be a transforming technology. Achieving the President's HPCC goal is the "grand challenge" that must be met to sustain this kind of progress.

The Importance of the Federal Government's Role in High-Performance Computing

Sig Hecker

Siegfried S. Hecker is the Director of Los Alamos National Laboratory, Los Alamos, New Mexico. For more complete biographical information, please refer to Dr. Hecker's presentation in Session 1.

As I look at the importance of the federal government's role, I think I could sum it all up by saying that Uncle Sam has to be a smart businessman with a long-term outlook because his company's going to be around for a long time. Someone has to take that long-term outlook, and I think clearly that's where the federal government has to come in.

What I want to discuss today is specifically the role of Los Alamos National Laboratory (Los Alamos, or the Laboratory) and how we'd like to respond to the High Performance Computing Initiative. I think most of you know that what we are interested in first and foremost at Los Alamos is solutions. We have applications, and we want the best and the fastest computers in order to be able to do our jobs better, cheaper, and faster and, hopefully, to be able to do things that we had not been able to do before.

Our perspective comes from the fact that we want to do applications; we're users of this computing environment. So we've always considered it imperative to be at the forefront of computing capability.

Let me review briefly the important roles that I think we've played in the past and then tell you what we'd like to do in the future. It's certainly fair to say that we've played the role of the user—a sophisticated, demanding user. In that role we have interfaced and worked very, very closely with the computing vendors over the years, starting early on with IBM and then going to Control Data Corporation, to Cray Research, Inc., to Thinking Machines Corporation, and then of course all along with Sun Microsystems, Inc., Digital Equipment Corporation, and so forth.

Out of necessity, in a number of cases we've also played the role of the inventor—in the development of the MANIAC, for instance, right after development of the ENIAC. Our people felt that we had to actually create the capabilities to be able to solve the problems that we had.

Later on, we invented things such as the common file system. The high-performance parallel interface, better known as HIPPI, is also a Los Alamos invention. That's the sort of product that's come about because we're continually pushed by the users for this sort of capability.

New algorithms to solve problems better and smarter are needed. So things like the lattice gas techniques for computational fluid dynamics were basically invented here by one of our people, along with some French collaborators. Also, we are very proud of the fact that we helped to get three of the four NSF supercomputer centers on line by working very closely with them early on to make certain that they learned from our experiences.

We also introduced companies like General Motors to supercomputing before they bought their computers in 1984. We were working with them and running the Kiva code for combustion modeling. As Norm Morse points out (see Session 10), we have 8000 users. At least half of them are from outside the Laboratory.

The role that we've played has been made possible by our feeling that we have to be the best in the defense business. Particularly in our mainline business, nuclear weapons design, we felt we needed those capabilities because the problems were so computationally intense, so complex, and so difficult to test experimentally. We were fortunate for many, many years that first the Atomic Energy Commission and then the Department of Energy (DOE) had the sort of enlightened management to give us the go-ahead to stay at the forefront and, most importantly, to give us the money to keep buying the Crays and Thinking Machines and all of those good machines.

What we have proposed for the Laboratory is an expanded national charter under the auspices of this High Performance Computing Initiative. First of all, our charter has already significantly expanded beyond nuclear weapons R&D, which represents only about a third of our activities. The remaining two-thirds is a lot of other defense-related activities and many civilian activities.

Today, in terms of applications, we worry about some of the same grand challenges that you worry about, such as environmentally related problems—for instance, the question of global climate change. The Human Genome Initiative is basically an effort that started at Los Alamos and at Lawrence Livermore National Laboratory because we have the computational horsepower to look at how one might map the 3 billion base pairs that exist on your DNA. We also have other very interesting challenges in problems like designing a free-electron laser essentially from scratch with supercomputers.

In response to congressional legislation earlier this year, I outlined a concept called Collaborative R&D Centers that I'd like to see established at Los Alamos or at least supported at places like Los Alamos. There are several aspects of this proposed center I would like to mention. For one thing, we'd like to make certain that we keep the U.S. at the leading edge of computational capabilities. For another, we intend to make the high-performance computing environment available to greater numbers of people in business, industry, and so forth.

But there are five particular things I'd like to see centers like this do. First of all, continue this very close collaboration with vendors. For instance, at Los Alamos we're doing that now, not only with Cray but also with IBM, Thinking Machines, and many others.

Second, continue to work, perhaps even closer, with universities to make sure that we're able to inject the new ideas into the future computing environment. An example of that might be the work we've done with people like Al Despain at the University of Southern California (see Session 4) as to how one takes the lattice gas concepts and constructs a computational architecture to take advantage of that particular algorithm. Despain has thought about how to take one million chips and construct them in such a fashion that you optimize the problem-solving capabilities.

As part of this collaboration with universities, we could provide a mechanism for greater support, through DOE and Los Alamos, of graduate students doing on-campus research, with provisions for work at the Laboratory, itself. We do a lot with graduate students now. In fact, we have about 400 graduate students here during the course of a summer in many, many disciplines. I think in the area of computational sciences, we will really boost student participation.

The third aspect is to have a significant industrial user program to work even more closely with U.S. industry—not only to make available to them supercomputing but also to promote appreciation of what supercomputing and computational modeling can do for their business. So we'd like to have a much greater outreach to U.S. industry. I agree with the comments of other presenters that supercomputing in industry is very much underutilized. I think one can do much better.

The fourth aspect would be to help develop enabling technologies for tomorrow's innovations in computing—technologies such as photolithography (or at least the next generations of photolithography), superconducting microelectronics, optical computers, neural networks, and so forth. In the Strategic Defense Initiative (SDI) program, where we've done a lot of laser development, we'd like to provide essentially the next "light bulb" for photolithography—that "light bulb" being a free-electron laser we've developed for SDI applications.

The benefit of the free-electron laser is that you can do projection lithography, that you can take the power loss because you start with extremely high power, and that you can tune to wavelength. We think that we can probably develop a free-electron laser, tune it down to the order of 10 nanometers, and get feature sizes down to 0.1 micron—perhaps 0.05 microns—with that sort of a light bulb. It would take a significant development to do that, but we think it's certainly possible. We're working right now with a number of industrial companies to see what the interest level is so that we might be able to get beyond what we think can be done with X-ray synchrotron proximity lithography. It's that type of technology development that, again, I think would be a very important feature of what laboratories like ours could do.

The fifth aspect would be a general-user program to make certain that we introduce, as much as is possible and feasible, some of these capabilities to the local communities, schools, businesses, and so forth. This collaborative R&D would have a central organization in a place like Los Alamos, but there would be many arms hooked together with very-high-speed networks. It would also be cost-shared with industry. The way that I see this working, the government invests its money in us to provide this capability. Industry invests its money and shows its interest by providing its own people to interact with us.

These are just a few remarks on what I think a laboratory like Los Alamos can do to make certain that this country stays at the leading edge of computing and that high-power computing is made available to a broader range of users in this country.

Legislative and Congressional Actions on High-Performance Computing and Communications

Paul G. Huray

Paul G. Huray is Senior Vice President for Research at the University of South Carolina at Columbia and a consultant for the President's Office of Science and Technology Policy. From 1986 to 1990, he served with the Federal Coordinating Committee on Science, Engineering, and Technology, chairing that body's Committee on Computer Research and Applications, which issued An R&D Strategy for HPC *and* The Federal HPC Program. *He has assisted in the development of Manufacturing Technology Centers for the National Institute of Standards and Technology.*

I thought I'd begin this presentation by making a few comments about the participation of the U.S. Congress in the High Performance Computing and Communications (HPCC) Initiative. Table 1 shows a chronology of the legislative actions on HPCC. There is nothing in particular that I want to bring to your attention in this figure except the great number of events that have occurred since August 1986. In July 1990, there were several pieces of legislation on the floor of the Senate: S-1067, S-1976, the Senate Armed Services Authorization, and a couple of complementary House bills. Let me outline the legislative situation as of this writing (August 1990).

At the recent meeting of the President's Council of Advisors on Science and Technology, I tried to guess where the legislation was headed; Table 2

Table 1. Legislative History of HPCC Initiative

Date	Designation	Action
Aug 86	PL 99-383	NSF Authorization: OSTP network report
Nov 87	An R&D Strategy for HPC	FCCSET: systems, software, NRN, human resources
Aug 88	Sen. Sci. (CS&T)	Hearing: "Computer Networks and HPC"
Oct 88	S.2918	"The National HPC Technology Act": Strategy + AI, Inf. Sci., and budget
May 89	S.1067	"The HPC Act of 1989"
Jun 89	H. SR&T (SS&T)	Hearing: "U.S. Supercomputer Industry"
Jun 89	Sen. CS&T	Hearing: "S.1067—NREN"
Jun 89	Gore roundtable	Off-the-record: network carriers
Jul 89	Sen. CS&T	Hearing: "S.1067—Visualization and Software"
Aug 89	H.R.3131	"The National HPC Technology Act of 1989"
Sep 89	*The Federal HPC Program*	Implementation plan: DARPA, DOE, NASA, NSF
Sep 89	Sen. Sci. (CS&T)	Hearing: "S.1067—Advanced Computing and Data Management"
Oct 89	H. SR&T (SS&T)	Hearing: "HPC"
Oct 89	H. Telecom. (E&C)	Hearing: "Networks of the Future"
Nov 89	S.1976	"The DOE HPC Act"
Mar 90	Sen. En. R&D (E)	Hearing: "S.1976"
Apr 90	S.1067, amended	"The HPC Act of 1990" (same as *The Federal HPC Program*)
Jun 90	S.1976, amended	Puts NREN under DOE
Jul 90	H.R.5072	"The American Technology Preeminence Act": DOC Authorization, ATP, S/W amend., OSTP
Jul 90	Sen. Armed Serv.	Authorizes $30 M for DARPA in FY 91
Jul 90	Gore roundtable	Off-the-record: HPC users

Table 2. Legislative Prognosis on HPCC Initiative

Action	Best Guess
Conference Compromise	S.1067, S.1976, Armed Services authorization will pass. Committee report will resolve Senate NREN issue. H.R.5072 or H.R.3131 will pass. House will accept full Senate bill.
Consolidation	Bill could attract ornaments or become a subset of other S&T legislation.
Appropriations	Budget committee will single out HPCC for a line item with a "special place in the FY 91 budget," but appropriations will fall short of authorization because of general budget constraints.
Education	Senate hearings will be held to place HPCC in an educational context. Human interfaces (sound, interactive graphics, multimedia) will be considered in view of previous failure of computer-aided instruction.
Business	Senate hearings will be held to consider the value of HPCC to the manufacturing environment and to corporate network services.

shows what could happen. S-1067 and S-1976 have passed out of committee and are on the floor for compromise. Some of the people attending this conference are participating in the compromise process.

There are some issues to be resolved, especially issues related to the National Research and Education Network (NREN), but it is our belief that these pieces of Senate legislation will pass in some form and that the complementary legislation will also pass in the House, although the language is quite different in some respects. However, we have some assurance that the House will accept the Senate bill when it's eventually compromised.

There are a couple of other dangers, however, in the consolidation process. What might happen to these pieces of legislation? As indicated in Table 2, they could attract ornaments from other activities, or they could become a subset of some other science and technology legislation. When I asked Senator Albert Gore of Tennessee what he thought might happen, he said that apart from issues related to abortion and the Panama Canal, he couldn't imagine what might be added.

The appropriations process, of course, is the key activity. The bills that are on the floor of the House and Senate now are just authorizations. But when it comes budget time—time to actually cut those dollars—we believe that HPCC will have a special status. The point is that there are still plenty of opportunities for this initiative to fall apart in the Congress. In fact, some ideologues would like that to happen. I think, on the other hand, that we will see a few HPCC additions in the FY 1991 budget produced by Congress, and certainly there will be a special place in the FY 1992 budget coming out of the Executive Office.

Table 2 also indicates that more hearings will be held on HPCC. I realize that probably half the people in this audience have participated in these hearings, but I would guess that the next hearings would be associated with education, probably with the network and its usefulness as a tool for education. I would also guess that the following hearing would concentrate on business and how the manufacturing sector will benefit from a national network. Because we were asked to discuss specific topics, I'm going to address the rest of my remarks to the usefulness of the HPCC Initiative to the business sector.

To start, I want to note a very important key word in the title of the program (see Figure 1), the word "federal." The word "federal" has allowed this initiative to go forward without becoming industrial policy. At one point, that word was "national," but we realized we were potentially running into trouble because we wanted to retain an emphasis on economic competitiveness without crossing the political line into industrial policy.

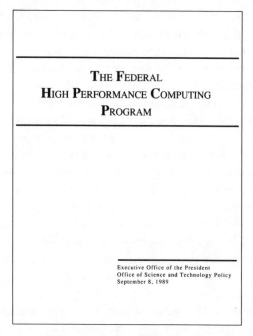

Figure 1. Cover of the report issued by the Office of Science and Technology Policy.

Three goals were stated in the development of the High Performance Computing Initiative, as the HPCC Initiative is also known:
- maintain and extend U.S. leadership in high-performance computing, especially by encouraging U.S. sources of production;
- encourage innovation through diffusion and assimilation into the science and engineering communities; and
- support U.S. economic competitiveness and productivity through greater utilization of networked high-performance computing in analysis, design, and manufacturing.

These goals are getting close to the political line of industrial policy. It is the third of these goals I want to focus on during my presentation.

I think everyone who participated in this process understood the relevance of this initiative to the productivity of the country. In order to examine that potential, I decided to take a look at what might become a more extended timetable than the three phases of the national network that are listed in HPCC Initiative documents.

As you probably know, phases 1, 2, and 3 of NREN are addressed in the document that covers growth in performance of the network and growth in the number of institutions. We're probably already into NREN-2 in terms of some of those parameters. But many of us believe that the big payoffs will occur when business begins to participate in the national network and gains access to high-performance computing. Figure 2 suggests that large numbers of institutions could become part of the network once the business sector participates.

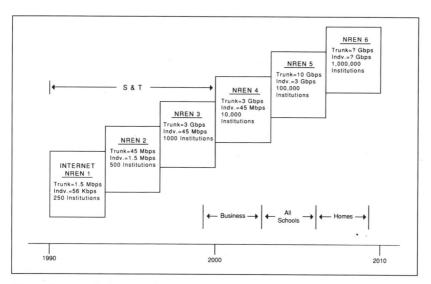

Figure 2. Extended timetable for a National Research and Education Network (NREN).

How will that happen? One way is through a program that the National Institute of Standards and Technology (NIST) runs. It is a program that I'm very familiar with because it takes place partly in South Carolina.

In 1988 Congress passed the Omnibus Trade and Competitiveness Act, as shown in Table 3. That act was very controversial because it had all kinds of issues that it dealt with—unfair trade, antidumping, foreign investments, export control, intellectual property. But one of the activities it established was the development of regional centers for the transfer of manufacturing technology, and this activity is intimately connected with the national network.

As shown in Table 4, there are currently three centers that participate in this manufacturing technology center program. One is at Rensselaer Polytechnic Institute in Troy, New York, one is in Cleveland at the Cleveland Area Manufacturing Program, and one is at the University of South Carolina. The South Carolina-based initiative provides a process for delivering technology to the work force; and this is quite different than the normal Ph.D. education process. We're delivering technology to small and medium-size companies, initially only in South Carolina but later throughout the 14 southeastern states in a political consortium called the Southern Growth Policies Board.

This technology involves fairly straightforward computation, that is, workstation-level activity. But in some cases the technology involves numerically intensive computing. Table 5 shows the kind of technologies that we're delivering in South Carolina. The small-to-medium-size companies range in capabilities from a blacksmith's shop up to a very sophisticated business. But we're having terrific impact. Nationwide, the number of computer scientists has increased by two orders of magnitude since 1983, to 60,000 in 1989. In 1989, we trained 10,000 workers in the technologies shown in Table 6 as part of this program. Those are the workers trained in one state by one of three manufacturing technology centers—a number that is expected to expand to five in 1991. This will be a model for a technology extension program for the whole United States, in which many people in the work force will have access not just to workstations but to high-performance computing, as well.

As an example, Figure 3 shows the network that we're currently using to distribute technology electronically. This is a state-funded network aided by Digital Equipment Corporation. This is an example of a local initiative that is fitting into the national program. There are 26 institutions participating in this consortium, mostly technical colleges whose instructors have previously trained many of the work force in discrete parts manufacturing activities around the state.

What Now? 585

Table 3. Omnibus Trade and Competiveness Act of 1988

- Tools to open foreign markets and help U.S. exporters (examples):
  - Equitable Trade Policy
  - Antidumping Measures
  - Foreign Investment
  - Export Control
  - Intellectual Property Protection

- Initiative to boost U.S. industry in world markets (NIST responsibilities)
  - Assist State Technology Programs
  - Implement Advanced Technology Program
  - Establish Clearing-House for State Technology Programs
  - Establish Regional Centers for Transfer of Manufacturing Technology

Table 4. Characteristic Descriptions of the Existing NIST Manufacturing Technology Centers (MTCs)

Northeastern MTC at Rensselaer Polytechnic Institute:
Focuses on hardening of federal software for commercialization.
Features excellent engineering and enjoys support of industry vendors.
"The Shrink-Wrap Center."

Great Lakes MTC at Cleveland Area Manufacturing Program:
Focuses on real-time measurement and metal-cutting machinery.
Promotes a large toolmaking base in the Cleveland area.
"The Manufacturing Resource Facility."

Southeastern MTC at the University of South Carolina:
Focuses on workforce training through existing technical institutions.
Addresses company needs, including training, in 14 southeastern states.
"The Delivery Center."

Table 5. SMTC Technical Emphasis

Computer-Aided Design	Robotics
Computer-Aided Manufacturing	Metrology
Computer-Aided Engineering	Integration Cells
Numerically Controlled Machines	Inventory Control
Advanced Machine Tools	Quality Control

Table 6. SMTC: General Successes

- Training approximately 10,000 workers in
  - CAD
  - CAM
  - CAE
  - Piping Design
  - Geometric Dimensioning/Tolerancing
  - Info Windows (IBM)
  - Prog Logic CON
  - SPC
  - TQC
  - ZENIX
- Implementing manufacturing-company-needs assessment
- Establishing technical colleges network
- Establishing center of competence for manufacturing technology
- Transferring numerous technologies from Oak Ridge National Laboratory and Digital Equipment Corporation

We estimate that more than 80% of the commerce in South Carolina is within 25 miles of one of these institutions, so linking them into a corporate network is going to be reasonably straightforward. We've initiated that linking process, and we believe we're going to bring about a cultural change with the help of a manufacturing network.

One of the things we are doing, for example, is bringing up an electronic bulletin board, which is essentially going to be a bidding board on which we have electronic specifications for subcontracts from the Department of Defense or large corporations. We will let those small corporations call the bulletin board and see what's up today for bid, with a deadline perhaps at three o'clock. The small businessman will look at the electronic specifications for a particular subcontract, which might be a fairly large data set displayed on a high-quality workstation. The businessman will ask himself if he can manufacture that item in a numerical fashion quickly for the subcontract bid. In one scenario, before the three o'clock deadline comes, the businessman will bid on the project, the bid will be let at five o'clock, the manufacturing takes place that night, and delivery happens the next morning. I think that's not an unrealistic vision of a future manufacturing infrastructure aided by a national network.

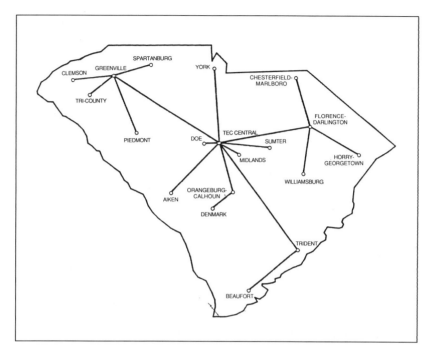

Figure 3. South Carolina State Technical Colleges' wide-area network plan.

The companies that are currently participating in our program are, for the most part, using 1950s technology, and they are just now coming into a competitive environment. Unfortunately, they don't even know they're competing in most cases. But we can see such a network program extending to other states in the southeastern United States, as shown in Figure 4. The plan for NIST is to clone the manufacturing technology centers throughout other states in the country. We can imagine eventually 20,000 small- to medium-sized companies participating in such a program—20,000 businesses with employees of perhaps, in the case of small businesses, 50 persons or less.

We need to remember that these small corporations produce the majority of the balance of trade for the United States. Seventy-five per cent of our manufactured balance of trade comes from small- or medium-sized companies; these are the people we need to impact. I have described a mechanism to initiate such a program. I believe that this program can address the infrastructure of manufacturing and productivity in the country, as well as accomplish some of the very sophisticated projects we are doing in academia and in our national laboratories.

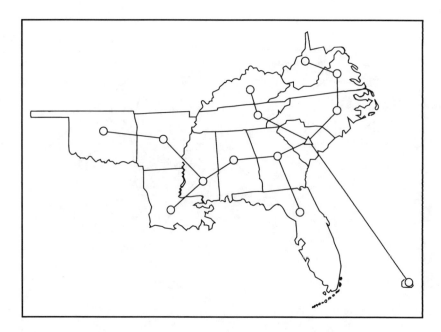

Figure 4. Southeastern Manufacturing Technology Network.

The Federal Role as Early Customer

David B. Nelson

David B. Nelson is Executive Director in the Office of Energy Research, U.S. Department of Energy. For more complete biographical information, please see his presentation earlier in this session.

Federal agencies—and I'll use Department of Energy (DOE) as an example because I'm most familiar with it—have for many years played an important role as an early customer and often as a first customer for emerging high-performance computing systems. This is true not only in federal laboratories but also in universities with federal funding support, so the federal role is diffused fairly widely.

Let's talk about some of the characteristics of the good early customer.

First, the customer must be sophisticated enough to understand his needs and to communicate them to vendors, often at an early stage when products are just being formulated.

Second, the customer must work closely with the vendors, often with cooperative agreements, in order to incorporate the needs of the customers into the design phase of a new product.

Third, the customer is almost always a very early buyer of prototypes and sometimes even makes early performance payments before prototypes are built to provide capital and feedback to the vendor.

Fourth, the early buyer must be willing to accept the problems of an immature product and must invest some of his effort to work with the vendor in order to feed back those problems and to correct them. That has historically been a very important role for the early good buyer.

Fifth, the early buyer has to be willing to add value to the product for his own use, especially by developing software. Hopefully, that added value can be incorporated into the vendor's product line, whether it is software or hardware.

And sixth, the early buyer should be able to offer a predictable market for follow-on sales. If that predictable market is zero, the vendor needs to know so he doesn't tailor the product to the early buyer. If the early buyer has a predictable follow-on market that is not zero, the vendor needs to be able to figure that into his plans. For small companies this can be extremely important for attracting capital and maintaining the ability to continue the product development.

The High Performance Computing Initiative authorizes an early-customer role for the federal government. The six characteristics of a good customer can be translated into five requirements for success of the High Performance Computing Initiative.

First, there must be a perceived fair process that spreads early customer business among qualified vendors. As we learn from Larry Tarbell's description of Project THOTH (Session 12) at the National Security Agency, competition may or may not be helpful. In fact, it has been our experience in DOE that when we are in a very early-buy situation, there is usually no opportunity for competition. But that doesn't mean that fairness goes out the window, because various vendors will be vying for this early buyer attention. They ought to. And by some mechanism, the spreading of the business among the agencies and among the user organizations needs to take place so that the vendors can perceive that it's a fair process.

Second, the early buyer agency or using organization must provide a patient and tolerant environment, and the agency sponsors need to understand this. Furthermore, the users of the product in the early-buy environment need to exercise restraint in either bad-mouthing an immature product or, equally bad, in making demands for early production use of this immature product. We have seen instances in agencies in the past where this aspect was not well understood, and mixed expectations were created as to what this early buy was supposed to accomplish.

Third, these early buys must be placed into sophisticated environments. Those agencies and those organizations that are going to participate in this aspect of the initiative need to be able to provide sophisticated environments. The last thing that a vendor needs is naive users. He accommodates those later on when his product matures, but when he is first bringing his product to market, he needs the expertise of the experienced and qualified user.

Fourth, early buys should be placed in an open environment, if possible. The information as to how the computer is performing should be fed back at the appropriate time, not only to the vendor but also to the rest of the user community. And the environment should involve outside users so as to spread the message. The organization that is participating in the early buy or the prototype evaluation needs to have both a willingness and an ability to share its experience with other organizations, with other agencies, and with the broader marketplace. Clearly, there's a tension between the avoidance of bad-mouthing and the sharing of the news when it's appropriate.

Finally, we need to have patient capital—the federal agencies must be patient investors of advanced technology and human capital. They must be willing to invest today with an understanding that this product may not return useful work for several years. They must be willing to find users to try out the new computer and feed back information to the vendor without being held to long-term programmatic accomplishments. In many cases, these sophisticated institutions are leading-edge scientific environments, and scientists must make tradeoffs. Is a researcher going to (1) publish a paper using today's technology or (2) risk one, two, or three years of hard work on a promising but immature technology?

As we put the High Performance Computing Initiative together, we kept in mind all of these aspects, and we hope we have identified ways to achieve the goals that I laid out. However, I don't want to leave you with the idea that this is easy. We, all of us, have many masters. And many of these masters do not understand some of the things that I've said. Education here will help, as in many other places.

A View from the Quarter-Deck at the National Security Agency

Admiral William Studeman

Admiral William O. Studeman, U.S. Navy, currently serves as Deputy Director of Central Intelligence at the CIA. Until April 1992, he was the Director of the National Security Agency (NSA). His training in the field of military intelligence began subsequent to his graduation in 1962 with a B.A. in history from the University of the South, Sewanee, Tennessee. Thereafter, he accumulated a series of impressive academic credentials in his chosen specialty, taking postgraduate degrees at the Fleet Operational Intelligence Training Center, Pacific, in Pearl Harbor, Hawaii (1963); the Defense Intelligence School in Washington, DC (1967); the Naval War College in Providence, Rhode Island (1973); and the National War College in Washington, DC (1981). He also holds an M.A. in Public and International Affairs from George Washington University in Washington, DC (1973).

The Admiral's early tours of duty were based in the Pacific, where he initially served as an Air Intelligence Officer and, during the Vietnam conflict, as an Operational Intelligence Officer, deploying as the Command Staff of the Amphibious Task Force, U.S. Seventh Fleet. Later assignments posted him to such duty stations as the U.S. Sixth Fleet Antisubmarine Warfare Force, Naples, Italy; the Defense Intelligence Agency station in Iran; the Fleet Ocean Surveillance Information Center, Norfolk, Virginia; and the U.S. Sixth Fleet Command,

> Gaeta, Italy. His duties immediately preceding his appointment as NSA Director included assignments as Commanding Officer of the Navy Operational Intelligence Center (1984–85) and Director of Naval Intelligence (1985–88), both in Washington.

The National Security Agency (NSA) possesses an enlightened, harmonious, heterogeneous computing environment, probably the most sophisticated such environment anywhere. In a sense, it is the largest general-purpose supercomputing environment in the world. Its flexibility in using microcomputers to supplement specially adapted high-power workstations and state-of-the-art supercomputers goes beyond anything that could have been imagined just a few years ago. The investment represents many billions of taxpayer dollars. A major portion of that investment is in supercomputing research, particularly in massively parallel areas, and in the applications that we use routinely on our networking systems. Because of special applications like these, and because of multilevel security requirements, NSA is obliged to operate its own microchip factory. Even if Congress declared tomorrow that the world had become safe enough to halt all defense work at the microelectronics lab, one could still justify the lab's existence on the basis of a whole range of nondefense applications related to the national interest.

So the computing environment at NSA is a complete computing environment. Of course, the focus is narrowly defined. Raw performance is an overriding concern, especially as it applies to signal processing, code-breaking, code-making, and the operations of a very complicated, time-sensitive dissemination architecture that requires rapid turnaround of collected and processed intelligence. In turn, that intelligence must be sent back to a very demanding set of users, many of whom function day-to-day in critical—sometimes life-threatening—situations.

One must admit to harboring concern about certain aspects of the federal High Performance Computing Initiative. Most especially, the emphasis must be on raw performance. Japanese competition is worrisome. If Japanese machines should someday come to outperform American machines, then NSA would have to at least consider acquiring Japanese machines for its operations.

Let me digress for a moment. You know, we operate both in the context of the offense and the defense. Consider, for instance, Stealth technology as it relates to battle space. Stealth benefits your offensive posture by expanding your battle space, and it works to the detriment of your enemy's defensive posture by shrinking his battle space. Translate battle space into what I call elapsed time: Stealth minimizes elapsed time when you are on the offense and maximizes it when you are on the defense.

I raise this rather complicated matter because I want to address the issue of export control. Some in government continue holding the line on exporting sensitive technology—especially to potentially hostile states. And I confess that this position makes some of us very unpopular in certain quarters, particularly among industries that hope to expand markets abroad for American high-tech products. I have heard a great many appeals for easing such controls, and I want to assure you, relief is on the way. Take communications security products: there now remains only a very thin tier of this technology that is not exportable. The same situation will soon prevail for computer technology; only a thin tier will remain controlled.

It most certainly is not NSA's policy to obstruct expansion of overseas markets for U.S. goods. Ninety-five per cent or more of the applications for export are approved. NSA is keenly aware that our vital national interests are intimately bound up with the need to promote robust high-tech industries in this country, and it recognizes that increased exports are essential to the health of those industries. Among the potential customers for high-tech American products are the Third World nations, who, it is clear, require supercomputing capability if they are to realize their very legitimate aspirations for development.

And it is just as clear that the forces of darkness are still abroad on this planet. In dealing with those forces, export controls buy time, and NSA is very much in the business of buying time. When the occasion demands, you must delay, obfuscate, shrink the enemy's battle space, so to speak.

It is no secret that NSA and American business have a community of interest. We write contracts totaling hundreds of millions—even billions—of dollars, which redound to the benefit of your balance sheets. Often, those contracts are written to buy back technology that we have previously transferred to you, something that never gets reflected on NSA's balance sheet. Still, in terms of expertise and customer relations, the human resources that have been built up over the years dealing with American business is a very precious form of capital. And that capital will become even more precious to NSA in the immediate future as the overall defense budget shrinks drastically (disastrously, in my view). If NSA is to continue carrying out its essential functions, it will more and more have to invest in human resources.

This may be a good time to bring up industrial policy. "Government-business cooperation" might be a better way to put it. I'm embargoed from saying "industrial policy," actually. It's not a popular term with the current administration. Not that the administration or anyone else in Washington is insensitive to the needs of business. Quite the opposite. We are more sensitive than ever to the interdependence of national security and a healthy industrial base, a key component of which is the supercomputer sector. If we analyze that interdependence, we get the sense that the dynamic at work in this country is not ideal for promoting a competitive edge. The Japanese have had some success with vertical integration, or maybe they've taken it a bit too far. Regardless, we can study their tactics and those of other foreign competitors and adapt what we learn to our unique business culture without fundamentally shifting away from how we have done things in the past.

What is NSA's role in fostering government-business cooperation, particularly with respect to the supercomputing sector? Clearly, it's the same role already mentioned by other speakers at this conference. That role is to be a good customer in the future, just as it has been in the past. NSA has always been first or second in line as buyers. Los Alamos National Laboratory took delivery on the first Cray Research, Inc., supercomputer; NSA took delivery on the second and spent a lot of money ensuring that Cray had the support capital needed to go forward in the early days. We at NSA are doing the same thing right now for Steve Chen and the Supercomputer Systems, Inc., machine. NSA feels a strong obligation to continue in the role of early customer for leading-edge hardware, to push the applications of software, and to be imaginative in its efforts to fully incorporate classical supercomputing with the massively parallel environment.

One area in which NSA is certain to devote more resources is cryptography problems that prove less tractable when broken down into discrete parallel pieces than they are when attacked in a massively parallel application. Greater investment and greater momentum in this line of research is a high priority.

Producing computer-literate people: through the years, NSA has nurtured a great pool of expertise that the entire nation draws upon. Many of you in this very audience acquired your credentials in computing at NSA, and you have gone out, like disciples, to become teachers, researchers, and advocates. Of course, NSA maintains a strong internship program, with the object of developing and recruiting new talent. Many people are not aware that we have also mounted a very active education program involving students in the Maryland school system, grades K through 12. Further, NSA brings in teachers from surrounding areas on summer sabbaticals and provides training in math and computer science that will improve classroom instruction. All of these efforts are designed to address directly the concerns for American education voiced by the Secretary of the Department of Energy, Admiral James D. Watkins (retired).

Thank you for providing me the opportunity to get out of Washington for a while and attend this conference here in the high desert. And thank you for bearing with me during this rambling presentation. What I hoped to accomplish in these remarks was to share with you a view from the quarter-deck—a perspective on those topics of mutual interest to the military-intelligence community and the high-performance computing community.

Supercomputers and Three-Year-Olds

Al Trivelpiece

Alvin W. Trivelpiece is a Vice President of Martin Marietta Energy Systems, Inc., and the Director of Oak Ridge National Laboratory. He received his B.S. from California Polytechnic State College-San Luis Obispo, and his M.S. and Ph.D. in electrical engineering from Caltech. During his professional career, Dr. Trivelpiece has been a Professor of Electrical Engineering at the University of Maryland-College Park, Vice President of Engineering and Research at Maxwell Laboratories, Corporate Vice President at Scientific Applications Incorporated, Director of the Office of Energy Research for the Department of Energy, and Executive Officer of the American Association for the Advancement of Science.

I always enjoy coming to the Southwest, as many of you do. There is something that people who live here sort of forget, and that is that when you step outside at night and you look up, you see a lot of stars. They're very clear and bright, just as if they are painted on the sky.

Ancient humans must have done the same thing. That is, after they skinned the last sabertooth of the day, they looked up at the sky to see all those stars. Among those people there must have been a few who were as smart as Richard Feynman. Some of them probably had the same IQ that Feynman had. It must have been very frustrating for them because they lacked the tools to answer the questions that must have occurred to them. What things we now recognize as planets, moons, meteors, and comets were phenomena they could only wonder about. But eventually some of them persuaded colleagues to join them and build Stonehenge (early "big science") and other kinds of observatories. These were some of the early tools.

As time has gone by, tools have included ships that permitted exploring what was on the other side of the water, transport that enabled travelers to cross deserts, and eventually, vehicles that allowed humans to go into space and deep into the ocean. Tools are also those things that permit the intellectual exercises that involve going into the interior of the atom, its nucleus, and the subparts thereof.

All of that has been made available by, in a sense, one of the most enduring human characteristics, and that is curiosity. Beyond the need to survive, I think, what drives us more than any other single thing is curiosity. But curiosity can't be satisfied. Ancient man couldn't figure out what was on the inside of an atom, but with the right kind of an accelerator, you can. However, as the tools have come along, they've led to new questions that weren't previously asked.

You, the computer manufacturers, are in the process of developing some truly spectacular tools. Where is all that going to go? One of the things that you are going to need is customers. So I want to just talk for a few minutes about one of the customer bases that you're going to have to pay attention to. I ask you to imagine that it's the year 2000 and that in 1997 a child was born. The child is now three years old. I pick a three-year-old because I think that when you're three years old, that's the last time in your life that you're a true intellectual. The reason I believe this is because at that particular stage a three-year-old asks a question "Why?" for no other reason than the desire to know the answer. Anybody who has been the parent of a three-year-old has put up with a zillion of these questions. I won't try to recite the string; you all know them.

That curiosity, however, is fragile. Something rather unfortunate seems to happen to that curiosity as the child gets a little older. Children reach the third grade, and now you direct them to draw a picture of a fireman, a policeman, a scientist. What do you get? A lady named Shirley Malcolm, who works at the American Association for the Advancement of Science, has a remarkable collection of pictures by third graders who were asked to draw a scientist. What do these pictures look like? Let me just tell you, these are usually people you wouldn't want to be. These are bad-looking, Einstein-like critters who are doing bad things to other people, the environment, or animals. That's your customer base. Incidentally, there's another set of pictures that I've never seen, but they're supposed to be from a similar study done in the Soviet Union, and what those children drew were people being picked up in limos and driven to their dachas.

Most of you have heard about the education demographics in the United States. Eighty-five per cent of the work force between now and the year 2000 is going to be minorities and women; they have not traditionally chosen to pursue careers in science and technology. We have as a result a rather interesting and, I think, a serious problem.

Now, how might you go about fixing that? Well, maybe one way is for every three-year-old to get a terminal and at government expense, get access to a global network. The whole world's information, literally, would be available. This can be done by the year 2000. If this were to occur, what would the classroom of the 21st century look like? I believe it starts with a three-year-old, a three-year-old who gets access to an information base that permits, in that very peculiar way that a three-year-old goes about things, skipping from one thing to another—language, animals, mathematics, sex, whatever. Three-year-olds have a curiosity that just simply doesn't know any particular bounds. Somehow the system that we currently have converts that curiosity into an absolute hostility toward intellectual pursuits. This seems to occur between the third year and third grade.

So, I believe that the distinction between home and classroom is probably going to be very much blurred. Home and classroom will not look significantly different. The terminals may be at home, the terminals may be in schools, they may be very cheap, and they may be ubiquitous. And the question is, how will they be used?

What about the parents between now and then? I suspect that the parents of these children, born in 1997 and three years old in the year 2000, are going to be very poorly equipped. I don't know what we can do about that. But I have a feeling that if you get the right tools in the hands of these three-year-olds in the year 2000, a lot of the customer base that you are counting on will eventually be available. Remember that you, the computer developers and vendors, have a vision. But, unless you do something to help educate the people needed to take advantage of that vision, the vision simply will not exist. So you have a serious problem in this regard.

Think also of children who are disabled in some way—blind, dyslexic, autistic, or deaf. You can make available to these children in their homes or schools through the information bases an ability to overcome whatever disabilities they might have. High-performance computing might provide one means to help in a broad-based campaign to overcome large collections of disabilities.

Thus, I think one of the questions is a rhetorical question. Rather than leaving you with an answer, I leave you with some questions. Because high-performance computing is going to have an impact on the classroom of the 21st century, you have to ask, what is that impact going to be? And how are you going to prepare for it?

NASA's Use of High-Performance Computers: Past, Present, and Future

Vice Admiral Richard H. Truly

Richard H. Truly, Vice Admiral, U.S. Navy (retired), was until 1992 the Administrator of the National Aeronautics and Space Administration. He has a bachelor's degree in aeronautical engineering from the Georgia Institute of Technology, Atlanta. He was the first astronaut to head the nation's civilian space agency. In 1977, he was pilot for one of the two-man crews that flew the 747/Space Shuttle Enterprise approach-and-landing test flights. He served as back-up pilot for STS-1, the first orbital test of the Shuttle, and was pilot of STS-2, the first time a spacecraft had been reused. He was Commander of the Space Shuttle Challenger (STS-8) in August–September 1983. As a naval aviator, test pilot, and astronaut, the Vice Admiral has logged over 7500 hours in numerous military and civilian jet aircraft.

I am delighted to be here at Los Alamos, even for a few hours, for three big reasons. First, to demonstrate by my presence that NASA means it when we say that our support for high-performance computing, and particularly the High Performance Computing Initiative (HPCI), is strong, and it's going to stay that way. Second, to tell you that we're proud of NASA's support over the last five years that led to this initiative. Third, to get out of Washington.

NASA needs high-performance computing to do its job. Let me begin by telling you that some of our current missions and most of our future missions absolutely depend upon the power and the capability that supercomputers can and are providing.

The Hubble Space Telescope, despite what you read, in the next several weeks is going to make a long series of observations that are going to create great interest among scientists and the public, alike. The Hubble will look out to the stars, but the value of the data it brings back can only be understood by programs that can be run on very powerful computers. Within two or three years, when we go back and bring the Hubble up to its originally intended performance, I can assure you that that mission is going to produce everything that we said it would.

There's a space shuttle sitting on the pad that's going to launch in about a week. Its launch will make ten perfect flights since the Challenger accident. This would have been impossible if a supercomputer at Langley Research Center had not been able to analyze the structural performance of the fuel joint that caused the Challenger accident—a problem we did not understand at the time of the accident.

As I speak, 26 light minutes away, the Magellan spacecraft is moving around the planet Venus. It has a very capable, perfectly operating, synthetic-aperture side-looking radar that we've already demonstrated. Magellan is under our control, and we're bringing data back to understand a problem that we've had with it. However, I must tell you the problem is in an on-board computer, apparently.

The reason that we need a side-looking radar to understand the planet Venus is that today, as we sit enjoying this beautiful weather out here, it is raining sulfuric acid on Venus through an atmosphere that produces a surface temperature of something just under 1000 degrees Fahrenheit. To understand that planetary atmosphere and to see the surface, we need supercomputers to interpret the data Magellan sends us. Supercomputers not only allow us to explore the planets through a robot but also will help us understand our own earth.

NASA is in a leadership business, and I think the HPCI is a leadership initiative. NASA is in a visionary business, and I think the HPCI is a visionary program. NASA very much is in a practical business, a day-to-day practical business, and NASA believes that the HPCI is a practical program where federal agencies can get together, along with cooperation from you, and solve some of these disparate problems. The 1992 budget has not been submitted to the OMB, but I assure you that when it is, NASA's support for the HPCI will be there.

Very briefly, our role in the HPCI is to take on the daunting and difficult task of coordinating the federal agencies on the software side and on algorithm developments. Our major applications areas in the

Initiative fall into three areas. First, in computational aeronautical sciences, I'm proud to say, NASA's relationship with the aircraft industry and the aeronautical research establishment over the years is possibly the best example of a cooperative government/private-industry effort. The supercomputers that we use in that effort are only tools, but they are necessary to make sure that the aircraft that sit at airports around the world in future years continue to be from the Boeing Company and McDonnell Douglas and that they continue their record of high returns to our trade balance.

The second major area in applications is in the earth sciences and in the space sciences. For instance, conference participants were given a demonstration of a visualization of the Los Angeles Basin, and it showed us the difficulties in understanding the earth, the land, the ice, the oceans, and the atmosphere. There's no way that our Planet Earth Initiative, or the Earth Observing System that is the centerpiece of it, can be architected without large use of supercomputing power. We recognize that, and that's why the computational planning part of that program is on the front end, and that's also why the largest single item in the budget is no longer the spacecraft, itself, but the analysis and computational systems. And developing those systems will probably turn out to be the most difficult part of the job.

The third applications area is in exploration, both remote today and manned in the future—to the planets and beyond, with robots and people.

Another major area that I ought to mention as our part of the HPCI is that of educating the next generation. In our base program, which I'll speak to in just a minute, we have about seven institution-sited or university-sited centers of excellence, and we intend to double that with our share of the HPCI funds we intend to propose, first to the President and then to Congress.

I should point out that the initiative already sits on a $50 million per year NASA research base in high-performance computing, principally targeted toward scientific modeling, aeronautical research modeling, and networking, both within NASA and outside.

In closing, let me make an observation about what I've seen from interacting with conference participants and from touring Los Alamos National Laboratory. I've noticed that this area is something like the space business. Every single person knows how to run it, but no two people can agree how to run it.

I believe, as Admiral Studeman said earlier in this session, that we absolutely need high performance, but we also need the entire range of work that the companies represented here can provide. We cannot do our job without supercomputing performance.

As far as priorities go, let me just say that in the little over a year that I've been the NASA administrator, I've had a lot of meetings over in the White House, particularly with Dr. Bromley. (I think that you and others in the science and research community ought to thank your lucky stars that Allan Bromley is the Science Advisor to the President.) Among the various topics that I recall—in conversations with small groups, talking about where the federal government and the nation should go—two subjects stand out. The first is in high-performance computing. The second is a subject which, frankly, stands higher in my priority list and is my first love, and that is math and science education. NASA's education programs—I can't miss this opportunity, as I never miss one, to tell you about our many great education programs—have three main thrusts. First, we try to capture at the earliest possible age young kids and make them comfortable with mathematics and science so that later they are willing to accept it. Second, we try to take those young people and channel more of them into careers in mathematics and science. Third, we try to enhance the tools—particularly in the information systems and computers—that we give to the teachers to bring these young people along.

In short, NASA intends to continue to be part of the solution, not part of the problem.

A Leadership Role for the Department of Commerce

Robert White

Robert M. White was nominated by President Bush to serve as the first Department of Commerce Undersecretary for Technology. His nomination was confirmed by the U.S. Senate on April 5, 1990.

Dr. White directs the Department of Commerce Technology Administration, which is the focal point in the federal government for assisting U.S. industry in improving its productivity, technology, and innovation to compete more effectively in global markets. In particular, the Administration works with industry to eliminate legislative and regulatory barriers to technology commercialization and to encourage adoption of modern technology management practices in technology-based businesses.

In addition to his role as Under Secretary for Technology, Dr. White serves on the President's National Critical Technologies Panel and the National Academy of Science's Roundtable on Government/University/Industry Research. Before joining the Administration, Dr. White was Vice President of the Microelectronics and Computer Technology Corporation (MCC), the computer industry consortium, where he directed the Advanced Computing Technology program. Dr. White served as Control Data Corporation's Chief Technical Officer and Vice President for Research and Engineering before he joined MCC.

> *In 1989, Dr. White was named a member of the National Academy of Engineering for his contributions to the field of magnetic engineering. He is also a Fellow of the American Physical Society and the Institute of Electrical and Electronics Engineers. In 1980 he received the Alexander von Humboldt Prize from the Federal Republic of Germany. Dr. White holds a B.S. in physics from MIT and a Ph.D. in physics from Stanford University.*

During Eugene Wong's presentation in this session, he named the eight major players in the High Performance Computing Initiative. You may recall that the Department of Commerce is not among them. Of course, we do have activities at the National Institute of Standards and Technology (NIST) that will relate to some of the standardization efforts. And certainly the National Oceanic and Atmospheric Administration, within the Department of Commerce, is going to be a user of high-performance equipment. But more generally, Commerce has an important role to play, and that is what I'd like to discuss.

Near the close of his remarks, Gene listed the ways in which the federal government might help, and I was intrigued by the one that was at the very bottom. You all probably don't remember, but it was leadership. And so I want to talk a little bit about the role that the government can play in leadership. And to do that, I sort of thought of some of the attributes of leadership that we often think about with regard to actual leaders. I want to try to apply them, if you like, to an organization, particularly the Department of Commerce.

One of the first things I think that leadership involves is vision, the conveying of a vision. I think in the past—in the past few years and even in this past week—we've heard a lot of discouraging information, a lot of discouraging data, a lot of discouraging comparisons to competitors. I think Commerce certainly is guilty of pessimism when they publish a lot of their data. We're all guilty of focusing too readily on the gains of our competitors.

I think our vision—the vision I want to see Commerce adopt—is one of a resilient, innovative and competitive nation, a positive vision, one that can look to the future.

One of the elements of competitiveness has to do with manufacturing, particularly manufacturing quality products. And one of the ways in which Commerce plays a leadership role in manufacturing is that we manage the Malcolm Baldrige Award, which basically promotes quality improvement. This program is only a few years old. During the first half of 1990, we had requests for over 100,000 guideline booklets. I don't know who's going to handle all those applications when they come in.

I think the exciting thing about those guidelines is that they double as a handbook for total quality management. So whether or not you apply for the Baldrige Award or just read the book, you're bound to benefit in terms of quality.

With regard to manufacturing itself, Paul Huray in this session has already mentioned the regional manufacturing centers. Paul emphasized the important fact that this is really the beginning of a very important network. Within Commerce we are also promoting another manufacturing effort, which is the shared manufacturing centers. These are manufacturing centers that are by and large funded by state and local governments, but they are available for small companies to utilize and try new equipment, try new approaches, and perhaps even do prototype runs on things.

And finally NIST, as many of you know, has actually a major automation effort under way that involves many collaborations with industry and federal agencies.

One of the other attributes of leadership is the role of catalyzing, coordinating, and generally focusing. One of the most important assets of the role that Commerce plays is that of convening, the power of convening, and in this way we have access to a lot of data that we can make available to you, hopefully in a useful way. We do maintain clearing-house efforts. We have a database now that has all the state and local technology efforts catalogued for easy accessibility.

Another attribute of a leader, or leadership, is that of empowering those who are part of the organization. In the Department of Commerce, the way that we empower industry, hopefully, is by removing barriers. When it became clear a few years ago that antitrust laws were inhibiting our ability to compete on a global scale, we worked with others to pass the Cooperative Research Act, which has so far made possible several hundred consortia throughout the country. And there is now in Congress a bill to allow cooperative production. I often think that if this bill existed a few years ago, the manufacturing line of the Engineering Technology Associates System could have been formulated as a cooperative effort to share the cost in that very far-thinking effort. And so, if this act passes, companies will certainly be able to work together to benefit from economies of scale.

The federal government also now offers exclusive licenses to technology, particularly that developed within the government laboratories.

And in fact, we have within Commerce a large policy organization that welcomes suggestions from you on removing barriers. In some of our discussions, we heard of some things we do that sound dumb. We'd like to identify those things. We'd like you to come and tell us when you think the federal government is doing something foolish. We can try to change that—change the policy, change the laws. That's one of our functions.

And finally, that brings me to technology policy. Many agencies, as you know, have identified their "critical technologies." The Department of Defense has done so. The Department of Commerce published something that they called "Emerging Technologies."

And as a result of legislation initiated by Senator Jeff Bingaman of New Mexico (a Session 1 presenter), we are now assembling a list of national critical technologies. The thought now is that there will be maybe 30 technologies listed and that the top 10 will be identified.

We also have under way through the Federal Coordinating Committee on Science, Engineering, and Technology an across-the-board exercise to actually inventory all the federal laboratories with regard to those national critical technologies.

And what we hope to do as a result of all of that is to bring together the relevant industrial and federal lab players in these critical technologies—much in the way in which you're brought together here to consider high-performance computing—to talk about priorities and establish a strategic plan—a five-year plan or longer.

The Technology Administration that we have in Congress has been given the power to award grants to industry for technology commercialization. I'm talking about the new Advanced Technology Program. Currently it's very small and very controversial, but it has the potential to become very large.

One of the important elements of this program is that it does require matching funds, and so active involvement by industry is certain. You often hear that this program will support precompetitive and generic technologies, which are acceptable words in Washington now. Precompetitive, to me, means something that several companies are willing to work together on. Anytime you have a consortium effort, almost by definition what they're working on is precompetitive.

All of these programs within the Technology Administration, and the Technology Administration, itself, may be new, but I'm highly optimistic that they will have a major impact.

Farewell

Senator Pete Domenici

Pete V. Domenici of New Mexico has served as a U.S. Senator since 1972. A central figure in the federal budget process, he is the ranking Republican on the Senate Budget Committee and was his party's Senate Coordinator at the President's 1990 Budget Summit. As a leader in formulating government science and technology policy, he authored the original technology-transfer bill, enacted in 1989. This ground-breaking piece of legislation strengthens the relationship between the national laboratories and the private sector. He also played a key role in the drafting of the 1990 Clean Air Act. The Senator, a native New Mexican, has garnered numerous awards during his distinguished political career, including the 1988 Outstanding Performance in Congress Award from the National League of Cities, the 1989 Public Service Award from the Society for American Archeology, and the 1990 Energy Leadership Award from Americans for Energy Independence.

I have been proud to consider myself a representative of the science community during my years in the Senate, and I have been proud to count among my constituents such institutions as the national laboratories, the National Science Foundation (NSF), and the National Aeronautics and Space Administration (NASA). So let me express, by way of opening these remarks, my profound admiration for the principles and the promise that the members of this audience embody.

You have all heard enough during these proceedings about jurisdictional disputes in Washington. You have heard how these disputes get bound up with policy decisions concerning the scope and accessibility of the supercomputer environment being developed in America—how these disputes affect whether we discuss supercomputers, per se, or just increased computing capacity of a high-sensitivity nature. But these issues—the substantive ones—are too important to see mired in turf battles, not only back in Washington but here at this conference, as well. To those of you arguing whether it's NSF versus NASA versus Department of Energy, let me suggest that we have to pull together if the optimum computing environment is to be established.

But I want to turn from the more immediate questions to a more fundamental one. And I will preface what I am about to say by emphasizing that I have no formal scientific training apart from my 1954 degree from the University of New Mexico, which prepared me to teach junior-high math and chemistry. Everything I have subsequently learned about high-performance computing I have learned by interacting informally with people like you. At least I can say I had the best teachers in the world.

It is obvious that this is the age of scientific breakthrough. There has never been anything quite like this phenomenon. We will see more breakthroughs in the next 20 to 30 years, and you know that. Computers are the reason. When you add computing capabilities to the human mind, you exponentially increase the mind's capacity to solve problems, to tease out the truth, to get to the very bottom of things. That's exciting!

So exciting, in fact, that the realization of what American science is poised to accomplish played a major role in my decision to run for the Senate again in 1990. I wanted to be in on those accomplishments. American science is a key component in the emerging primacy of American ideals of governance in a world where democracy is "breaking out" all over. Oh, what a wonderful challenge!

Of course, while the primacy of our ideals may not be contested, our commercial competitiveness most certainly is. In the years just ahead, the United States is going to be conducting business in what is already a global marketplace—trying to keep its gross national product growing predictably and steadily in the company of other nations that want to do the very same thing. That's nothing new. Since World War II, other nations have been learning from us, and some have even pulled ahead of us in everything except science. In that field, we are still the envy of the world.

I submit to you that few things are more crucial to maintaining our overall competitive edge than exploiting our lead in computing. Some members of Congress are focusing on the importance of computers and computing networks in education and academia. They have my wholehearted support. You all know of my continuing commitment to science education. But I, for one, tend to focus on the bottom line: how American business will realize the potential of computers in general and supercomputers in particular. What is it, after all, that permits us to do all the good things we do if not the strength of our economy? Unless business and industry are in on the ground floor, the dividends of supercomputing will not soon accrue in people's daily lives, where it really counts.

Let me tell you a little bit about doing science in New Mexico. Back in the early 1980s, believe it or not, a recommendation was made by a group of business leaders from all over the state, along with scientists from Los Alamos National Laboratory and Sandia National Laboratories, that we establish a network in New Mexico to tie our major laboratories, our academic institutions, and our business community together. We now have such an entity, called Technet.

We encouraged Ma Bell to accelerate putting a new cable through the state; we told them that if they didn't do it, we would. And sure enough, they decided they ought to do it. That helped. We got in three or four years ahead of time so that we could rent a piece of it.

Now, it is not a supercomputer network, but it's linking a broad base of users together, and the potential for growth in the service is enormous.

You know, everyone says that government ought to be compassionate. Yes, I too think government ought to be compassionate. I think the best way for government to be compassionate is to make sure that the American economy is growing steadily, with low inflation. That's the most compassionate activity and the most compassionate goal of government. When the economy works, between 65 and 80 per cent of the American people are taken care of day by day. They do their own thing, they have jobs, business succeeds, business makes money, it grows, it invests. That's probably government's most important function— marshaling resources to guarantee sustained productivity.

Every individual in this room and every institution represented here, as players in the proposed supercomputing network, have an opportunity to be compassionate because you have an opportunity to dramatically improve the lives of the people of this great nation. You can do it by improving access to education, certainly. But mostly, in my opinion, you can do it by keeping your eye on the bottom line and doing whatever is necessary and realistic to help industry benefit from high-performance computing.

You all know that there are people in Congress who are attracted to supercomputers because they're high tech—they're, well, neat. The truth is, we will have succeeded in our goal of improving the lives of the American people when supercomputers are finally seen as mundane, when they're no longer high tech or neat because they have become a commonplace in the factory, in the school, and in the laboratory.

It's a great experiment on which we are embarking. To succeed, we'll have to get the federal government to work closely with the private sector, although there are some who will instantly object. I will not be one of those. I think it's good, solid synergism that is apt to produce far better results than if those entities were going at it alone. It would be a shame if we, as the world leader in this technology, could not make a marriage of government and business work to improve the lives of our people in a measurable way.

The rest of the world will not wait around to see how our experiment works out before they jump in. Our competitors understand perfectly well the kind of prize that will go to whoever wins the R&D and marketing race. The United States can take credit for inventing this technology, and we are the acknowledged leaders in it. I say the prize is ours—but that we'll lose it if we drop the ball.

We Americans have a lot to be proud of as we survey a world moving steadily toward democracy and capitalism. Our values and our vision have prevailed. Now we must ensure that the economic system the rest of the world wants to model continues vibrant, growing, and prosperous. The network contemplated by those of us gathered here, the supercomputing community, would most certainly contribute to the ongoing success of that system. I hope we're equal to the task. I know we are. So let's get on with it.

Contributors

Readers are encouraged to contact the conference contributors directly in regard to questions about particular papers in these proceedings. The contributors are listed below.

Fran Allen
IBM Thomas J. Watson Research Center
P. O. Box 704
Yorktown Heights, NY 10598

Ben Barker
BBN Advanced Computers Inc.
70 Fawcett Street
Cambridge, MA 02138

Richard Bassin
c/o nCUBE Corporation
919 East Hillsdale Boulevard
Foster City, CA 94404

Fernand Bedard
National Security Agency
9800 Savage Road
Ft. Meade, MD 20755-6000

Gordon Bell
450 Old Oak Court
Los Altos, CA 94022

Contributors

R. E. Benner
Sandia National Laboratories
P. O. Box 5800, Org. 1424
Albuquerque, NM 87185

Senator Jeff Bingaman
United States Senate
524 Senator Hart Office Building
Washington, DC 20510-3102

David L. Black
Open Software Foundation Research Institute
1 Cambridge Center
Cambridge, MA 02142

Erich Bloch
Council on Competitiveness
900 17th Street, NW, Suite 1050
Washington, DC 20006

Barry Boehm
Defense Advanced Research Projects Agency
Software and Intelligent Systems Technology Office
3701 North Fairfax Drive
Arlington, VA 22203-1714

Robert Borchers
Lawrence Livermore National Laboratory
University Relations Office
P. O. Box 808, L-414
Livermore, CA 94551

Joe Brandenburg
Intel Supercomputer Systems Division
15201 NW Greenbrier Parkway
Beaverton, OR 97006

Al Brenner
Supercomputing Research Center
17100 Science Drive
Bowie, MD 20715-4300

Contributors 617

Jerry Brost
Cray Research, Inc.
900 Lowater Road
Chippewa Falls, WI 54729

Charles Brownstein
National Science Foundation
1800 G Street, NW
Washington, DC 20550

Duncan Buell
Supercomputing Research Center
17100 Science Drive
Bowie, MD 20715-4300

Dick Clayton
Thinking Machines Corporation
245 1st Street
Cambridge, MA 02142

Robert Cooper
Atlantic Aerospace Electronics Corporation
6404 Ivy Lane, Suite 300
Greenbelt, MD 20770-1406

George Cotter
National Security Agency
9800 Savage Road
Ft. Meade, MD 20755-6000

Harvey Cragon
University of Texas
Department of Electrical Engineering and Computer Science
Austin, TX 78712

Neil Davenport
c/o Cray Computer Corporation
1110 Bayfield Drive
Colorado Springs, CO 80906

Les Davis
Cray Research, Inc.
Harry Runkel Engineering Building
1168 Industrial Boulevard
Chippewa Falls, WI 54729-4511

Alvin Despain
University of Southern California
Department of Electrical Engineering Systems/Advanced Computer Architecture Laboratory
University Park, EEB 131
Los Angeles, CA 90089-2561

Senator Pete Domenici
United States Senate
427 Senator Dirksen Office Building
Washington, DC 20510-3101

Jack Dongarra
University of Tennessee
Department of Computer Science
104 Ayers Hall
Knoxville, TN 37996-1301

Bob Ewald
Cray Research, Inc.
Development Building
900 Lowater Road
Chippewa Falls, WI 54729

Lincoln Faurer
LDF, Inc.
1438 Brookhaven Drive
McLean, VA 22101

Dave Forslund
Los Alamos National Laboratory
MS B287
Los Alamos, NM 87545

Henry Fuchs
University of North Carolina
CB-3175, Sitterson Hall
Chapel Hill, NC 27599

Sam Fuller
Digital Equipment Corporation
146 Main Street
Maynard, MA 01754-2571

Myron Ginsberg
EDS Advanced Computing Center
GM Research and Environmental Staff
P. O. Box 905
Warren, MI 48090-9055

Seymour Goodman
University of Arizona
College of Business and Public Administration
McClellend Hall
Tucson, AZ 85721

Sheryl L. Handler
Thinking Machines Corporation
245 1st Street
Cambridge, MA 02142

Michael Harrison
University of California
Computer Science Division
Berkeley, CA 94720

Sig Hecker
Los Alamos National Laboratory
MS A100
Los Alamos, NM 87545

Robert Hermann
United Technologies Corporation
United Technologies Building
Hartford, CT 06101

Alan Huang
AT&T Bell Laboratories
Crawfords Corner Road
Holmdel, NJ 07733

Paul Huray
University of South Carolina
106 Osborne Administration Building
Columbia, SC 29208

Ken Iobst
Supercomputing Research Center
17100 Science Drive
Bowie, MD 20715

Bill Joy
Sun Microsystems, Inc.
2550 Garcia Avenue
Mountain View, CA 94043

Bob Kahn
Corporation for National Research Initiatives
1895 Preston White Drive, Suite 100
Reston, VA 22091

Ken Kennedy
Rice University
Center for Research on Parallel Computation
P. O. Box 1892
Houston, TX 77251

David Kuck
University of Illinois
Center for Supercomputing Research and Development
305 Talbot Lab
104 South Wright Street
Urbana, IL 61801

H. T. Kung
Carnegie Mellon University
School of Computer Science
Pittsburgh, PA 15213-3890

George Lindamood
Gartner Group, Inc.
56 Top Gallant Road
Stamford, CT 06902

C. L. Liu
University of Illinois-Urbana/Champaign
1304 West Springfield Avenue
Urbana, IL 61801

Alan McAdams
Cornell University
Johnson Graduate School of Management
Malott Hall, Room 515
Ithaca, NY 14853-4201

Oliver McBryan
University of Colorado
Campus Box 430
Boulder, CO 80309-0430

Ralph Merkle
Xerox Research Center
Xerox Parc
3333 Coyote Hill Road
Palo Alto, CA 94304

George Michael
Lawrence Livermore National Laboratory
P. O. Box 808, L-306
Livermore, CA 94551

Emilio Millán
University of Illinois-Urbana/Champaign
Department of Computer Science
1304 West Springfield Avenue
Urbana, IL 61801

Gary Montry
Southwest Software
11812 Persimmon, NE
Albuquerque, NM 87111

Norm Morse
Los Alamos National Laboratory
MS B255
Los Alamos, NM 87545

David B. Nelson
Office of Scientific Computing, ER-7
Department of Energy
Washington, DC 20585

Kenneth W. Neves
Boeing Computer Services
P. O. Box 24346
Mail Stop 7L-25
Seattle, WA 98124

Vic Peterson
National Aeronautics and Space Administration
Ames Research Center
MS 200-2
Moffett Field, CA 94035

John Pinkston
National Security Agency
9800 Savage Road
Ft. Meade, MD 20755-6000

Justin Rattner
Intel Supercomputer Systems Division
15201 NW Greenbrier Parkway
Beaverton, OR 97006

Gian-Carlo Rota
Massachusetts Institute of Technology
Department of Mathematics, 2-351
77 Massachusetts Avenue
Cambridge, MA 02139-4307

Patric Savage
Shell Development Company
P. O. Box 481
Houston, TX 77001-0481

Paul Schneck
Supercomputing Research Center
17100 Science Drive
Bowie, MD 20715-4300

Jack T. Schwartz
New York University
Courant Institute of Mathematical Sciences
251 Mercer Street
New York, NY 10012

Bob Selden
Los Alamos National Laboratory
MS A113
Los Alamos, NM 87545

Howard E. Simmons
E. I. du Pont de Nemours and Company
1007 Market Street
Wilmington, DE 19898

Larry Smarr
University of Illinois-Urbana/Champaign
National Center for Supercomputing Applications
Computing Applications Building
605 East Springfield Avenue
Champaign, IL 61820

Burton Smith
Tera Computer Company
400 North 34th Street, Suite 300
Seattle, WA 98103

Kermith Speierman
National Security Agency
Attn: ADDR(T)
9800 Savage Road
Ft. Meade, MD 20755

George Spix
Supercomputer Systems, Inc.
1414 West Hamilton Avenue
Eau Claire, WI 54701

Stephen Squires
Defense Advanced Research Projects Agency
Computing Systems Technology Office
3701 North Fairfax Drive
Arlington, VA 22203-1714

Herbert E. Striner
4979 Battery Lane
Bethesda, MD 20814

Admiral William Studeman
Central Intelligence Agency
Washington, DC 20505

Larry Tarbell
National Security Agency
9800 Savage Road
Ft. Meade, MD 20755-6000

William L. "Buck" Thompson
Los Alamos National Laboratory
MS A109
Los Alamos, NM 87545

Lloyd Thorndyke
DataMax, Inc.
2800 East Old Shokapee Road
Bloomington, MN 55425

Howard Thrailkill
FPS Computing
3601 SW Murray Boulevard
Beaverton, OR 97005

Al Trivelpiece
Oak Ridge National Laboratory
P. O. Box 2008
Oak Ridge, TN 37831-6255

Vice Admiral Richard H. Truly
6116 Berlee Drive
Alexandria, VA 22312

Tony Vacca
Cray Research, Inc.
Harry Runkel Engineering Building
1168 Industrial Boulevard
Chippewa Falls, WI 54729

Steven J. Wallach
CONVEX Computer Corporation
P. O. Box 833851
Richardson, TX 75083-3851

David Wehrly
ForeFronts Computational Technologies, Inc.
P.O. Box 307
Shokan, NY 12481-0307

Peter Weinberger
Software and Systems Research Center
AT&T Bell Laboratories
PO Box 636
Murray Hill, NJ 07974-0636

Robert White
Department of Commerce
14th Street and Constitution Avenue, NW
Washington, DC 20230

Karl-Heinz A. Winkler
Los Alamos National Laboratory
MS B260
Los Alamos, NM 87545

Peter Wolcott
University of Arizona
College of Business and Public Administration
McClellend Hall
Tucson, AZ 85721

Eugene Wong
Office of Science and Technology Policy
Executive Office of the President
17th Street and Pennsylvania Avenue, NW
Washington, DC 20506